Dryland Rivers

Dryland Rivers

Hydrology and Geomorphology of Semi-arid Channels

Edited by

L.J. BULL
University of Durham, UK

and

M.J. KIRKBY
University of Leeds, UK

JOHN WILEY & SONS, LTD

Other Wiley Editorial Offices

John Wiley & Sons, Inc., 605 Third Avenue,
New York, NY 10158-0012, USA

WILEY-VCH Verlag GmbH, Pappelallee 3,
D-69469 Weinheim, Germany

John Wiley & Sons (Australia) Ltd, 33 Park Road, Milton,
Queensland 4064, Australia

John Wiley & Sons (Asia) Pte Ltd, 2 Clementi Loop #02-01,
Jin Xing Distripark, Singapore 129809

John Wiley & Sons (Canada) Ltd, 22 Worcester Road,
Rexdale, Ontario M9W 1L1, Canada

Library of Congress Cataloging-in-Publication Data
Dryland Rivers: Hydrology and Geomorphology of Semi-arid Channels
edited by L.J. Bull and M.J. Kirkby.
    p. cm.
  Includes bibliographical references (p.).
  ISBN 0-471-49123-3 (alk. paper)
  1. Arid regions—Mediterranean Region. 2. River channels—Mediterranean Region.
I. Bull, L.J. (Louise J.) II. Kirkby, M.J.

  GB618.68.M43 D79 2002
  551.41′5′091822—dc21   2001045410

British Library Cataloguing in Publication Data
A catalogue record for this book is available from the British Library

ISBN 0-471-49123-3

Typeset in 9/11 pt Times by C.K.M. Typesetting, Salisbury, Wiltshire.
Printed and bound in Great Britain by TJ International, Padstow, Cornwall.
This book is printed on acid-free paper responsibly manufactured from sustainable forestry,
in which at least two trees are planted for each one used for paper production.

Contents

List of Contributors

Dr Francisco Alonso-Sarría — Department of Physical Geography, Laboratory of Geomorphology, University of Murcia, Campus 'La Merced', E-30001, Murcia, Spain

Professor Keith Beven — Institute of Environmental and Natural Sciences, Lancaster University, Lancaster, LA1 4YQ, UK

Dr Louise J. Bull — Department of Geography, University of Durham, Science Laboratories, South Road, Durham, DH1 3LE, UK

Dr Carmelo Conesa-García — Department of Physical Geography, Laboratory of Geomorphology, University of Murcia, Campus 'La Merced', E-30001, Murcia, Spain

Dr Francesc Gallart — Estación Experimental de Zonas Aridas (CSIC), General Segura, 1, 04001 Almeria, Spain

Dr Adrian M. Harvey — Department of Geography, University of Liverpool, PO Box 147, Liverpool, L69 3BX, UK

Dr Janet Hooke — Department of Geography, University of Portsmouth, UK

Dr Mike J. Kirkby — School of Geography, University of Leeds, Leeds, LS2 9JT, UK

Dr A. David Knighton — Department of Geography, University of Sheffield, Sheffield, S10 2TN, UK

Dr Roberto Lázaro — Estación Experimental de Zonas Aridas (CSIC), General Segura, 1, 04001 Almeria, Spain

Dr Francisco López-Bermúdez — Department of Physical Geography, Laboratory of Geomorphology, University of Murcia, Campus 'La Merced', E-30001, Murcia, Spain

Dr Jenny Mant — Department of Geography, University of Portsmouth, UK

Dr J. Nachtergaele — Laboratory for Experimental Geomorphology, K.U. Leuven, Redingenstraat 16, 3000 Leuven, Belgium

Dr Gerald C. Nanson — School of Geosciences, University of Wollongong, New South Wales 2522, Australia

Dr D. Oostwoud Wijdenes — Laboratory for Experimental Geomorphology, K.U. Leuven, Redingenstraat 16, 3000 Leuven, Belgium

Dr J. Poesen — Laboratory for Experimental Geomorphology, K.U. Leuven, Redingenstraat 16, 3000 Leuven, Belgium

Dr Joan Puigdefàbregas   Estación Experimental de Zonas Aridas (CSIC), General
                          Segura, 1, 04001 Almeria, Spain

Dr Ian Reid   Department of Geography, Loughborough University,
                         Loughborough, Leicestershire, LE11 3TU, UK

Dr Roy Richardson   Philip Williams & Associates, Ltd. (PWA), San Francisco
                         Bay Area Office, 770 Tamalpais Drive, Suite 401, Corte
                         Madera, CA 94925, USA

Dr Julie Shannon   Department of Geography, Kings College London, Strand,
                         London, WC2R 2LS, UK

Dr Albert Solé   Estación Experimental de Zonas Aridas (CSIC), General
                         Segura, 1, 04001 Almeria, Spain

Dr John Thornes   Department of Geography, Kings College London, Strand,
                         London, WC2R 2LS, UK

Dr Stephen Tooth   Institute of Geography and Earth Sciences, University of
                         Wales, Aberystwyth, SY23 3DB, UK

Dr L. Vandekerckhove   Laboratory for Experimental Geomorphology, K.U.
                         Leuven, Redingenstraat 16, 3000 Leuven, Belgium

Dr B. van Wesemael   Laboratory for Experimental Geomorphology, K.U.
                         Leuven, Redingenstraat 16, 3000 Leuven, Belgium

Dr G. Verstraeten   Laboratory for Experimental Geomorphology, K.U.
                         Leuven, Redingenstraat 16, 3000 Leuven, Belgium

Preface

Drylands cover approximately half of the Earth's land surface, but only 20% of its population. Warm drylands provide an environment in which solar energy is plentiful, but water is scarce. It is in semi-arid drylands that there is evidence of some of the earliest development of agriculture associated with management of water, first through water-harvesting techniques and later with increasingly sophisticated irrigation schemes. Water was spread from large rivers flowing through arid areas in Mesopotamia and Egypt, and from groundwater in fans in Iran and elsewhere in the Middle East. With the spread of irrigation water, salinisation of topsoil has become a growing limitation on agriculture, and one that is difficult to reverse. Nevertheless the further extension of efficient cultivation into dryland areas remains one of the major areas of potential increase in world food production, and scarcity of water in dryland areas is a growing source of international conflict.

Understanding and sustainable management of dryland rivers is therefore becoming increasingly important. In this book it is shown how and why dryland rivers differ from the more widely studied rivers of temperate regions. This understanding of their special features provides a basis for forecasting their hydrology, sediment transport and geomorphology.

The book intends to present a discussion of previous research into dryland river systems, discuss some of the latest insights, and sum up our current understanding. Some of the research presented was initiated under the EU-funded MEDALUS project (Project 4) but this is augmented by key individuals and research initiated under other funding sources. The book is divided into four parts: Part I presents a general overview of research into dryland rivers; Part II relates to catchment processes in dryland rivers; Part III deals with channel network expansion; and Part IV discusses flooding in ephemeral channels.

In the introductory part, Chapter 1 provides a brief introduction to the key characteristics of dryland rivers and raises some issues in terms of key concepts in understanding the landscape. In Chapter 2 Nanson et al. sum up the current global perspective of research into dryland rivers, and expand on key aspects raised in the introduction by focusing on preconceptions and misconceptions regarding main climatic drivers, hydrology, sensitivity to change, channel geometry, sediment yield and evidence for aeolian dune and river interaction.

In Part II, which discusses catchment processes in drylands, analyses runoff production, sediment transport, channel dynamics and sediment deposition. In Chapter 3 Beven investigates runoff generation, including a discussion on different perceptual models such as the point, infiltration, plot, subsurface and catchment models. Modelling at the point and catchment scales are reported in detail and are supported by case studies taken from Australia and the Mediterranean basin. In Chapter 4 Reid moves on to discuss issues regarding sediment dynamics. Aspects covered include different forms of transport in different streams and implications for trends in sediment yield in dryland river systems, and examples from a wide range of systems are discussed. Shannon et al., in Chapter 5, focus on modelling flow and sediment transport under the conditions peculiar to ephemeral channels using a catchment in southeast Spain as a test case. The focus then shifts in Chapter 6 to understanding the morpho-dynamics of dryland river systems. Hooke et al. propose a conceptual model of change, and discuss the impacts of floods over different time scales, ranging from the event to decades. The final chapter in this section, by Harvey (Chapter 7), analyses depositional environments, focuses on the influence of fans on coupling or buffering dryland river systems, presents examples of fan morphology and discusses fan aggradation and dissection.

Part III, on channel network expansion, analyses gullies, channel heads and badlands. In Chapter 8 Poesen et al. examine gullies as an important source of sediment in drylands and debate different methods of classification. The chapter then moves on to a discussion of topographic thresholds and predicting soil loss and sediment production. In Chapter 9 Bull and Kirkby examine previous research on channel heads, discuss the dominant processes and debate the theories surrounding channel initiation. The relationship of the channel head to the wider catchment is also explored. In Chapter 10 Gallart et al. investigate badland systems. Geological controls, geomorphological processes and the relationship between vegetation and geomorphology are all discussed, before a climatic classification of badlands is presented.

Chapters 11 and 12 (Part IV) are used to explore flooding using specific examples of dryland systems from the Mediterranean basin. Chapter 11 by López-Bermúdez et al. investigates the magnitude and frequency of flooding in the Iberian peninsula, while Alonso-Sarría et al. in Chapter 12 examine synoptic conditions that result in these floods.

The greater emphasis on the Mediterranean region than other dryland areas in this book provides some focus on the dryland systems to which this book most fully applies. The particular climatic regime is almost certainly less relevant than other geographical aspects of the Mediterranean. Most notably, the region is one of continuing tectonic activity, with a long coastline, giving rise to a mountainous terrain containing mainly small drainage basins. There is therefore a concentration on catchments of up to 5000 km^2, and many of the studies refer to upland catchments of 1–100 km^2. The book is therefore directed less towards river engineers than to catchment managers, foresters and water supply engineers who are responsible for understanding and planning land use, soil erosion and upstream control of floods.

We would like to acknowledge the wide range of advice we have received, and tried to incorporate into the balance of the book. We hope and intend that, by taking this advice, the book goes far beyond our individual experiences of dryland rivers.

Louise Bull and Mike Kirkby

Part I   An Overview of Dryland Rivers

1 Dryland River Characteristics and Concepts

LOUISE J. BULL[1] AND MIKE J. KIRKBY[2]
[1] *Department of Geography, University of Durham, UK*
[2] *School of Geography, University of Leeds, UK*

1.1 DRYLANDS AND DRYLAND RIVERS

The essence of a warm dryland environment is its sparse vegetation cover resulting from aridity. Vegetation cover shows dryland adaptations when the rainfall is less than the potential evapotranspiration for all or part of the year, creating a permanent or seasonal soil moisture deficit. Many dryland areas show a strong seasonal variation in moisture deficit, or a seasonal alternation between deficit and surplus. One important seasonal dryland regime is the Mediterranean environment. Its definition follows that of the 'Mediterranean climate' adopted by the Study Group on Erosion and Desertification in Regions of Mediterranean-type Climate (MED) from the International Geographical Union (established 1994) and discussed in Conacher and Sala (1998). Summer drought is taken as the distinguishing characteristic of mediterranean type environments, but regions tend to have similar climates, vegetation and landscape forms (Conacher and Sala, 1998). Climates are characterised by hot, dry summers and cool, wet seasons although variability results in periods of seasonal drought and torrential rains. Relief, aspect and altitude modify local climates.

Drylands as a whole include arid, semi-arid and dry subhumid regions. They form a highly significant global environment, occupying 50% of the land area, and supporting 20% of the world's population (UNEP, 1992; Middleton and Thomas, 1997). In this book we have chosen to concentrate on warm drylands, which share aridity and sparse and unevenly distributed vegetation. They are most widespread around 30° North and South latitudes, but can occur locally elsewhere in rain shadow areas.

Arid and semi-arid climates produce a characteristic balance of hillslope and channel processes which give dryland rivers their special features. Many arid climates are associated with intense rainstorms which, over sparsely vegetated surfaces, generate locally high rates of overland flow runoff that lead to hillslope erosion by wash processes. Runoff tends to be patchy however, and much of it re-infiltrates before reaching a channel. Little subsurface flow is available for solute removal, so that soils tend to weather only slowly, and younger dryland areas are characteristically coarse grained with little formation of clay minerals. The products of weathering tend to remain in situ, leading to the gradual accumulation of the more soluble components as duricrusts and evaporites in suitable locations (Figure 1.1).

Drylands occur in a range of tectonic settings. The Mediterranean region is associated with active tectonics, so that the area shows widespread Quaternary uplift, with many erodible late

Dryland Rivers: Hydrology and Geomorphology of Semi-arid Channels. Edited by L.J. Bull and M.J. Kirkby.
© 2002 John Wiley & Sons, Ltd.

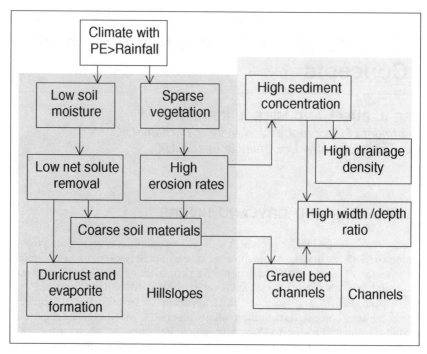

Figure 1.1 Flow diagram showing dominant geomorphological processes and impacts in dryland environments

Tertiary deposits exposed. These characteristics are shared by dryland areas around the Pacific rim, for example in California and Chile, but other major dryland areas, particularly in Africa and Australia, occur in cratonic settings where weathering has been continuous throughout most of the Tertiary and surface soils are now strongly weathered. In some cratonic areas which are now drylands, plate movement and climate change have also exposed the surface to a wider range of climatic conditions during their evolution.

The balance of hillslope impacts may be seen by plotting rainfall intensity (expressed as mean rainfall per rain-day) against percentage vegetation crown cover, on a monthly basis (Figure 1.2). Maximum erosion occurs when high intensities fall on sparsely vegetated surfaces. In a highly seasonal dryland climate with dry summers and wet winters, like the Mediterranean, vegetation cover usually lags behind rainfall by about a month, so that heavy autumn rains can fall on poorly vegetated surfaces. Where the summer is wet and the winter dry, vegetation cover varies less between seasons, and erosion is less severe; but both are more severe than in a more equable climate.

The most obvious and important property of most dryland rivers is that flow is ephemeral, occurring only for a short period during and after rainstorms, unless they have their source outside the arid area (allogenic rivers like the Nile or the Colorado). Hence the relative importance of many fluvial processes, especially the magnitude and frequency of their operation, differs considerably from more humid regions (Graf, 1988; Thornes, 1994). At the start of a storm, the advance of flood waves is limited by channel infiltration except where there is sealing by fines. The frequency distribution of discharges is therefore very different from humid rivers. This can be seen in the form of the curves (Figure 1.3), or summarised in the coefficient of

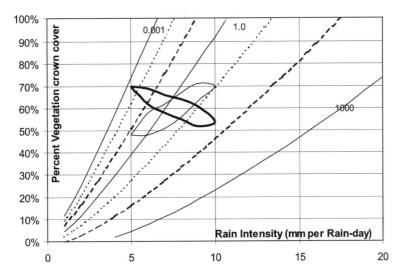

Figure 1.2 A schematic plot of vegetation cover against rainfall intensity, expressed as mean rain per rain-day. Smooth curves and values indicate relative erosion rates by surface wash. The loops show idealised types of annual cycle for seasonal dryland climates. The heavier curve shows a Mediterranean type of climate with dry summers and cool winters, with extreme seasonal variations in erosion. The lighter curve shows seasonality with wet summers and dry winters, with more muted seasonal erosion contrasts. In both cases the seasonal regime generates more erosion than an equable climate with the same average values

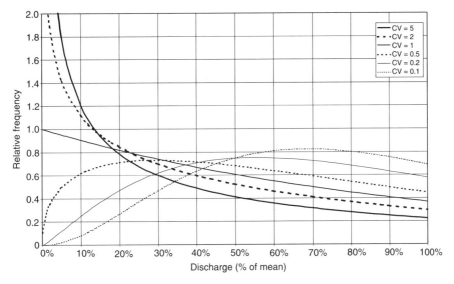

Figure 1.3 A schematic plot showing flood frequency regimes as gamma functions, related to their coefficients of variation (CV). CV > 1 is associated with dryland regimes, and CV < 1 with humid regimes. These distributions are all unimodal, but monsoonal and other seasonal climates may also give bimodal distributions

variation of discharge, which tends to be greater than 1.0 for dryland rivers, and smaller for temperate rivers. High erosion rates and limited runoff give high sediment concentrations in rivers, leading to closely spaced channels and a high drainage density. In headwater areas this may lead to gullying and/or badland development. Downstream, the high sediment concentration and coarse grain sizes give rise to channels with high width–depth ratios. Although intense storms in all climates tend to have a limited areal extent, average storm intensity tends to fall off more rapidly with area for dryland climates than for temperate climates. Total flood discharge therefore increases only slowly, or in some cases decreases downstream, and this pattern is reflected in the channel geometry. Although none of these characteristics is unique to drylands, and there are some exceptions, dryland rivers commonly share many characteristics which place them outside the normal range of temperate rivers, and merit separate study.

Other issues that vary substantially from humid temperate regions include characteristics of rainfall, flow, bedload transport and flooding. Rainfall is highly spatially and temporally variable, with variability possibly increasing with aridity (Bell, 1979). Convective thunderstorms are frequently less than 10–14 km in diameter (Renard and Keppel, 1966; Diskin and Lane, 1972) and result in locally highly concentrated rainfalls (Sharon, 1972; Thornes, 1994; Chapter 12). In humid-temperate regions small patches of extreme precipitation have been recorded, but these events are masked by perennially flowing rivers. Due to the lack of flow in drylands, the limited extent of high-intensity rainstorms is much more evident and more important for producing runoff and flow. The energy of individual raindrops is also high due to the high-intensity rainfall hitting bare ground (Thornes, 1994). Rapid runoff is dominated by Hortonian overland flow and is again highly spatially variable, responding to surface properties such as crusting, stoniness, vegetation and micro-topography (Chapter 3). Sediment is therefore readily entrained and results in highly dissected hillslopes with high drainage densities, although networks can be poorly connected (Chapter 9). Vegetation has a huge influence on both runoff and sediment transport.

Studies of dryland rivers have been dominated by investigations into flash floods, single peak events, multiple peak events and seasonal floods (Graf, 1988). Although conditions vary dramatically across drylands, several key characteristics can be identified that result in variations from humid-temperate rivers including flow hydraulics, low runoff, and decreases in discharge downstream due to transmission losses. The flow hydraulics of dryland rivers are possibly more complex than temperate humid streams due to more complex roughness elements in the channel. Flow hydraulics and roughness are strongly influenced by channel bottom vegetation. Roughness thus varies within individual flood events as flow depth increases, but also between events as vegetation grows and dies. The differences between humid-temperate and dryland rivers are analysed in more detail in Chapter 2 of this volume, and the impacts of vegetation on channel morphology are discussed in Chapter 6.

Runoff in dryland catchments tends to be low, due to low annual rainfall, and there is also high interannual variability in rainfall. This means that often catchment area cannot be used as a reliable surrogate for estimating runoff in drylands (Tooth, 2000). The ratio of large to small flows also results in a highly skewed flood frequency distribution. Dryland channels are highly sensitive to the effects of large floods due to limited resistance and the long-lasting effects (due to limited floods available to modify the channel)(see Chapter 11 for more details).

Decreases in discharge downstream are due to transmission losses that result from infiltration of the floodwaters into unconsolidated alluvium that tends to form dryland channel boundaries. Magnitudes and rates of transmission losses are highly variable and depend on storm size, position of storm, storm tracking, hydrograph volume and duration and channel characteristics. The most important impact of transmission losses on dryland channels is that flow is rarely continuous throughout the whole catchment (Chapter 5). Thornes (1977) described three phases of flow: asynchronous, axial and fully integrated. Asynchronous flow is where tributaries flow but the main trunk of the channel remains dry; axial flow is when flow occurs only in the trunk stream and not in the tributaries; and fully integrated flow is when the whole catchment flows.

Sediment transport within dryland rivers is characteristically high. This reflects catchment processes that produce readily mobilised sediment, delivery of this sediment from sparsely vegetated hillslopes and the availability of sediment within the channel beds themselves (see Chapter 4). Bedload transport results in scour and fill of dryland channels, most of which tend to be sand- or gravel-bed channels (Leopold et al., 1966; Foley, 1978; Schick et al., 1987; Hassan, 1990). Studies have suggested that similar amounts of material are eroded and deposited during floods, which results in a highly mobile channel bed, although this does not leave much evidence of minor channel change. This supports the supposition that dryland rivers have poor armour layers that may encourage equal mobility. Suspended sediment transport tends to result from entrainment of fines in overland flow, as well as contributions from bank erosion and bed scour. Suspended sediment concentrations tend to be high at low discharges, although no hysteresis has been recorded. Suspended sediment transport is controlled by hydraulic forces and is also influenced by chemistry of local sediments.

Because of the high ratio of sediment to water (compared to temperate rivers), channels are irregular and there is a high occurrence of depositional features such as alluvial fans. Dryland rivers tend to be braided systems with high width–depth ratios and low sinuosity. Fans are common between tributaries and the main channel due to selected areas of the catchment producing runoff during different events (see Chapter 7). They are especially frequent in basins that tend to produce asynchronous flow. Dryland rivers may also terminate in a fan and do not necessarily form part of an increasingly larger channel network.

1.2 CONCEPTUAL THEMES IN RELATION TO DRYLAND RIVERS

Drylands are in a delicate state of balance that can easily be upset, but research has shown that most dryland rivers have the following distinct characteristics:

1. The rivers are significant agents of erosion and deposition and are effective causes of landform development (Frostick and Reid, 1987; Reid and Frostick, 1997).
2. There are long histories of well-preserved fluvial deposits, which allows palaeoenvironmental construction (Merrifield, 1987; Reid and Frostick, 1997).
3. Water resource issues are of major importance.

Langbein and Schumm (1958) related effective precipitation to sediment yield and proposed that maximum sediment yield is produced with an effective precipitation of 300 mm yr^{-1}, which lies squarely within the dryland environments, although additional data have since blurred this relationship (Chapter 2). Changes to any factors influencing runoff and sediment production can therefore have a great influence on dryland environments. Wolman and Gerson (1978) noted the rapid increase in channel width with drainage area, which is probably related to the high drainage densities that are sustained by rapid runoff. When the width–area relationship is thought about in conjunction with the high sediment loads, it emphasises the sensitivity of dryland systems. Thus large events have the potential to move large quantities of sediment, dramatically alter the channel morphology and disrupt within-channel vegetation.

Sensitivity concerns the likelihood that a given change in the controls of a system or the forces applied to the system will produce a sensible, recognisable, sustained but complex response (Brunsden, 2001). For rivers this has been interpreted in a number of different ways. Firstly, as the ratio of disturbing to resisting forces (Brunsden and Thornes, 1979); secondly, the propensity for change and the capacity of the river system to absorb change (Graf, 1979); and, lastly, in terms of thresholds, above which change may occur if the disturbing force is sufficiently large (Schumm, 1991). Robust landforms have the ability to absorb change with only modest adjustment, while responsive landforms are those which undergo a fundamental and persistent change (Werrity and Leys, 2001). Using these definitions, dryland rivers may not actually be classified as sensitive, responsive systems. Even though drylands undergo major, catastrophic flooding, the channel systems themselves do not change their overall character. In general, dryland rivers remain as very wide, braided, gravel-bed rivers, despite changes in within-channel deposits and aggradation or dissection of fan deposits at junctions. Therefore drylands may be located in sensitive climatic zones, but the river systems themselves tend to have the propensity to absorb change and are relatively unresponsive. However, land use and climate changes have the ability to trigger a major change in flood magnitude and frequency, although this may not alter the overall appearance of the dryland river systems.

It has been suggested that dryland rivers are in disequilibrium. Channel form represents the effects of the last major flood and hence complete adjustment of form to process is inhibited. Richards (1982) proposed that 'because recovery periods exceed the recurrence interval of channel-changing floods, equilibrium between channel form and channel forming discharge is inconceivable'. Most discussions of disequilibrium have focused on climate as the cause; however, in some areas, such as the Mediterranean, tectonics have also played a major role in shaping the landscape. In such locations the lag between uplift and channel response to attain equilibrium may be even longer than the lag between a major flood and reworking of channel deposits to obtain equilibrium.

However, what do river systems in disequilibrium look like? The more obvious answers to this question would refer to dryland morphology, with rivers adjusted to catastrophic flood events, basins characterised by incised channels with a readily available supply of coarse sediment from hillslopes and through the channel network (Chapter 4). As an example, for some channels in southeast Spain the channel-forming discharge is likely to be in the region of a 1-in-10-year event, with annual events having little impact. Floods that are larger than this produce major channel change, but the 1-in-10-year event is able to rework the channel. In a temperate river system we would expect the channel-forming discharge to be much more frequent, perhaps an annual event, although larger floods may produce major changes in channel morphology. The relatively frequent channel-forming discharge then modifies the effects of the large flood. But how does this link to the equilibrium of a system? Both systems respond to events of different magnitude and frequency, but the time-scales of channel modification and the lag between the disturbing event and response is greater for dryland channels. Just because dryland systems are adjusted to a less frequent event, does it mean they are not in equilibrium?

Equilibrium can also be debated by considering the hillslope characteristics and the availability of sediment. Upland river channels in humid-temperate regions (which are generally presumed to be in equilibrium), tend to have a landscape generally composed of rounded, convex hillslopes, although with a dense cover of vegetation, be that grass or forest. We also have few details of bed characteristics except for small streams. Hence we do not know how much sediment is awaiting transport at the channel bed. The characteristics that we would assume indicate a system not in equilibrium, actually refer to the sediment transport regime (supply or transport limited), and may not be substantially different between different types of fluvial system. It is also unclear how much bed degradation occurs during catastrophic floods due to highly mobile channel beds that infill areas of scour during flood recession. It is therefore difficult to know the rates of downcutting of dryland rivers during single events and even over a period of a few years. For longer time periods stratigraphic evidence and dating methods assist in estimating long-term rates.

As mentioned above, sediment transport tends to be transport limited rather than supply limited. When dryland rivers are in flood they are much more efficient erosional agents than perennial systems (Reid and Frostick, 1997). This is due to the availability of material for erosion on poorly vegetated slopes, and the coarse material available on the channel bed. As a result, suspended sediment concentrations and bedload transport rates reach record levels (Reid and Frostick, 1997). So how does the sediment regime of a system relate to equilibrium? To answer this we need to consider different temporal scales. Over long time scales of cyclic time when dryland systems are responding to changes in tectonics or climate, the fluvial system is an open system undergoing continual change to a varying imposed potential energy caused by changes in relief and/or climate. This relates to Schumm's (1991) definition of decay equilibrium as sediment is removed from hillslopes and transported downstream during large, infrequent flood events. At the other end of the spectrum, dryland rivers may be in a static equilibrium as they experience no, or limited, change between major flood events. If sediment transport regime is linked to disequilibrium, it is more likely to be over an intermediate timescale and relate to dynamic equilibrium as proposed by Schumm (1991). Dynamic equilibrium refers to variability about a changing mean, but in dryland systems the 'mean' is very difficult to define because of high temporal and spatial variability. During discussions of dynamic equilibrium it is generally assumed that change within the system is always in the same direction. For dryland rivers this is not necessarily the case. With so much available sediment it is difficult to establish whether a system is predominantly cutting or filling. Between major flood events the system may switch between the two, but with our current understanding it is difficult to identify a longer-term trend.

Another key theme in drylands is the relationship between the runoff thresholds and the dominant event size. Runoff thresholds in drylands tend to be relatively low, such as 5–10 mm for areas of marl and 20 mm for areas of mica-schist in southeast Spain (Bull et al., 2000). Factors encouraging low runoff thresholds include local geology, sparse vegetation, soil chemistry, high slopes and thin soils. If these thresholds are greater than the mean rainfall during a storm event, runoff and ephemeral channel flow become rare. For small areas, local runoff can lead to intense erosion and the formation of gully systems (Chapter 8), but runoff tends to decrease strongly with increasing area. Floods from larger areas (>100 km^2) generally only occur during the rare rainfall events that cover a substantial fraction of the catchment area.

1.3 IMPLICATIONS FOR LOCAL POPULATIONS

If drylands are defined as areas where potential evapotranspiration is greater than rainfall, then it follows that these regions experience water shortages. Local populations therefore cannot rely on surface water or groundwater to recharge locally and need to store any temporary runoff and to exploit water transfer schemes. Annual summer drought is common and is a problem in its own right, but is also a cause of other issues, especially as the demand for water increases from agriculture, industry and expanding urban uses. Severe problems result when economic development is not matched by developments in water resources. A number of factors exacerbate water scarcity or availability to the point of crises. These include prolonged periods of drought, increasing population demands, issues of sovereignty where water transcends regional or national boundaries, problems of upstream interception (e.g. dams and reservoirs), land degradation, or water pollution (Conacher and Sala, 1998).

Problems of land degradation include accelerated erosion, flooding, and vegetation loss. Many dryland regions are experiencing unsustainable rates of soil loss, although this is a highly spatial variable (Sala et al., 1991). Abandoned fields are becoming an important part of upland dryland landscapes due to land use change, and once plant colonisation is disturbed, sheetwash and erosion are important and can lead to rilling and higher rates of soil erosion (García-Ruíz et al., 1991). In Portugal, intensive use and recent changes in land use are responsible for 30% of soils being under high risk from erosion, 54% under medium risk and 15% under low erosion risk (Commission of European Communities, 1992).

Soils exhibit many forms of degradation other than physical loss, including salinisation, acidification, structural decline, water repellency, declining fertility and pollution by chemical residues (Conacher and Sala, 1998). The main contamination sources are industrial and mining wastes, and agricultural pesticides and fertilisers (Conacher and Sala, 1998). Land degradation affects the viability of natural ecosystems as well as agricultural productivity in terms of yield losses and loss of income. It also imposes a range of direct costs including costs of repair, prevention, damage to ecosystems, adverse effects on human well-being and health and rural depopulation.

By definition drylands are characterised by strong contrasts and hence can experience devastating floods. Both the volumes of water and the sediment transported provide hazards. The floodwave itself can take villages by surprise due to the characteristics of dryland areas discussed above. There are many reports of the speed of flood development and loss of human life (e.g. Conesa-García, 1995). Floods accelerate erosion processes, cause the loss of soil and the aggradation of gravel-bed rivers (Conacher and Sala, 1998). Sediment produced from soil erosion and landsliding is transported downstream and deposited as ridges in braided channels, or as fans at the confluence of tributaries and trunk streams. Sediment may therefore be transported downstream as waves related to large flood events, and this not only represents hazardous conditions for cultivated fields close to the main channel but also results in the sedimentation of dams and reservoirs. Removal of native vegetation also disrupts the hydrological cycle and leads to increased runoff, which through positive feedback results in increased flooding, erosion, sediment delivery and habitat changes downstream.

Most problems of land degradation have been at least partly caused by the removal, degradation or replacement of the vegetative cover. In the Mediterranean basin little indigenous vegetation remains due to the long period of human settlement. Mixed oak forests have been replaced by matorral on steep hillslopes and by cultivated dryland farming on less steep areas. The primary roles of vegetation are to prevent wind erosion by reducing wind speed near the surface and to reduce the impact velocity of raindrops. Although extremes can be found, the effect is generally modest, and is not limited to trees (Grove and Rackham, 2001). The critical factor is ploughing, and not necessarily deforestation (Grove and Rackham, 2001). Ploughing removes the protective cover of vegetation beneath tree crops and over areas not used for agriculture. This leaves the soil bare and open to the influence of wind erosion, splash erosion and erosion by overland flow. Neglect of established terraces may also lead to accelerated erosion, as gullies cut through terrace walls and release the sediment stored behind them.

Grove and Rackham (2001) propose four more recent activities that promote erosion. Although this research is based on the Mediterranean, it tends to apply to most dryland regions.

1. Digging false terraces out of a hillside and leaving them to stand or fall unsupported. This may be carried out to plant crops or to revegetate hillsides.

2. Bulldozing land and leaving it bare for months or years.
3. Enlarging fields by removing cross-slope boundaries and increasing slope length.
4. Modern road-making by digging roads out of hillsides, leaving scarp and talus unsupported, and compensating for instability with greater width.

Hence the bulldozer is becoming an important tool for modifying landscapes in dryland environments and has initiated a novel direction for dryland research.

Desertification is also an issue in drylands, and has been highlighted as a key problem since the 1970s following the UN Conference in 1977 (UNCOD). The target set by this conference was the eradication of desertification by 2000, and since this time it has been viewed as one of the most pressing issues affecting human kind. UNEP (1992) developed a common definition of desertification as 'land degradation in arid, semi-arid and dry subhumid areas, resulting from various factors including climatic variability and human activities'. Land degradation was seen as 'the reduction or loss of the biological or economic productivity caused by land use change, from a physical process, or a combination of the two' (Thomas, 1997). Desertification is therefore seen as a process that is triggered by changes in climate and socio-economic boundary conditions of dryland systems. Triggers include overgrazing, deforestation, precarious agriculture, uncertain rainfall, river flooding, depletion of surface water, and depletion of ground water, while the extent of desertification is determined by rainfall patterns, soil morphology, soil pedology, vegetation, and land use. The role of dryland rivers in the process of desertification is poorly understood and it is vital that we develop our understanding of landscape sensitivity and erosion to enable improved understanding and management.

1.4 THE FUTURE

Within dryland river research there remains a major stumbling block, namely a lack of good quality data. This is due to the infrequent nature of events and the problems of establishing monitoring networks over large areas where rainfall and runoff are highly variable. This has meant that much research has been carried out by a select number of research groups, and has focused on relatively few research locations. This is addressed in more detail in Chapter 2. With a greater range of data for dryland areas hopefully the accepted trends that have been discussed in previous reviews of dryland rivers will be augmented and developed.

Connectivity within dryland rivers is an important area where relatively little is known. Ideas of connectivity incorporate coupling between tributaries and trunk streams (Thornes, 1976), local coupling between hillslopes leading directly into channels (Harvey, this volume), and catchment-wide connectivity between fields and local tributaries to the main trunk stream (e.g. Wainwright, 1996; Bull et al., 2000). Research in this area can be divided into understanding runoff production and variation in discharge throughout dryland river systems (Yair and Lavee, 1982; Bull et al., 2000), or can be focused on aspects of sediment sources and travel distances (Hassan et al., 1991; Schick and Lekach, 1993). Our understanding needs to be significantly developed before the widespread application of models to predict water and sediment flow occurs without recognising these difficulties (e.g. Morgan et al., 1998; Chapter 5).

Associated with this is the issue of scale. Much research into dryland rivers has been carried out at the plot scale and only a few investigations have been executed at the catchment scale. This is especially true for studies of runoff, soil erosion such as interrill and rill erosion (e.g. Parsons et al., 1996; 1997) and the effects of vegetation on runoff and sediment transport (e.g. Imeson and Verstraeten, 1989; van Wesemael et al., 2000). An important direction for future research is to determine if, and how, knowledge at the local scale can be scaled up to the catchment. This is particularly relevant for understanding how changes in land use affect flood magnitude and frequency, and also for resource planning for local populations. We have also been more skilled at elucidating processes at the event time scale rather than for longer periods (e.g. Wainwright, 1996; Bull et al., 2000; Torri et al., 2000). This is also related to the logistics of data collection, but the evolution of dryland landscapes and how this influences runoff and sediment production is a huge gap in our knowledge. At the plot scale a number of advances have been made (e.g. Cammeraat and Imeson, 1999), but again we need to scale this up to be more widely applicable.

We also need to develop our current understanding of interactions between different aspects of dryland catchments. Connectivity between slopes and channels is only one aspect of this problem, which also includes the effect of vegetation on runoff (at plot and catchment scales) (e.g. Wainwright et al., 2000), interactions between bedload and channel morphology (e.g. Reid et al., 1998; Hassan et al., 1999; Chapter 6) and interactions between rare floods and within-channel vegetation (e.g. Brookes et al., 2000; Hooke and Mant, 2000). Again the issue of scale is important here, with logistics making it more possible to improve our understanding at the plot and event time scale. However, to comprehend large-scale changes related to variations in climate and land use we need to elucidate interactions between all processes at all scales. This type of research has been undertaken in some avenues such as investigations of rills and gullies (e.g. Oostwoud Wijdenes et al., 1999), but interactions of water and sediment between gullies and dryland river processes now need to be appraised (see Chapters 8 and 10).

As mentioned earlier in this chapter a recent phenomenon in dryland regions is bulldozing natural landscapes predominantly for road-building and agriculture. Road-building tends to affect small areas and produce local landslides and sources of sediment. Bulldozing for agriculture affects much larger areas because it involves large sites being flattened for intensive farming, or areas of gullies and badlands being infilled to again produce areas of flat land. Evidence for these changes include gully systems literally disappearing and pylons and telegraph poles standing on mounds of sediment above the flattened field surface. As yet we have little idea of the long-term effects of altering landscape morphology in this way. In terms of infilling gullies we do not know (i) whether the dumped material will bond with the previous crusted surfaces; (ii) whether the dumped and compacted material has been left with slope angles that will encourage stability at the edge of the areas affected; or (iii) whether large infrequent storms can erode and cut back down into the infilled areas. In terms of flattening large areas, we do not know how this will influence runoff production and frequency of floods. The answers to these questions may well depend on the exact location of the altered area with respect to runoff production areas and the river system. Since our understanding of some of these issues remains in its infancy, this gives us another line of inquiry to explore.

Bulldozing may be a relatively new phenomenon in drylands but historically we have interfered with ephemeral systems by building reservoirs and check dams. These structures exert a strong control on the connectivity of drylands, transmission of water and sediment downstream and possibly groundwater recharge. Research has used reservoirs to investigate sediment yields from areas feeding into the reservoir (e.g. Oostwoud Wijdenes et al., 2000), but we still know little about the effects on the system at the catchment scale. Check dams also raise issues for future research. Ideally they delay (at worst) or stop (at best) the movement of water and sediment during flood events. However, upstream of check dams, ponds of sediment develop, while below check dams dryland channels can experience severe erosion. As check dams gradually fill up with sediment their effect on the dryland river systems becomes less. We do not generally know the useful life of a check dam, although many appear to fill after only a few storms, nor can we be sure how the steps in channel profiles modify sediment transport and flood protection. One suggestion is that check dams are effective in small events, but become useless during large events and can even exacerbate flooding, most strikingly where failure affects one or more in a sequence of check dams, e.g. the Arás basin, Spain.

A related issue is the maintenance of terraces. Terraces have been successfully managed in many dryland regions to assist in farming, water harvesting and stabilising land. However, in some areas these are not now being maintained and the effect of this on dryland landscape development is still unknown. Terraces adjacent to dryland rivers may be especially problematic. As the maintenance of these terraces declines several problems arise including increasing the connectivity of hillslopes and channels, increasing erosion and reducing stability of the actual terraces. Hence large floods may actually be exacerbated, which is undesirable for local farmers and villages. Management of our drylands is therefore important and we need to improve our understanding of the impacts of human intervention in these areas to be aware of the impact on processes and landscape development.

REFERENCES

Bell, F.C. 1979. Precipitation. In D.W. Goodall and R.A. Perry (Eds) *Arid Land Ecosystems*. Cambridge University Press, Cambridge, 373–393.

Brookes, C.J., Hooke, J.M. and Mant, J.M. 2000. Modelling vegetation interactions with channel flow in river valleys of the Mediterranean region. *Catena*, 40, 93–118.

Brunsden, D. 2001. A critical assessment of the sensitivity concept in geomorphology. *Catena*, 42, 99–123.

Brunsden, D. and Thornes, J.B. 1979. Landscape sensitivity and change. *Transactions of the Institute of British Geographers*, 4, 463–484.

Bull, L.J., Kirkby, M.J., Shannon, J. and Hooke, J.M. 2000. The variation in estimated discharge in relation to the location of storm cells in SE Spain. *Catena*, 38(3), 191–209.

Cammeraat, L.H. and Imeson, A.C. 1999. The significance of soil-vegetation patterns following land abandonment and fire in Spain. *Catena*, 37, 107–127.

Commission of European Communities, 1992. *CORINE – Soil Erosion Risk and Land Resources in the Southern Regions of the European Community*. EUR 13233, Office for the publications of the European Community.

Conacher, A.J. and Sala, M. 1998. *Land Degradation in Mediterranean Environments of the World: Nature and Extent, Causes and Solutions*. Wiley, Chichester, 491pp.

Conesa-García, C. 1995. Torrential flow frequency and morphological adjustments of ephemeral channels in south-east Spain. In E.J. Hickin (Ed.) *River Geomorphology*. Wiley, Chichester, 169–192.

Diskin, M.H. and Lane, L.J. 1972. A basinwide stochastic model of ephemeral stream runoff in southeastern Arizona. *International Association Scientific Hydrologists Bulletin*, 17, 61–76.

Foley, M.G. 1978. Scour and fill in steep, sand bed ephemeral channels. *Geological Society of America Bulletin*, 103, 504–511.

Frostick, L.E. and Reid, I. 1987. *Desert Sediments: Ancient and Modern*. Geological Society Special Publication, 35. Blackwell, Oxford.

García-Ruíz, J.M., Lasanta, T. and Martínez, R. 1991. Erosion in abandoned fields, what is the problem? In M. Sala et al (Eds) *Soil Erosion Studies in Spain*. Geoforma Ediciones, Logroño.

Graf, W.L. 1979. Mining and channel response. *Annals of the Association of American Geographers*, 69, 262–275.

Graf, W.L. 1988. *Fluvial Processes in Dryland Rivers*. Springer-Verlag, Berlin.

Grove, A.T. and Rackham, O. 2001. *The Nature of Mediterranean Europe; An Ecological History*. Yale University Press, London, 383pp.

Hassan, M.A. 1990. Observations of desert flood bores. *Earth Surface Processes and Landforms*, 15, 481–486.

Hassan, M.A., Church, M. and Schick, A.P. 1991. Distance of movement of coarse particles in gravel bed streams. *Water Resources Research*, 27, 503–511.

Hassan, M.A., Schick, A.P. and Shaw, P.A. 1999. The transport of gravel in an ephemeral sandbed river. *Earth Surface Processes and Landforms*, 24, 623–640.

Hooke, J.M. and Mant, J.M. 2000. Geomorphological impacts of a flood event on ephemeral channels in south east Spain. *Geomorphology*, 34, 163–180.

Imeson, A.C. and Verstraeten, J.M. 1989. The microaggradation and erodibility of some semi-arid and Mediterranean soils. *Catena*, 16, 11–24.

Langbein, W.B. and Schumm, S.A. 1958. Yield of sediment in relation to mean annual precipitation. *Transactions of the American Geophysical Union*, 55, 91–107.

Leopold, L.B., Emmet, W.W. and Myrick, R.M. 1966. Channel and hillslope processes in a semi-arid area, New Mexico. *United States Geological Survey Professional Paper*, 352-G.

Merrifield, P.M. 1987. Mapping desert alluvial deposits from satellite imagery. In L. Berkofsy and M.G. Wurtele (Eds) *Progress in Desert Research*. Rowan & Littlefield, New Jersey, 198–216.

Middleton, N.J. and Thomas, D.S.G. 1997. *World Atlas of Desertification*, 2nd Edn. UNEP/Edward Arnold, London.

Morgan, R.P.C., Quinton, J.N., Smith, R.E., Govers, G., Poesen, J., Auerswald, K., Chisci, G., Torri, D. and Styczen, M.E. 1998. The European Soil Erosion Model (EUROSEM): a dynamics approach for predicting sediment transport from fields and small catchments. *Earth Surface Processes and Landforms*, 23, 527–544.

Oostwoud Wijdenes, D., Poesen, J., Vandekerckhove, L. and Ghesquiere, M. 2000. Spatial distribution of gully head activity and sediment supply along an ephemeral channel in a Mediterranean environment. *Catena*, 39, 147–167.

Oostwoud Wijdenes, D., Poesen, J., Vandekerckhove, L., Nachtergaele, J. and De Baerdemaeker, J. 1999. Gully head morphology and implications for gully development on abandoned fields in a semi-arid environment, Sierra de Gata, south east Spain. *Earth Surface Processes and Landforms*, 24, 585–603.

Parsons, A.J., Abrahams, A.D. and Wainwright, J. 1996. Responses of interrill runoff and erosion rates to vegetation change in southern Arizona. *Geomorphology*, 14, 311–317.

Parsons, A.J., Wainwright, J., Abrahams, A.D. and Simanton, J.R. 1997. Distributed dynamic modelling of interrill overland flow. *Hydrological Processes*, 11, 1833–1859.

Reid, I. and Frostick, L.E. 1997. Channel form, flows and sediments in deserts. In D.S.G. Thomas (Ed.) *Arid Zone Geomorphology: Process, Form and Change in Drylands*. Wiley, Chichester, 205–229.

Reid, I., Laronne, J.B. and Powell, D.M. 1998. Flash-flood dynamics of desert gravel-bed streams. *Hydrological Processes*, 12, 543–557.

Renard, K.G. and Keppel. R.V. 1966. Hydrographs of ephemeral streams in the Southwest. *Proceedings of the American Society of Civil Engineers, Journal of the Hydraulics Division*, 92 (HY2), 33–52.

Richards, K.S. 1982. *Rivers: Form and Process in Alluvial Channels*. Methuen, London.

Sala, M., Rubio, J.L. and García-Ruíz, J.M. (Eds) 1991. *Soil Erosion Studies in Spain*. Geoforma Ediciones, Logroño.

Schick, A.P. and Lekach, J. 1993. An evaluation of two 10-year sediment budgets, Nahal-Yael, Israel. *Physical Geography*, 14, 225–238.

Schick, A.P., Lekach, J. and Hassan, M.A. 1987. Bedload transport in desert floods: observations in the Negev. In C.R. Thorne, J.C. Bathurst and R.D. Hey (Eds) *Sediment Transport in Gravel Bed Rivers*. Wiley, Chichester, 617–636.

Schumm, S.A. 1991. *To Interpret the Earth: Ten Ways to be Wrong*. Cambridge University Press, Cambridge.

Sharon, D. 1972. The spottiness of rainfall in a desert area. *Journal of Hydrology*, 17, 161–175.

Thomas, D.S.G. 1997. Science and the desertification debate. *Journal of Arid Environments*, 37, 599–608.

Thornes, J.B. 1976. *Semi-arid erosional systems*. Occasional Paper 7. London School of Economics, London.

Thornes, J.B. 1977. Channel changes in ephemeral streams: observations, problems and models. In K. Gregory (Ed.) *River Channel Changes*. Wiley, Chichester, 317–335.

Thornes, J.B. 1994. Catchment and channel hydrology. In A.D. Abrahams and A.J. Parsons (Eds) *Geomorphology of Desert Environments*. Chapman and Hall, London, 257–287.

Tooth, S. 2000. Process, form and change in dryland rivers: a review of recent research. *Earth Science Reviews*, 51, 67–107.

Torri, D., Regues, D., Pellegrini, S. and Bazzoffi, P. 2000. Within-storm soil surface dynamics and erosive effects of rainstorms. *Catena*, 40, 14–25.

UNEP 1992. *World Atlas of Desertification*. Edward Arnold, London.

Van Wesemael, B., Mulligan, M. and Poesen, J. 2000. Spatial patterns of soil water balance on intensively cultivated hillslopes in a semi-arid environment – the impact of rock fragments and soil thickness. *Hydrological Processes*, 14, 1811–1828.

Wainwright, J. 1996. Infiltration, runoff and erosion characteristics of agricultural land in extreme storm events, SE France. *Catena*, 26, 27–47.

Wainwright, J., Parsons, A.J. and Abrahams, A.D. 2000. Plot-scale studies of vegetation, overland flow and erosion interactions: case studies from Arizona and New Mexico. *Hydrological Processes*, 14, 2921–2943.

Werrity, A. and Leys, K. 2001. The sensitivity of Scottish rivers and upland valley floors to recent environmental change. *Catena*, 42, 251–273.

Wolman, M.G. and Gerson, R. 1978. Relative scales of time and effectiveness of climate watershed geomorphology. *Earth Surface Processes and Landforms*, 3, 189–208.

Yair, A. and Lavee, H. 1982. Factors affecting the spatial variability of runoff generation over arid hillslopes, southern Israel. *Israel Journal of Earth Sciences*, 31, 133–143.

2 A Global Perspective on Dryland Rivers: Perceptions, Misconceptions and Distinctions

GERALD C. NANSON,[1] STEPHEN TOOTH[2] AND A. DAVID KNIGHTON[3]
[1] *School of Geosciences, University of Wollongong, Australia*
[2] *Institute of Geography and Earth Sciences, University of Wales, UK*
[3] *Department of Geography, University of Sheffield, UK*

2.1 INTRODUCTION

The past several decades have seen increasing interest in the river systems that drain the world's extensive hyper-arid, arid, semi-arid and subhumid regions. These regions are collectively termed 'drylands' and cover nearly 50% of the world's land area and support nearly 20% of its population (UNEP, 1992; Middleton and Thomas, 1997). Despite this increasing interest in dryland rivers, some of our understanding of dryland fluvial processes has been derived from general studies of rivers in more humid regions and then transferred to dryland settings, while knowledge gained specifically on dryland rivers has come from relatively few studies in a limited range of settings. As a result, our present understanding of dryland rivers is reminiscent of the apocryphal three blind men who, on individually touching just an elephant's trunk, side or tail, could not agree on whether they were investigating a tree, a wall or a snake! (Chorley, 1965). There is a real risk that our comprehension of dryland rivers will be generalised from a limited number of studies in specific environments and reflective of only part of the story.

This chapter assesses our present perceptions of dryland rivers, draws attention to a number of situations where misconceptions have arisen from incomplete or inappropriate evidence, and attempts to identify some of the distinctive features of rivers in drylands. The focus is on dryland river hydrology and channel forms and processes, and does not address extensive and often related research into alluvial fans, hillslopes or pediments. The specific aims here are three-fold: (1) to identify the diversity of river hydrological and geomorphological forms and processes in drylands; (2) to correct some misconceptions and oversimplifications that have arisen from attempts to generalise information on dryland rivers; and (3) to consider whether dryland rivers can be regarded as a group truly distinctive from rivers in other environmental settings. In order to achieve these aims we have found it necessary to emphasise our own work conducted in central Australia, particularly in the Channel Country of western Queensland, because so little research has been focused on very large, low-gradient rivers characterised by abundant fine sediment load. The Australian work contrasts with evidence from other drylands, mostly in steeper, smaller, coarse-load rivers. Even so, in a global context this analysis remains inadequate, for there is relatively little evidence in the scientific literature of dryland

Dryland Rivers: Hydrology and Geomorphology of Semi-arid Channels. Edited by L.J. Bull and M.J. Kirkby.
© 2002 John Wiley & Sons, Ltd.

fluvial processes from extensive areas of the Middle East, central and east Asia, South America, the Sahara and southern Africa. At the very least, this emphasis on antipodean examples highlights the diversity of dryland systems and reduces their chance of becoming typecast in a single category.

2.2 DEVELOPMENT OF RESEARCH INTO DRYLAND RIVERS

Notable early work on dryland rivers was conducted in the American southwest by earth scientists such as G.K. Gilbert, K. Bryan, L.B. Leopold, J.P. Miller and S.A. Schumm. Their findings on sediment transport, badlands, arroyos and channel changes – summarised in the classic fluvial geomorphology text by Leopold et al. (1964) – not surprisingly reflect dryland regions in the American southwest. The work was dominated by studies of relatively small rivers draining upland catchments, often where arroyo or gully formation had been of particular concern. W.L. Graf followed these studies with further research in the American southwest, and in 1988 published the only text specifically devoted to fluvial processes in dryland rivers. Graf's book incorporated work by A.P. Schick, J.B. Thornes and others in relatively small river catchments in Israel and Spain, and at much the same time I. Reid and L. Frostick were extending the geographical range of dryland river research by studying small rivers in northern Kenya. Not as widely known by geomorphologists, but respected by sedimentologists, were the studies of sand-bed rivers in central Australia by G.E. Williams. By the early 1990s, a growing body of literature suggested that dryland rivers have typically low sinuosity, wide and shallow channels and transport relatively unconsolidated sediment. These rivers were considered to be commonly braided, subject to scour-and-fill, and largely in disequilibrium (see reviews by Graf, 1988; Cooke et al., 1993; Thornes, 1994a, 1994b). As this chapter will show, however, subsequent work has demonstrated that dryland rivers are not so circumscribed but are characterised by a considerable range of conditions and forms.

2.3 DRYLAND RIVERS: DIVERSE, DISTINCTIVE OR UNIQUE?

While many more individual studies must be undertaken in a wide variety of settings before a truly global understanding of dryland rivers is achieved, it is possible from the existing literature to recognise some regional similarities as well as contrasts. However, so far there are no landform features that can be described as truly unique to dryland rivers, with the possible exception of waterholes that characterise some low-gradient, muddy systems (see Section 2.6). As Knighton and Nanson (1997) emphasised, there is in reality a continuum of fluvial environmental conditions from hyper-arid to very humid, although they stress that climate is just one criterion influencing river systems. Their review drew attention to the diversity of dryland rivers, yet, along with others researchers (e.g. Rodier, 1985; Graf, 1988; Reid and Frostick, 1994, 1997; Cooke et al., 1993; Thornes, 1994a, 1994b; Nanson and Tooth, 1999), they recognised the convenience of regarding dryland rivers as a definable group, just as other workers have tended to treat tropical (e.g. Gupta, 1993, 1995) or periglacial (McEwen and Matthews, 1998) rivers as definable groups.

The science of rivers consists of studies of hydrology, geomorphology, sedimentology, water quality and biology; however, only certain aspects of the first three are considered here. Comprehensive reviews of dryland river hydrology and geomorphology *per se* can be found in Thornes (1994a, 1994b), Knighton and Nanson (1997), Reid and Frostick (1997) and Tooth (2000a). Assessments of the use of fluvial evidence for palaeoenvironmental reconstructions in drylands have been undertaken by Reid (1994) and Nanson and Tooth (1999). These sources provide greater detail and information complementary to that presented here.

Approaches that attempt to categorise dryland rivers as a distinctive or even unique group are reminiscent of a major philosophical school in geomorphology, popular in Europe in the mid-twentieth century, known as 'climatic geomorphology' (e.g. Tricart and Cailleux, 1972; Peltier, 1975; Budel, 1980). The central thesis of this school was that the Earth's surface evolved towards a series of distinctive 'morphoclimatic regions', each one characterised by landforms reflecting regional, climate-driven processes. Such a prescriptive approach gradually fell out of favour when geomorphologists began to appreciate the extent of Cainozoic climate change. This resulted in a recognition that certain landforms in a given climatic region may be relic, or complex palimpsests partly inherited from previous climates, and that there can be equifinality in landforms evolving under different climates (Chorley et al., 1984; Summerfield, 1991). Rivers, however, adjust far more rapidly and completely to climate change than do whole landscapes. It is possible, therefore, that if climate does influence river type, dryland rivers and their associated landforms and sedimentary successions could adjust rapidly to reflect essentially contemporary climatic conditions, although the possibility of equifinality remains.

Any attempt to categorise dryland rivers as a distinctive group thus remains an intriguing but complex problem. To do so requires both an appreciation of the controls and characteristics of dryland rivers and a recognition of the extent to which these controls and characteristics are unique to dryland situations.

2.4 CLIMATE

In any assessment of dryland rivers as a distinctive group, an appreciation of the role of climate is paramount, as it is the main determinant of rainfall and runoff. Drylands as a whole are characterised by high (but variable) degrees of aridity, with aridity being promoted by various factors. Most of the world's drylands are located in the subtropics where they are under the influence of dry, stable, high-pressure air masses that inhibit the penetration of rain-bearing weather systems for a significant part of the year. On the west coast of the southern hemisphere continents (South America, Australia, southern Africa), low rainfall is exacerbated by cold ocean currents which limit evaporation and inland moisture penetration. Other drylands are located in rain shadows formed by topographic barriers (e.g. deserts in the American south-west, Tibetan plateau deserts) or in the centre of large continental land masses (e.g. central Eurasian deserts). Annual precipitation in most drylands is not only low but also highly variable, with interannual variabilities often greater than 80% (Goudie, 1987; Cooke et al., 1993), and many drylands are subject to decadal-scale cycles of above and below average rainfall (e.g. Allan, 1985, 1988; Tyson and Preston-Whyte, 2000).

Dryland rainfall is often described as being derived mainly from localised convectional storms, but there is evidence to the contrary across many drylands. For instance, in western New South Wales, Australia, of all rainfall events exceeding 5 mm, <10% are derived from localised convective storms (Cordery et al., 1983), with most rainfall being derived from a variety of general (often frontal) storms. Similarly, in northern and central Australia, extensive rainfall is usually derived from incursions of moist air from the Indo-Pacific Basin that result from prolonged southward shifts of the monsoon trough. These incursions commonly are associated with La Niña conditions (Kotwicki and Isdale, 1991; Kotwicki and Allan, 1998).

Even under conditions where annual rainfall is relatively high, aridity can result from high evaporation. Australia is the world's driest continent (with the exception of Antarctica), yet it has an average annual rainfall of 470 mm and nowhere is annual rainfall less than about

(a)

(b)

Figure 2.1 Oblique aerial photographs illustrating the broad similarity of (a) the sand-bedded Plenty River in the arid centre of Australia and (b) the sand-bedded Gilbert River in the monsoon tropics of the Gulf of Carpentaria, northern Australia

120 mm. However, average potential evaporation across the continent is 3250 mm, creating a continental average potential moisture deficit of 2780 mm. Even in northern Australia, where annual summer monsoon rainfall ranges from 500 to 1500 mm, annual potential evaporation varies from 2000 to 2600 mm. This highly seasonal rainfall runs off quickly, with much of the remainder evaporating during the dry season, so that the rivers in the monsoon tropics are dry for months at a time. These rivers are commonly wide, single-thread or anabranching sand-bed channels, more characteristic of rivers near the arid centre of the continent (see Section 2.8 below) than those of equivalent annual runoff in cooler and less strongly seasonal southern Australia (Figure 2.1).

Clearly, average climatic parameters are often not a particularly useful guide for defining the environments that are characterised by dryland rivers. Short-term variability of rainfall and runoff is more important than long-term means, and strongly seasonal monsoonal climates with high evaporation can produce landscapes with a strong dryland imprint. Climate affects more than just surface moisture availability, as it can directly and indirectly influence weathering processes and rates, associated river sediment loads, and the nature of riparian vegetation, all of which can then influence river form and process in diverse ways (see below). When these climatically influenced variables are combined with those that are independent of contemporary climate yet can also influence river form and process, such as physiography, geological structure and lithology, tectonism and palaeoclimate, it is not difficult to see why the classification of river type based on climate alone becomes a complex issue. Any characterisation of dryland rivers as a group must reflect both climate and the range of other variables with which climate may or may not interact.

2.5 RIVER HYDROLOGY

In association with climate, it is their hydrology that might be expected to set dryland rivers apart from rivers in more humid environments. However, dryland rivers themselves can be divided into two fundamentally different hydrological types. Allogenic (or exotic) rivers are sourced almost entirely from outside dryland areas, classic examples being the Euphrates River which receives almost 90% of its flow from the mountains of Turkey, and the Nile River which is supplied largely from the east African and Ethiopian highlands (Adamson, 1982; Beaumont, 1989). Such allogenic rivers commonly sustain perennial flow. Endogenic rivers are sourced almost entirely from within dryland environments and are therefore more diagnostic of truly dryland hydrological conditions. Owing to low precipitation, high evaporation rates, and depressed groundwater tables, endogenic dryland rivers are usually ephemeral or intermittent. For reasons of scale, many very large (>1000 km long) dryland rivers are commonly partly allogenic, whereas small ones are almost entirely endogenic.

In addition to such gross hydrological differences between dryland rivers, properties of floods also vary widely. Graf (1988) identified four types of flood in dryland rivers: flash floods, single-peak events, multiple-peak events and seasonal floods. While these are not mutually exclusive, flash floods and single-peak events are most common in endogenic rivers, and multiple-peak events and sustained seasonal floods most common in allogenic rivers. Despite the abundant evidence for variable hydrological conditions in dryland rivers, it is the flash flood hydrographs with their steep rising limbs, sharp peaks and steep recession limbs that have come to be regarded as typical of dryland rivers. Steep rising limbs occur in response to intense rainfall and the dominance of overland flow in runoff generation, while steep recession limbs reflect the short-lived nature of runoff and large transmission losses.

However, flash floods characterise small, steep, upland or piedmont channels. Larger, lower gradient, endogenic rivers can be very slow to respond to heavy rainfall, sometimes giving months of warning for a changing stage. For instance, floods in the Channel Country of western Queensland, Australia, commonly take 2–3 months to travel from the headwaters, through the low gradient (0.0001–0.0002 m m^{-1}) middle reaches, and to their terminus at Lake Eyre, should they even reach that point (Kotwicki, 1986).

From a study of flow data from dryland rivers on six continents, McMahon (1979) has shown that there is a more rapid downstream decline in mean peak discharge per unit area than in humid catchments. However, in dryland rivers actual peak discharges can vary in a complex way downstream. Along Cooper Creek in Australia, flood peaks typically increase in volume downstream in the headwaters because of tributary additions, but decrease in size thereafter because of transmission losses and much fewer tributaries. This is true even for infrequent, high-magnitude events (Knighton and Nanson, in press a) (Figure 2.2).

Relative flood magnitudes are very much steeper in drylands compared to humid regions, as the slopes are often established by a few, very large flood events (Figure 2.3). Owing to the very

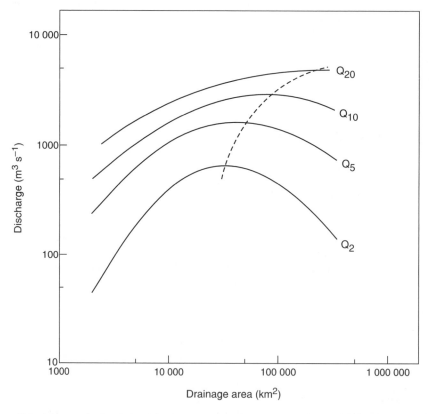

Figure 2.2 Log–quadratic relationships between the two-year (Q_2), five-year (Q_5), ten-year (Q_{10}) and twenty-year (Q_{20}) floods and drainage areas illustrating the complex relationship between drainage area and flood magnitude in the Channel Country rivers of Australia. The dashed line indicates the drainage area at which the transition from a positive to a negative relationship occurs for different magnitude events in this dryland system. After Knighton and Nanson (in press a)

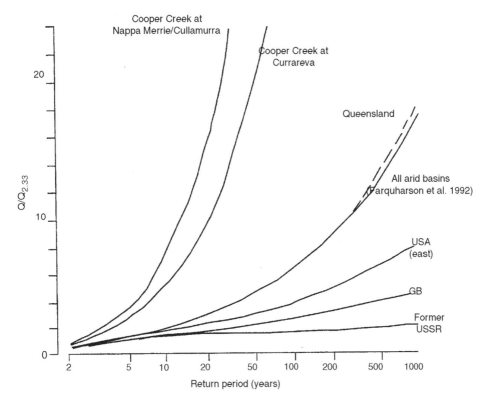

Figure 2.3 Regional flood frequency curves illustrating the large increases in relative flood magnitude $(Q/Q_{2.33})$ of dryland rivers compared to those in humid regions of the northern hemisphere. After Knighton and Nanson (1997)

similar form of flood frequency curves on endogenic dryland rivers, Farquharson et al. (1992) argue that they are representative of a single homogeneous region. Again, however, there are hydrological differences both within and between different drylands, such that Pilgrim et al. (1988) and McMahon et al. (1992) suggest that there is a greater diversity in hydrological characteristics within dryland regions than within humid ones. Such diversity can have a complex form. For instance, mean annual flow variability tends to be high in small upland catchments due to rapid runoff and limited storage, lower in medium-size catchments, but again high in very large catchments where flows must traverse great distances and infrequently penetrate far down stream. This is shown by data from the Lake Eyre basin (Figure 2.4). In a comparison of rivers in different drylands, McMahon et al. (1992) show that, for mean annual runoff, eastern Mediterranean rivers produce higher runoff per unit area of catchment than do Australian rivers, and North American dryland rivers tend to be much less variable than those elsewhere. In other words, flow variability in dryland rivers is itself variable, both within and between basins.

Cooper Creek in the Channel Country is one of the better-gauged large dryland rivers and consequently an analysis of its flow record is instructive. Upstream of Nappa Merrie the catchment area is 237 000 km^2, and during large floods flow expands up to 60 km wide across

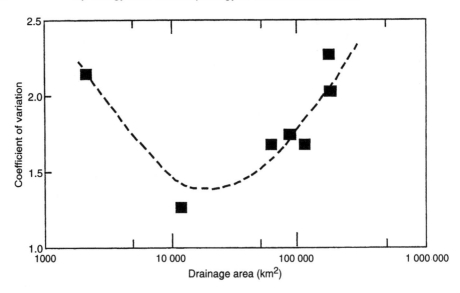

Figure 2.4   Coefficients of variation of annual flows in the Channel Country rivers of Australia, illustrating in this dryland system a strong non-linear relationship with drainage area. After Knighton and Nanson (in press a)

extensive floodplains in the middle reaches. As is not uncommon in dryland rivers, transmission losses are large. Between two gauging stations on the Cooper located ~420 km apart, transmission losses average 75–80%, but vary in a complex manner (Knighton and Nanson, 1994b). Flows with a total volume of less than 3 km³ do not usually traverse the full distance between the two stations, whereas larger flows which are confined to the primary channels are moderately efficient, transmitting about 20–50% of their discharge to the downstream station (Figure 2.5). Once flows exceed bankfull, however, there are total losses of about 80–90%, largely due to evaporation and infiltration on the wide floodplain. During the largest flood events when the floodplain becomes fully inundated and saturated, and the most direct floodways become fully active (Figure 2.6), transmission efficiency again increases sharply (Figure 2.5). Clearly, in contrast to the comparatively simple flash flood hydrographs, flood routeing across extensive dryland floodplains is a complex phenomenon dependent on gradients, sediment texture, antecedent soil moisture, soil porosity, available floodways, and channel and floodplain vegetation.

   To account for these diverse flood flow characteristics, Knighton and Nanson (in press a) advocate an event-based approach to dryland river hydrology and identify three distinctive types of event: single, multiple and compound. This approach is very suitable for ephemeral rivers because individual events are discrete, each being separated by a period of zero flow, thereby allowing a systematic comparison of different types of flood in a range of catchment sizes and dryland settings. Single events have a single peak discharge with no subsidiary peaks. Multiple events have more than one peak but on the rising limb the subsidiary peaks show a progressive increase in magnitude towards a well-defined maximum, whereas on the falling limb there is a progressive decrease in peak heights. Compound events consist of multiple peaks which are not symmetrically distributed by height around the highest peak. From an analysis of such events in small to very large catchments in the Channel Country, Knighton and Nanson (in press a) show that single events have significantly smaller magnitudes, shorter durations and

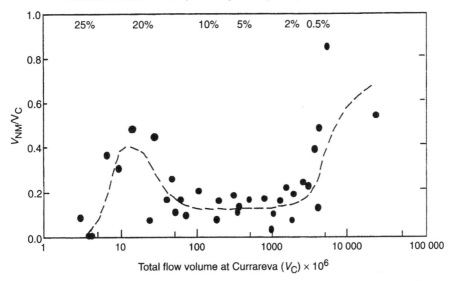

Duration of maximum daily discharge during flow period at Currareva

Figure 2.5   The total flow volume output at Nappa Merrie (V_{NM}) as a ratio of the total input at Currareva (V_C) plotted against the total input for a ~420-km reach of Cooper Creek. The sharp rise in the initial flow efficiency of the reach in the early stages while flow remains in bank contrasts with the marked drop in efficiency when flow extends overbank. Note that there is another marked increase in flow efficiency during very large floods (after Knighton and Nanson, 1994b)

Figure 2.6   An oblique aerial photograph showing overbank floodways crossing a main anastomosing channel on Cooper Creek, Australia. On sparsely vegetated dryland floodplains, downvalley overbank flow can be a major contribution to flood routeing. Flow is from top to bottom

more rapid time-to-peaks than do other types, whereas compound events have larger magnitudes, longer durations and slower time-to-peaks. Multiple events, although more akin to compound events than single ones overall, tend to occupy a middle ground, behaving more like single events at smaller runoff volumes and in smaller catchments, but more like compound ones at the other end of the scale. Generally, there is a progressive increase in magnitude, duration and time-to-peak, and a progressive decline in the level of predictability, from single to compound events. The extent to which this is true in other dryland catchments still needs to be assessed, but this event-based approach will enable direct comparisons between widely different dryland systems.

The speed of floodwaves travelling down small dryland rivers has been shown to be directly related to discharge, with large floods travelling most rapidly (Ben-Zvi et al., 1991; Sharma et al., 1994; Sharma and Murthy, 1998). This is largely true in the upper Cooper catchment (Figure 2.7(a)), but further downstream the speed of floodwaves varies non-linearly with increasing discharge (Figure 2.7(b)). Floodwave speed peaks at about two-thirds bankfull

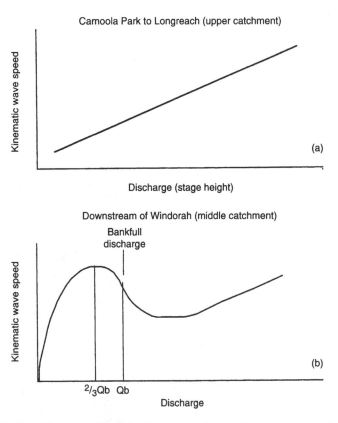

Figure 2.7 A schematic diagram illustrating the progressive log–linear increase in floodwave speed with discharge (estimated from stage height) in the upper reaches of the Thompson River (essentially upper Cooper Creek), and the suggested non-linear relationship between wave speed and discharge in the middle reaches of Cooper Creek downstream of Windorah. This illustrates the complexities of comparing smaller upstream with larger downstream dryland catchments. After Knighton and Nanson (in press b)

stage when the channel offers the least resistance, but speed declines as bankfull is approached, probably due to the increased resistance offered by riparian vegetation on the upper part of the banks. It continues to decline towards a minimum just above bankfull flow when channel–floodplain flow interactions reach a maximum (cf. Knight and Shiono, 1996) and high infiltration losses occur on the floodplain, but rises again as larger floods drown out these effects (Knighton and Nanson, in press a, in press b) in a manner similar to rivers in more humid regions. This pattern of change illustrates how, at one scale, dryland catchments may exhibit characteristics different from humid catchments, while at another they may be similar.

Clearly, there is a wide range of hydrological conditions in dryland rivers. Such diversity is reflected in terms of the almost continuous spectrum of flow occupancy, from rivers in hyper-arid deserts that sometimes remain dry for decades, through to allogenic rivers that sustain perennial flow (Figure 2.8). As with climate, therefore, this diversity makes it difficult to classify dryland rivers as a distinctive group on the basis of hydrology alone. Some hydrological characteristics are more common in dryland rivers than in rivers in other environments (e.g. large-scale transmission losses, large relative flood magnitudes), while other characteristics are shared with many rivers in more humid environments (e.g. non-linear variation of floodwave speed with stage, strong channel–floodplain flow interactions).

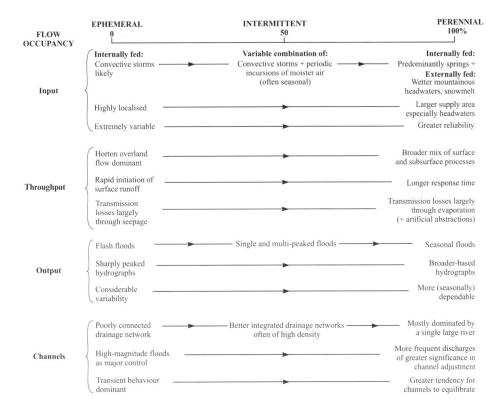

Figure 2.8 Diversity in hydrological input, throughput and output, and in channel characteristics in dryland rivers, expressed in terms of a linear flow occupancy scale. From Knighton and Nanson (1997)

2.6 CHANNEL GEOMETRY

As hydrologic regimes vary widely both within and between drylands, it might be expected that channel geometries would also vary widely. An early study of channel geometry in ephemeral rivers in New Mexico (Leopold and Miller, 1956) showed how width, depth and velocity all increase downstream with increasing catchment area (Figure 2.9(a)). Using data compiled from a variety of dryland rivers, Wolman and Gerson (1978) extended these findings to show how width initially increases rapidly until catchments reach an area of about $50-100 \text{ km}^2$, following which width stabilises at about $100-200 \text{ m}$, even for catchments up to thousands of square kilometres (Figure 2.9(b)). Wolman and Gerson (1978) argued that, owing to a common lack of riparian vegetation that otherwise provides bank protection and aids the processes of channel recovery, dryland rivers expand in width to accommodate the largest floods and, following these floods, do not readily reduce their width but remain in an expanded condition. They suggested that the zone of constant width in larger catchments (Figure 2.9(b)) reflects limited storm-cell sizes ($10-100 \text{ km}^2$) that rarely produce simultaneous runoff from larger catchments, the decreasing probability of bank erosion by low flows as channels grow wider, and increased infiltration losses that occur in wide, shallow channels.

These findings are frequently cited as typical of downstream changes in dryland channels (e.g. Reid and Frostick, 1997) but subsequent studies have revealed much greater diversity in channel geometry. Limited storm-cell sizes are apparent in only some areas and in other areas only partly drive the rivers (see Section 2.4 above), and riparian vegetation along some dryland rivers can provide very effective bank protection during even the largest floods (see Section 2.9 below) or aid in the process of channel recovery following destructive large floods. From experience in the Australian drylands, Mabbutt (1977) recognised that many rivers actually decrease in channel size downstream in response to transmission losses – a trend also shown in several subsequent studies in Australia and South Africa (e.g. Dunkerley, 1992; Tooth, 1999a, b; Tooth et al., in press) (Figure 2.9(c) and (e)). Channel size decreases downstream in some dryland rivers and this can be exacerbated by channel avulsions and the formation of splays (Figure 2.9(e)), and in extreme cases continuous channels cannot be maintained; instead, occasional flows spill across unchannelled 'floodouts' (Tooth, 1999a, b) (Figure 2.10). Floodouts are common to many channels in the drylands of Australia and southern Africa, and occur across a wide range of scales (c. $1-1000 \text{ km}^2$). In addition to downstream reductions in discharge, floodouts result from barriers to flow acting singly or in combination. These are aeolian barriers (e.g. where dunes have migrated across a river's course – see Section 2.11 below), hydrologic/alluvial barriers (e.g. where a trunk channel in flood results in ponding and sediment deposition on an adjacent tributary), and structural barriers (e.g. where resistant bedrock outcrops cross the path of a river). In certain instances, the unchannelled flows persist through the floodout and concentrate into small 'reforming channels', which sometimes form or join a larger river or which disappear in another floodout (Tooth, 1999a; Tooth et al., in press) (Figure 2.10).

Figure 2.9 (Right) The variable nature of channel geometry and hydraulic geometry changes in dryland rivers: (a) downstream hydraulic geometry changes in New Mexico rivers (Leopold and Miller, 1956); (b) downstream non-linear increases in channel width for a range of dryland rivers compared to linear increases in humid rivers (Wolman and Gerson, 1978); (c) downstream decline in channel width near the Barrier Ranges, western New South Wales (Dunkerley, 1992); (d) triaxial diagram of at-a-station hydraulic geometry exponents showing the lack of discrimination between humid and semi-arid rivers (Park, 1977); (e) downstream increases and decreases in channel width along the Woodforde River in central Australia. Downward-directed arrows indicate tributary inputs and upward-directed arrows indicate splay or distributary losses (Tooth, 2000b)

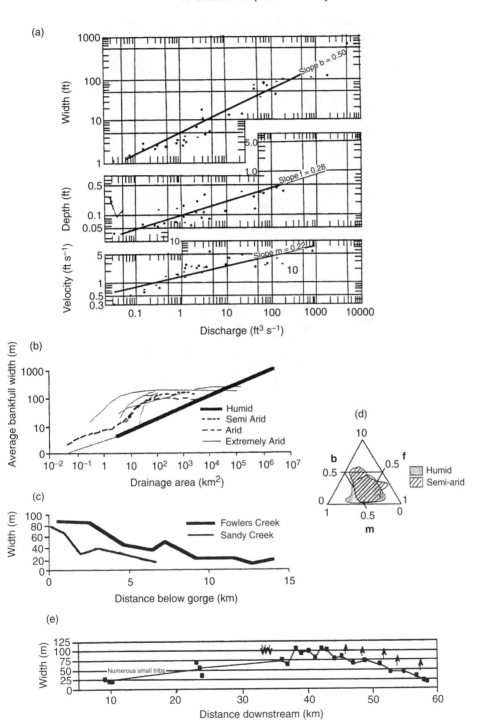

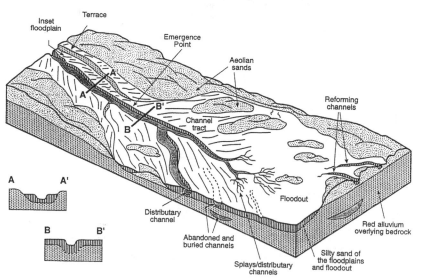

Figure 2.10 An oblique aerial view of the Sandover River floodout (flow away from the camera), and a block diagram illustrating a floodout and related landforms on the Northern Plains of central Australia. From Tooth (1999b)

While floodouts are characteristic of some dryland rivers, many alluvial fans in more humid settings are also characterised by the downstream disappearance of channelised flow, such that the process may not be entirely unique to drylands. From research to date, possibly the only large channel features that appear to be restricted to dryland or seasonally dry monsoonal settings are waterholes, self-scouring sections of channels or floodplains that maintain a perennial supply of water in otherwise ephemeral or intermittent systems. Although waterholes have a limited distribution overall, they have been described from a variety of physiographic settings in the Australian drylands, such as from rocky gorges and floodouts (Argue and Salter, 1977; Tooth, 1999a). The best studied examples, however, occur as relatively wide, deep sections of channel that characterise some of the low-gradient, muddy, anastomosing rivers in the Channel Country (Knighton and Nanson, 1994a, 2000) (Figure 2.11). They can also

(a)

(b)

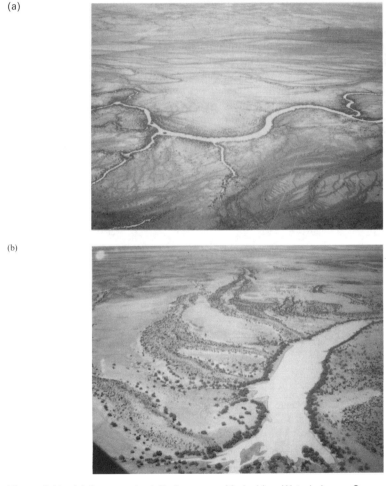

Figure 2.11 (a) Long and relatively narrow Meringhina Waterhole on Cooper Creek formed by coalescing anastomosing channels. Flow is from right to left. (b) A short, broad waterhole on the Diamantina River formed by flow constriction between obstructing dunes (left) and the valley side (right). Flow is from bottom to top right

occur as isolated features on muddy floodplains away from the main channels. Along Cooper Creek, Knighton and Nanson (1994a) identified three types: (i) Those formed between constricting aeolian dunes are relatively short, wide and shallow scour features that usually terminate as soon as the source of the constriction ceases (Figure 2.12). (ii) Those flanked on one side by the valley margin or a single dune that acts to focus flow energy. These are usually longer and deeper than those confined between dunes and most are part of the primary system of through-flowing channels. (iii) Those that are unrestricted but usually where there is the confluence of two or more anastomosing channels (Figure 2.12). In all cases, waterholes are very long-lived features that focus scour into a cohesive, muddy floodplain, commonly exhibiting contemporary splays of mud enriched with sand at their downstream ends. Their fixed position in the channel or floodplain helps to maintain both the stability of the multiple channel pattern and the continuity of flow in a hydrologically variable environment. This, along with their occurrence at points of flow convergence, is indicative of a contemporary origin with their dimensions in equilibrium with the present flow regime rather than as evidence of inheritance from earlier Quaternary flow conditions (Knighton and Nanson, 2000). Their ecological significance with the provision of permanent aquatic habitats in an otherwise arid landscape is particularly important.

Floodouts and waterholes aside, however, it is debatable whether on average the channel geometries of dryland rivers are distinctly different from rivers in more humid environments. Despite major variations in climate and hydrology from region to region, the hydraulic properties of running water and its influence on sediment transport hardly change, so rivers in different settings with similar sediment types still exhibit many similar geomorphic characteristics in response to changes in discharge and bank strength. Indeed, a summary of several

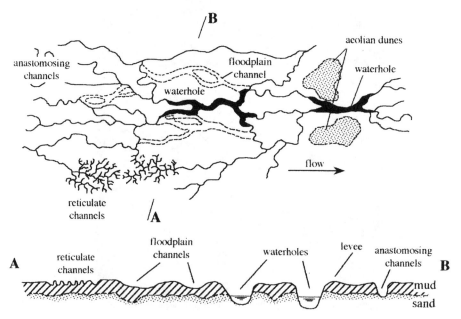

Figure 2.12 A schematic diagram of the Cooper Creek floodplain showing the interconnected nature of reticulate channels, floodplain channels (floodways), anastomosing channels and waterholes. Note the waterhole formed by coalescing anastomosing channels and that formed by flow constriction between aeolian dunes. From Tooth and Nanson (2000b)

at-a-station hydraulic geometry studies by Cooke et al. (1993, see their Table 11.1) shows no particular difference between ephemeral and perennial rivers, at least among those they selected for comparison, and tri-axial graphs show that the exponent sets for dryland and humid channels commonly overlap (Park, 1977) (Figure 2.9(d)). In the case of at-a-station hydraulic geometries, local effects like boundary materials and bank stability are usually largely independent of climate (for exceptions see Section 2.10 below), and tend to dominate over regional effects like climate. In the case of downstream hydraulic geometries, some trends such as downstream size decreases are perhaps more common in drylands, but they are not common to all dryland rivers and can also occur along rivers in more humid settings such as in southeastern Australia (e.g. Nanson and Young, 1981; Tooth and Nanson, 1995; Woodfull et al., 1996).

2.7 SEDIMENT TRANSPORT

Dryland regions typically support sparse, unevenly distributed or temporally variable hillslope vegetation covers, which is usually thought to promote additional hillslope sediment supply. Hillslope sediment yield influences the supply of sediment to valley floors and rivers, and measurements have shown that many dryland rivers transport large quantities of sediment during flood events, both as suspended and bedload (Sharma and Murthy, 1994; Laronne and Reid, 1993; Reid and Laronne, 1995; Reid, Chapter 4 in this volume). These high sediment loads occur partly because of the ready availability of sediment but also because infrequent flows greatly concentrate the period during which this available sediment is transported (Cooke et al., 1993). Suspended loads up to 100 g l^{-1} have been recorded in some dryland rivers (e.g. Reid, Chapter 4 in this volume) but such high concentrations of sediment usually cannot be sustained for long periods because available sediment becomes exhausted as runoff progresses. Sediment exhaustion can occur over a single flood cycle or perhaps over an entire flow season.

Interesting bedload measurements have been obtained from ephemeral rivers in Israel and show transport rates several orders of magnitude higher than those obtained for the same discharges in more humid rivers. For a given shear stress, these ephemeral rivers also appear to be much more efficient transporters of bedload than humid rivers (Laronne and Reid, 1993; Reid and Laronne, 1995; Reid et al., 1998; Reid, Chapter 4 in this volume). These differences have been attributed principally to the typically higher degree of armouring in humid rivers which tends to limit sediment availability at the onset of flood flows and the relative lack of bed armour in dryland rivers (Laronne and Reid, 1993; Reid and Laronne, 1995; Reid et al., 1998).

These studies offer insights as to the often highly sensitive nature of sediment transport in dryland rivers which may have implications for river response to future changes in runoff. At this stage, however, it is premature to extend the Israeli evidence to a wide range of dryland rivers. First, the measurements in Israel were made in relatively small, steep, upland reaches with gullies proximal to the channels that feed abundant sediment, conditions not necessarily representative of larger piedmont or lowland reaches. Second, the Israeli data were obtained during flash floods from sediment traps that filled during the first few minutes of a floodwave. The traps could not be reused to check for hysteresis or exhaustion effects at later times during the same flood, yet in dryland rivers flowing just above or on bedrock, or where flood durations are much longer than in the Israeli streams, bedload sediment may be exhausted during the passage of a floodwave.

Sediment load data from larger, lower gradient dryland catchments are rare. A detailed study over more than a decade along the continental-scale, low-gradient channels of Cooper Creek, however, shows sediment transport conditions at the opposite end of the range to those obtained from the Israeli rivers. The Cooper transports very little sand and no gravel (Nanson et al., 1986, 1988; Rust and Nanson, 1986, 1989), with much of its bedload in the middle reaches consisting of pelleted mud aggregates reworked from vertisols on the floodplains (Nanson et al., 1986). Individual mud pellets behave as low-density, fine sand-sized aggregates and form ripples and dunes (Maroulis and Nanson, 1996). Transport rates in the anastomosing channels and on the floodplain are not known in detail, but attempts to obtain bedload measurements in both using a Helly–Smith sampler during the 1990 and 1997 floods showed rates to be extremely low; in fact, less than the ability of the sampler to measure. Suspended load measurements were obtained for both these floods up to bankfull and range from 0.5 to 1.5 g l^{-1}. For a mud-dominated river system, these are extremely low values, and highlight the important control played by aggregates in limiting yields of muddy alluvium. Similarly, while Leopold and Miller (1956), Schick (1970) and Thornes (1976a) (see Cooke et al., 1993) report increases in sediment concentration downstream in small dryland rivers, it is difficult to see how this downstream increase could be maintained in large dryland rivers such as Cooper Creek.

The sharp contrasts between the sediment load measurements obtained by Laronne, Reid and others in Israeli ephemeral rivers (Laronne et al., 1994), and those obtained in the Channel Country serve to highlight the extreme variability of sediment transport characteristics in dryland rivers of vastly different size and gradient. If the impact of different intensive landuse practices on hillslope sediment yields is taken into account, going back over millennia in the case of some Mediterranean and Israeli catchments, it is clear that there are many factors within a dryland climate that can influence river sediment loads.

It is also open to question whether hillslope sediment yields and river sediment transport conditions are distinctly different in drylands as compared to other environments. In oft-cited research, Langbein and Schumm (1958) and Schumm (1965) suggested that, in small to medium catchments (<4000 km^2) (Cooke et al., 1993), sediment yields reach a maximum when mean annual effective precipitation is between 250 and 350 mm, and that yield falls off sharply towards areas of more or less effective precipitation (Figure 2.13(a)). Langbein and Schumm (1958) explained this pattern as resulting from limited runoff availability in drier areas which reduces sediment yield, while greater vegetation growth in wetter areas increases surface protection and reduces sediment yields. As with many other broad generalisations relating to distinctions between dryland and humid environments, however, this observation is now seen to be an over-simplification related to the limited evidence available at the time (Cooke et al., 1993). Reviews by Walling and Kleo (1979) and Walling and Webb (1983) of much larger global datasets show instead that annual sediment yields in many drylands are not distinctively high compared with yields from other more humid areas (Figure 2.13(b)). Similarly, many other aspects of sediment transport are not unique to dryland rivers and are shared with rivers in other environments. High suspended sediment loads typical of some dryland rivers are also common in many proglacial rivers (e.g. Gurnell, 1987). Seasonal declines in sediment availability, characteristic of some dryland rivers, also have been shown in dramatic form for both suspended and bedload in humid alpine environments (e.g. Nanson, 1974; Hayward, 1980). Furthermore, while many ephemeral rivers do not display well-developed armouring, but perennial rivers in humid regions typically do, these are tendencies only. There is probably no *a priori* reason why well-developed armour layers cannot develop in dryland rivers with more sustained or perennial flow regimes. Clearly, many more studies are needed before sound generalisations regarding the distinctive aspects of sediment transport in dryland rivers can emerge.

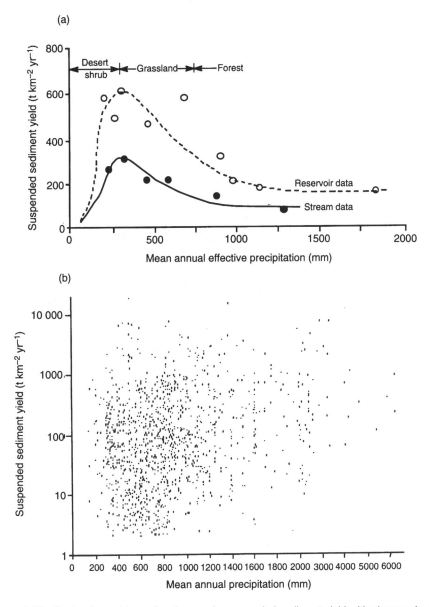

Figure 2.13 Contrasting evidence for changes in suspended sediment yield with changes in mean annual precipitation. (a) Langbein and Schumm (1958); (b) Walling and Kleo (1979)

2.8 RIVER PATTERNS AND FLOODPLAINS

From research conducted largely on arroyo-style rivers in the American southwest, a misconception arose that many dryland rivers do not have conventional floodplains (e.g. Graf, 1988, p. 217). This may have come about because of two other misconceptions: first, that meandering rivers are largely responsible for forming floodplains and that braided rivers rarely do; and, second, that braiding is the most common pattern in dryland areas (Slatyer and Mabbutt, 1964; Mabbutt, 1977; Graf, 1988). However, while traditional braided rivers *with floodplains* are common in some dryland settings, single-thread, planar bed channels are common in others (Tooth, 1997, 2000b) (Figure 2.1). Both appear to be a response to a relative abundance of gravelly or sandy bedload and sometimes relatively erodible banks, and are therefore more typical of upland or piedmont locations. In lowland dryland settings, where sediment is cohesive and where riparian vegetation forms stable banks, rivers flow across extensive floodplains forming braided floodways (Figure 2.14(a)), reticulate distributary channels (Figure 2.14(b)), anastomosing channels and meandering channels (Figure 2.14(c)). In sandy settings, ridge-form anabranching channels can also form (Figure 2.14(d)). In other words, a wide range of river patterns and floodplain types can be represented in dryland areas.

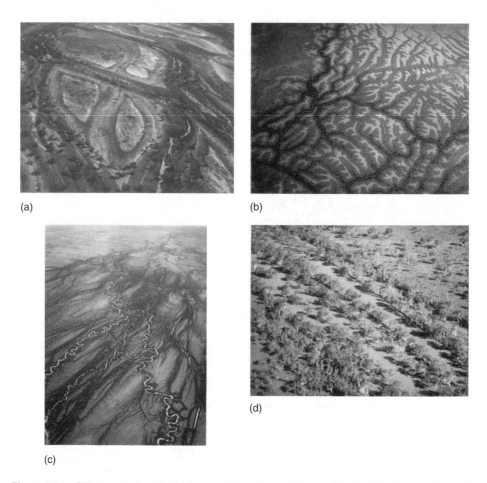

(a)

(b)

(c)

(d)

Figure 2.14   Oblique air photos of (a) braided floodways (Cooper Creek), (b) reticulate channels (Diamantina River), (c) meandering and anastomosing channels (Cooper Creek) and (d) ridge-form anabranching channels (Bundey River)

In semi-arid to subhumid eastern South Africa, meandering rivers are widespread, commonly occurring in relatively erodible sandstone or shale valleys upstream of highly resistant dolerite sills and dykes (Tooth et al., submitted). Wide, bordering floodplains marked by numerous oxbows and abandoned channels host substantial wetlands in these seasonally dry settings. In the Channel Country of Australia, Rust (1981) was the first to recognise anabranching as an important dryland river pattern, with later detailed descriptions of anabranching rivers in a wide variety of drylands being provided by Nanson et al. (1986, 1988), Rust and Nanson (1986), Schumann (1989), Knighton and Nanson (1994a, 1994b), Gibling et al. (1998), Makaske (1998), Wende and Nanson (1998), Taylor (1999), and Tooth and Nanson (1999). In the Channel Country, some wider and deeper sections of the anabranching channels and floodplain host waterholes (see Section 2.6). Distributary patterns have been described commonly for alluvial and terminal fans, but also for river systems subject to downstream declining discharges and gradients (Sullivan, 1976; Mabbutt, 1977; Tooth, 1999a, 1999b, 2000b).

These aerially extensive and diverse examples illustrate just how difficult it is to generalise globally about dryland river patterns and their associated floodplains. In Australia, where up to 80% of the continent is classified as arid or semi-arid, there are few examples of classical braided river patterns that occur in more than just local reaches, whereas single-thread, planar-bed rivers and anabranching are widespread. The preponderance of anabranching in Australia may have much to do with that continent's low gradients, fine-grained bank sediments and well-developed riparian vegetation (Nanson and Huang, 1999). While anabranching is not common in all drylands, it is clearly a misconception that braiding is the dominant river pattern in drylands as a whole. Furthermore, none of these river patterns is especially unique to drylands. The meanders visible in Figure 2.14(c) are not noticeably different from those in, say, the Canadian Arctic. Steep braided rivers in humid regions have many geometrical similarities with braided rivers in dryland regions. The anastomosing Rhine and Meuse Rivers in humid Europe exhibit pattern characteristics very similar to those of the anastomosing Niger River in the West African drylands (Makaske, 1998). Clearly, while different combinations of variables such as hydrology, gradient, sediment load and type, and vegetation can produce very different river patterns and floodplains in climatically-similar settings, similar fluvial forms can occur in widely different settings.

2.9 RIPARIAN VEGETATION

As recognised by Hickin (1984), the role played by vegetation is probably one of the least well understood major influences on river channel form and process, largely because of the difficulties in measurement and quantification of the influence. A particular problem is to establish the influence of vegetation on flow properties or on the erosional resistance of channel boundaries. Riparian vegetation along dryland rivers is commonly described as sparse (e.g. Leopold and Miller, 1956; Thornes, 1977; Clark and Davies, 1988) and as a result such rivers are often considered to be highly susceptible to major flood erosion from which it may take many years to recover (Wolman and Gerson, 1978). This is certainly true in some dryland environments where aridity is severe and surface flows infrequent, or where groundwater is too deep to be available to most vegetation. However, where surface flows are frequent enough to offer adequate irrigation, and where groundwater is maintained nearer the surface, bankline and in-channel growth of vegetation can be shown to be a major factor contributing to either channel instability (Hadley, 1961; Graf, 1978, 1981; Griffin et al., 1989) or channel stability (Kondolf and Curry, 1984, 1986; Tooth and Nanson, 1999), depending on the degree and nature of vegetation growth.

Some of the most detailed research on the impact of vegetation has been concerned with the invasion of *Tamarix* along rivers in the American southwest. *Tamarix* invasion displaced many native vegetation types and increased tree densities, which resulted in channels narrowing, becoming much more sinuous and sometimes avulsing (Hadley, 1961; Robinson, 1965; Graf, 1978, 1979, 1980, 1988; Blackburn et al., 1982). As *Tamarix* is an exotic, invasive species introduced from the Mediterranean, its interrelationship with dryland rivers in the United States is not especially useful in evaluating the role of natural vegetation in channel adjustment. By contrast, Australia is a continent that has been progressively drying since the middle Tertiary and as a consequence has many plant species that have evolved to take advantage of groundwater available beneath commonly dry river beds. While much of Australia supports sandy deserts, muddy alluvium also abounds, particularly along the rivers in the extensive Channel Country. Much of the continent is also of very low gradient and consequently slowly-moving floodwaves are not especially erosive. These floodwaves can provide prolonged inundation of riparian land (Knighton and Nanson, 1994a, 1994b and in press) after which the muddy alluvium retains some moisture. Species of *Eucalyptus, Melaleuca* and *Acacia* trees, and a variety of shrubs, forbs, sedges and grasses, have adapted to these conditions and provide a sometimes dense array of vegetation types along many Australian dryland rivers (Figure 2.15). The result is a continent where natural riparian vegetation has a profound impact on the character of its dryland rivers.

Numerous examples of the influence of riparian vegetation on Australian dryland channels have emerged over the last few decades. In a detailed study of densely treed ephemeral channels in New South Wales, Australia, Graeme and Dunkerley (1993) found that the roughness generated by trees in the channel approached that generated by the boundary itself. They also showed that roughness tended to increase with stage, unlike most humid environment channels where roughness generally decreases with increasing water depth, at least until overbank flows encounter the high resistance of bank-top and floodplain vegetation. Woodyer et al. (1979) found that teatrees (*Melaleuca linariifolia* Sm.) establish on the bed of muddy rivers during low-flow periods and initiate the formation of bank-attached benches. Wende and Nanson (1998) have shown how the growth of paperbark trees (*Melaleuca* spp.) is an important factor in the development of ridge-form anabranching channels in northwestern Australia, and Tooth and Nanson (1999) have described similar characteristics for rivers supporting river red gum trees (*Eucalyptus camaldulensis*) in more arid parts of central Australia. In this latter study, tributaries were shown to foster the development of ridge-form anabranches immediately downstream of their junctions by providing more frequent irrigation of the bed of the trunk river, which encourages the growth of trees. These trees offer considerable flow resistance (Graeme and Dunkerley, 1993) and are able to withstand inundation during the occasional floods, promoting sediment deposition on their downstream side. Over time, these sediment deposits build up in association with vegetation colonisation, eventually forming long, narrow ridges separating adjacent narrow anabranches. Ridge formation provides a location for trees above and separate from the sandy-floored anabranches (Figure 2.14(d)). These trees offer very substantial protection to banklines or ridges, with numerous large, flood-scarred trees providing evidence of long-term bank and ridge stability. In a related study of the role of inland teatrees (*Melaleuca glomerata*) in the formation of ridge-form anabranches in central Australia, Tooth and Nanson (2000a) describe a sequence of adjustment from: (i) trees colonising a flat river bed and causing high flow resistance; (ii) the development of numerous low-resistance, streamlined lemniscate islands colonised by trees; (iii) the alignment of islands to reduce obstruction; and (iv) the coalescing of islands to form well-defined ridges separating anabranches that are free of trees and offer minimum flow resistance (Figure 2.16).

Figure 2.15 An aerial and ground view of riparian vegetation in separate anabranches of the Diamantina River, Australia

Clearly, in some drylands there is a diverse range of complex interactions between rivers and associated riparian vegetation. Many aspects of dryland river–vegetation interactions are similar to rivers in more humid environments, where attention also has been given to the influence of in-channel and bankline vegetation on flow resistance, bank strength and channel morphology (e.g. Petryk and Bosmajian, 1975; Smith, 1976; Watts and Watts, 1990; Hey and Thorne, 1986; Gregory and Gurnell, 1988; Huang and Nanson, 1997, 1998). Subtle differences may still exist, however, for as Tooth and Nanson (2000a) suggest, the vegetation influence may

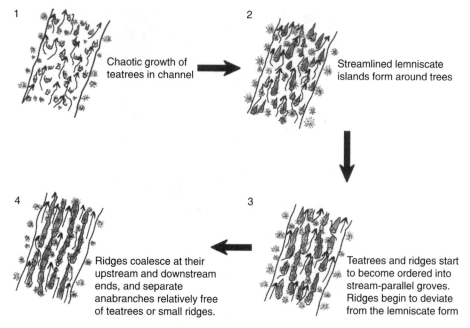

Figure 2.16 The sequence of island to ridge formation on the Marshall River in central Australia. After Tooth and Nanson (2000a)

differ between rivers with ephemeral or intermittent flow, and those with perennial flow. In channels that are dry year-round or seasonally, vegetation sometimes establishes on channel beds and can directly initiate the formation of various depositional features such as bars, benches, ridges or islands (e.g. Hadley, 1961; Graf, 1978; Woodyer et al., 1979; Pickup et al., 1988; Wende and Nanson, 1998; Tooth and Nanson, 1999). In contrast, as Fielding et al. (1997) have noted, vegetation in perennial rivers tends to respond to sedimentation, typically only colonising bars or islands that are sufficiently elevated to be protected from regular inundation (Nanson and Beach, 1977; Hupp, 1990; Hupp and Osterkamp, 1985, 1996).

2.10 ALLUVIAL INDURATION

Where dryland rivers experience seasonal flow regimes, such as in monsoonal or Mediterranean climates, there can be sufficient warmth and moisture for pronounced chemical alteration of bedrock and alluvium. Alteration is enhanced where the bedrock or alluvium is composed of relatively reactive lithologies, such as igneous rocks or carbonates. Chemical weathering releases certain substances in solution which are transported along moisture gradients and may be re-precipitated locally to form cemented or indurated horizons of ferricrete, silcrete, calcrete, gypcrete or halite, depending on local conditions of aridity and pH. Such induration, and in extreme cases lithification, of alluvium can produce hardened channel boundaries, islands and benches that greatly influence river forms and processes. In the upper Inland Niger Delta, for instance, there is evidence of substantial ferruginous and manganiferous induration of alluvium which greatly reduces lateral channel migration (Makaske, 1998).

Similarly, in northern Australia, many alluvial successions have become lithified during the late Quaternary as a result of ongoing cementation processes. On the Gilbert River in the Gulf of Carpentaria, a combination of thermoluminescence dates on alluvial sediment and uranium series dates on pedogenic minerals shows periods of alluviation to have been closely followed by induration by both calcrete and ferricrete (Nanson et al., 1991). Iron and manganese oxyhydroxides and calcite precipitation have been essentially contemporaneous in this wet–dry climate, in some cases with nodules of all three minerals intertwined. The Gilbert River is now partially confined between banks of its own irregularly lithified deposits and as a result flows in places within rock gorges up to 10 m deep and over waterfalls 4–6 m high that extend across the entire channel. There is evidence of knickpoint retreat with collapsed and tilted blocks of lithified alluvium downstream of these waterfalls (Figure 2.17). Flow over these lithified sediments has produced scalloped and fluted surfaces similar to those eroded by high-velocity flow over harder rock types.

In more arid settings, silcretes, calcretes and gypcretes can produce similar effects. In the rivers of the Okavango Delta and Makgadikgadi basin in Botswana, sandy bars and sheets cemented by silica and calcite provide resistant channel boundaries (McCarthy and Ellery, 1995; Shaw and Nash, 1998). Along the margins of the Dead Sea in Israel, halite horizons form extensive units that influence the form of river channels as they incise to keep pace with lowering base levels. Clearly, some dryland river systems are not free to adjust their geometry as can truly alluvial channels but instead are subjected to partial confinement within their own indurated or lithified alluvium. As can occur in bedrock (Heritage et al., 1999; Tooth and McCarthy, submitted), these rivers may incise to form shallow gorges, or their channels may relocate by avulsion to form anabranching or distributary patterns.

Figure 2.17 Indurated alluvium on the Gilbert River, Gulf of Carpentaria, Australia. Calcrete and ferricrete have indurated the channel floor through which a knickpoint has migrated upstream leaving small gorges and rapids. Note the figure for scale

While induration and lithification of alluvium can occur in other (particularly tropical) environments, it appears to be common in drylands where the dominance of evaporation over rainfall readily promotes re-precipitation of minerals in solution. Indurated alluvium is characteristic of some dryland rivers; however, balanced assessment of its spatial extent and importance relative to other variables influencing fluvial form and process awaits much more detailed investigation.

2.11 RIVER AND AEOLIAN DUNE INTERACTIONS

A distinctive characteristic of some dryland rivers is that they interact with adjacent aeolian dunefields. An example of river–aeolian interaction occurs where Cooper Creek emerges from the Innamincka Dome in South Australia to confront the northern-trending dunes of the Strzelecki Desert (Figure 2.18). Here, the irregular flood flows twist and turn through the

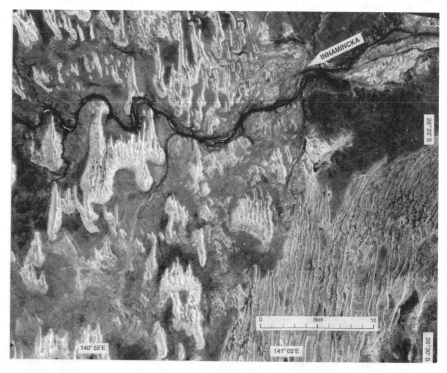

Figure 2.18 Cooper Creek emerging from the bedrock confines of the Innamincka Dome (the dark area north and south of the township of Innamincka) in South Australia. This area illustrates a long history of river and aeolian dune interaction (Wasson, 1983; Coleman, in prep.). The present course of Cooper Creek is the well-defined dark channel meandering to the west. In the past it flowed across much of the western half of the image, leaving a series of mid- to late-Quaternary palaeochannels and juxtaposed source-bordering or transverse dunes. During the last glacial maximum and Holocene, linear dunes have been deflated northwards from these, interacting with the present and previous channels of Cooper Creek. The extensive dunefield to the southeast was not derived from Quaternary alluvium (Wasson, 1983). Photography modified from a 1996 Landsat TM image, suppled by Santos Pty Ltd and South Australian Department of Mines and Energy

dunefield and a continuous channel disappears, becoming instead a series of floodbasins, disjunct floodplains and playas, before reforming further westward on its course to Lake Eyre (Wasson, 1983). Luminescence dating shows that during the late Quaternary, source-bordering transverse dunes were deflated from seasonally dry sections of river. During the last glacial maximum, southerly winds deflated sand from these transverse dunes, extending linear dunes northwards across the path of Cooper Creek (Figure 2.18), a process that has continued through the Holocene to the present (Coleman, in prep.). The present river is in places deflected around the northern tips of such dunes, while in other places is reworking them. The result is a situation where sand is deflated from the river bed to form large dunes, only to be eroded and reworked by the river during wetter phases (Figure 2.11(b)).

A similar situation has been described in some detail by Makaske (1998) for the upper Inland Niger Delta in central Mali. Here, Pleistocene and Holocene climate changes resulted in substantial interaction between the river and adjacent aeolian dunes, with blocking dune cordons later being breached by rivers. In the middle Holocene, as climate became wetter, stream powers increased but sand supply from the vegetating dunes declined, causing the rivers to incise. In the late Holocene, a drying climate and increased human activity again enhanced sand supply, causing the rivers to aggrade.

In the western Simpson Desert in Australia, Nanson et al. (1995) describe the Finke River truncating the northward-trending regional dunefield. At least three phases of source-bordering dune formation have been identified as resulting from deflation of the river and floodplain over the period of the last full glacial cycle (Figure 2.19). Dune orientations have recorded changes in wind direction over this period, with the latest Holocene dunes corresponding closely with the modern wind direction.

Other examples of river–aeolian interactions on a range of spatial and temporal scales have been provided from numerous drylands world wide, such as the Mojave Desert in California (e.g. Lancaster, 1997; Clarke and Rendell, 1998), the Namib Desert in Namibia (e.g. Ward and von Brunn, 1985; Lancaster and Teller, 1988), and the northern Kalahari in Botswana (e.g.

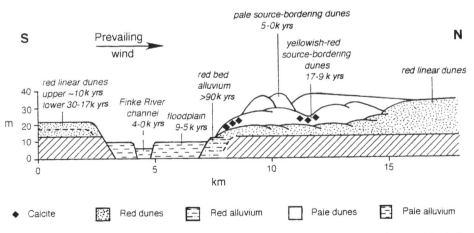

Figure 2.19 A schematic section across the Finke River valley and associated dunes in central Australia showing the stratigraphy and chronology of source-bordering dunes. Diagonal shading is bedrock. The regional (red) linear dunes north and south of the river are older than the last glacial maximum but have been reworked in places by wind. Since the last glacial maximum, pale source-bordering transverse dunes have been deflated northwards from Finke River alluvium, overwhelming the linear (red) dunes proximal to the northern valley margin (Nanson et al., 1995)

Thomas et al., 2000). While the nature of these interactions varies greatly depending on local circumstances, thus making generalisations difficult, aeolian processes are a feature common to many dryland rivers. This sets such rivers apart from those in more humid environments where a greater degree of soil development and vegetation cover limits aeolian activity.

2.12 SENSITIVITY OF DRYLAND RIVERS TO CHANGING ENVIRONMENT

If dryland rivers are strongly reflective of their particular environment (e.g. climate, hydrology, sediment loads, vegetation), then it might be expected that they will respond to changing environmental conditions. There is some disagreement, however, as to just how sensitive dryland rivers are to change. Hereford (1984) maintained that subtle changes in climate are important for bringing about significant changes in dryland rivers, and Knighton and Nanson (1997) also see dryland rivers as unusually sensitive to even small environmental perturbations. In contrast, Reid (1994, p. 571) has suggested that rivers in general and, by implication, dryland rivers '. . . *are comparatively insensitive to changes in climate unless these changes are substantial*'. From the divergent views given above, it is probable that dryland river sensitivity spans a considerable range (Nanson and Tooth, 1999). Dryland rivers which are confined within bed-rock gorges, stabilised by effective riparian vegetation, or restrained by indurated or very clay-rich alluvium, are probably relatively insensitive to all but extreme environmental changes, whereas rivers within erodible silty or sandy alluvium will probably respond much more readily to even modest changes.

Contrasts in the sensitivity of dryland rivers in responding to, and preserving, records of Quaternary environmental change have been clearly illustrated by Hereford (1984, 1986), Nanson et al. (1988, 1995), Tooth (1999b) and Bourke and Pickup (1999). These contrasts can be illustrated using three examples (for details see Nanson and Tooth, 1999). First, Hereford's studies are among the very few detailed examples of relationships between dryland climate change and alluvial history, albeit covering only a short period of a few decades. He showed that moderate- to high-energy dryland rivers in unconsolidated sandy to gravelly alluvium can be highly responsive to even relatively minor environmental changes, with flood-plain landforms and sediments providing detailed short-term records of climate and flow-regime change. Second, in central Australia near Alice Springs, the extent to which dryland rivers respond to climate change has been shown to depend on proximity to the headwater ranges, valley confinement and the resistance of alluvium to erosion (Bourke and Pickup, 1999; Tooth, 1999a, 1999b, 2000a). Some rivers have been relatively responsive to recent climate and flow-regime changes, with floodplains preserving detailed records of those changes, while other rivers have been much less responsive and provide very little evidence for the same changes. Third, the vast, low-gradient, mud-dominated rivers of the Channel Country appear to have been relatively unresponsive to recent Holocene climate and flow-regime changes but, partly as a consequence, have preserved significant long-term records of the major changes that affected the continent during the Pleistocene (Nanson et al., 1986, 1988; Rust and Nanson, 1986).

In short, the evidence for environmental change preserved by dryland rivers is a function both of the magnitude and character of those changes and of the river's ability to resist or respond to them (Nanson and Tooth, 1999). The ability to resist or respond to change in turn depends on the balance between available stream power and the erosional resistance of the channel and floodplain boundary (cf. Bull, 1979, 1991). In general, the piedmont reaches of a river system will often be the most sensitive to environmental change. Here there is usually sufficient stream power to modify valley fills of sand and gravel, whereas, further upstream, bedrock and very coarse alluvium may offer highly resistant boundaries, and further downstream stream powers may be lower and the alluvium relatively cohesive. This generalisation probably applies in both dryland and more humid settings, and it is therefore difficult to assess the relative sensitivity of rivers in different environments when so much variation in river response is a product of within-catchment conditions.

2.13 EQUILIBRIUM AND NON-EQUILIBRIUM CONDITIONS

The sensitivity of dryland rivers to changing environmental conditions is related to concepts of equilibrium and non-equilibrium in river behaviour. The concept of dynamic equilibrium discussed here is that conceived originally in geomorphology by Gilbert (1877) and recently defined by Tooth and Nanson (2000b, p. 186) as:

> . . . a condition in which, over a period of years, the relatively stable physical characteristics of a stream indicate that the controlling variables are balanced to provide just those conditions required for the transport of water and sediment from upstream, a system in which dynamic adjustments caused by negative feedback are implicit.

Early work on dryland rivers suggested that they tend towards approximate equilibrium conditions (e.g. Gilbert, 1877; Leopold and Miller, 1956; Leopold et al., 1964), with Slatyer and Mabbutt (1964, p. 24) stating explicitly that:

> Despite their ephemeral character, drainage channels in arid regions show the same tendency to adjust in equilibrium with hydraulic factors . . . and in fact, such channel adjustments are facilitated by the paucity of restraining vegetation . . .

However, owing to the propensity for major flood-related channel changes and long recovery times in *some* dryland rivers (Wolman and Gerson, 1978), it has subsequently been widely presumed that equilibrium conditions may rarely, if ever, occur (e.g. Graf, 1981, 1983; Warren, 1985; Bull, 1991; Cooke et al., 1993; Bourke and Pickup, 1999). As Graf (1988) suggested, difficulties in transferring concepts of equilibrium channel behaviour from humid areas where there is continuous flow and well-defined feedback processes occur because dryland rivers commonly do not provide such conditions. Cooke et al. (1993) and Knighton and Nanson (1997) considered that concepts such as equilibrium are less relevant in dryland rivers, and even today Hooke and Mant (Chapter 6 in this volume) state: 'It has long been accepted that ephemeral channels tend to be in a non-equilibrium state and are therefore unstable, with some propensity for sudden switches of characteristics . . . .' As Tooth and Nanson (2000b) show, however, the concept that dryland rivers are non-equilibrium systems is true for only some rivers, just as it is for some humid rivers. There is no *a priori* reason why all dryland rivers should exhibit non-equilibrium characteristics.

Small, high-gradient, upland reaches in dryland rivers transporting sand and gravel bedloads are typically characterised by short-lived flash floods that generate relatively high stream powers. In unconsolidated alluvium, and where bankline vegetation is sparse, such stream powers typically exceed bank erosional thresholds and channels are susceptible to dramatic changes during large floods (Schick, 1974; Hereford, 1986; Clark and Davies, 1988). Changes may also occur due to the exceedance of internal thresholds (Schumm, 1973). Such changes are often slow to reverse and, consequently, equilibrium with more frequently occurring lower magnitude flows is difficult or impossible to achieve. While such upland reaches thus may be characterised largely by non-equilibrium conditions, exceptions can occur where a small, frequently active channel in equilibrium with the short-term hydraulic regime occurs within a larger non-equilibrium channel that is the product of infrequent large floods (Rhoads, 1990).

Intermediate-size, moderate-gradient piedmont reaches in dryland rivers tend to be dominated by sandy bedloads and longer duration flows that generate more moderate stream powers (Tooth, 2000a, 2000b). Riparian vegetation may be better developed than in the uplands and may contribute greatly to increased bank strength. Here, equilibrium conditions can prevail for long periods of time. However, where bank erosional thresholds are low, and are exceeded by an extreme flood event or a succession of closely spaced more moderate events, channels can be susceptible to change in a non-equilibrium fashion (e.g. Schumm and Lichty, 1963; Burkham, 1972; Hereford, 1984).

Intermediate to large-, moderate- to low-gradient lowland reaches in dryland rivers are commonly transfer or depositional zones. Stream powers are low and the banks may be tree-lined, with the power of floods rarely (if ever) rising above the threshold for severe bank erosion. Hence, channel forms may be adjusted to higher frequency, lower magnitude flow and sediment discharge regimes rather than to individual large, rare events. As Tooth and Nanson (2000b) detail, there are numerous reaches of rivers in the Australian drylands that meet such conditions. Taking the four conditions that Richards (1982) suggests as diagnostic of equilibrium conditions in rivers, they show that: (1) channels have remained essentially stable despite occasional large floods; (2) sediment transport discontinuities, although present at a catchment scale in infilling basins or where flow declines downstream, are essentially insignificant in terms of disrupting equilibrium channel form and process in individual reaches; (3) there are strong correlations between channel form and process variables; and (4) dryland rivers can be adjusted to maximum sediment transport efficiency under conditions of low gradient, abundant within-channel vegetation, and declining downstream discharges. As an example, the muddy rivers in the Channel Country appear to have been exceptionally stable for many thousands of years despite frequent overbank flooding (Nanson and Tooth, 1999). Partly as a consequence, the rivers exhibit strong correlations between aspects of channel form and process. These include predictable relationships between waterhole dimensions and discharges, narrow and deep anastomosing channels for transporting water and limited suspended load at low flows and wide, shallow bedload-transporting floodways operative during overbank flows, and variable meander wavelengths that are scaled to channel geometry and bankfull discharge (Knighton and Nanson, 2000; Fagan, 2001). Strongly aggradational reaches in the Australian drylands are less stable, however, for here sediment transport discontinuities can be readily apparent (Tooth, 1999a, 1999b; Tooth and Nanson, 2000b).

While a lack of equilibrium between process and form may be characteristic of some dryland rivers, the identification of widespread equilibrium conditions in dryland rivers cited above requires recognition. At this stage, there is probably insufficient evidence to suggest whether dryland rivers *as a whole* are more likely to exhibit non-equilibrium conditions than rivers in more humid regions. As with rivers in other environments, accurate identification of equilibrium or non-equilibrium conditions in dryland rivers has important implications for their management. The presumption or misconception that all, or even most, dryland rivers are non-equilibrium systems could result in management strategies that are inappropriate in some contexts. Sudden changes in channel form or sediment transport conditions in non-equilibrium systems may have natural explanations, perhaps associated with periodic threshold adjustments, whereas those in equilibrium systems could be human-induced and diagnostic of a system undergoing severe disruption.

2.14   CONCLUSIONS

Research into dryland rivers is still in its infancy. While there have been numerous studies of rivers in the American southwest, parts of the Mediterranean and a small area of the Middle East, the emphasis has been mostly on small, relatively steep, coarse-load rivers in upland catchments. To some extent, this emphasis has been justified. In an analysis of the physiography of major northern hemisphere deserts, Clements et al. (1957) showed that mountains cover between 35 and 50% of the land area. The Mediterranean, for instance, which is a focus for many of the chapters in this book, is a region drained predominantly by steepland rivers (Macklin et al., 1995). By the same token, however, the analyses by Clements et al., and a similar analysis by Mabbutt (1971) which showed that mountains occupy less than 18% of the Australian deserts, indicate that a continued focus on upland areas will leave considerably more than 50% of the world's dryland regions poorly understood in fluvial terms. Many of the generalities expressed in reviews of dryland rivers reflect the physiographic and scale bias towards small, steep systems. Yet as Graf (1988, p. 56) has stated:

> Scale has important implications for the analysis of dryland rivers. What is learned of processes in small basins is not likely to be informative with regard to the operation of larger river systems. . . .

Indeed, as this chapter has shown, the generalities emerging from research in small, steep upland reaches regarding high sediment transport rates, the dominance of large floods as a control on channel morphology, and non-equilibrium conditions, often cannot be applied to larger, lower gradient, piedmont or lowland reaches. Dryland river characteristics vary across a wide spectrum of scales and in response to often subtle changes in regional climate, hydrology, sediment load and vegetation. Few features are unique to dryland rivers, with the possible exception of large waterholes and the tendency for trees to grow on channel beds. However, some characteristics are more common in dryland rivers compared with rivers in other environments (e.g. downstream transmission losses, downstream decreases in channel size, floodout formation, fluvial–aeolian interaction and alluvial induration) while others are widely shared with rivers in a range of climates (e.g. diversity of river patterns, role of vegetation in influencing river form and process). Similarly, while some dryland rivers may be closely adjusted to present-day aridity, others may exhibit characteristics partly inherited from more humid conditions in the past. As was found with the study of 'climatic geomorphology' in the mid-twentieth century, the search for one defining set of dryland river characteristics may be something of a chimera.

The way forward for understanding the sum total of dryland river form and process in most cases will be to recognise the important variations between upland (erosional), piedmont (transfer) and lowland (depositional) reaches (cf. Schumm, 1977; Pickup, 1986; Bull, 1991; Tooth, 1997). Degree of aridity, flow seasonality, physiographic setting, tectonic stability, geological structure and lithology, and vegetation type and density, are examples of other variables that contribute to dryland river diversity and need to be taken into account when comparing different reaches of the same river, different rivers in the same dryland, or different rivers in different drylands. Descriptions of turbid, viscous, flash floods in upland rivers draining small, tectonically active catchments contrast greatly with very low sediment loads in slowly moving floodwaters in lowland rivers draining continental-scale, tectonically-stable catchments. Wide, shallow, sandy cross-sections in piedmont settings present a dramatic difference to narrow, deep, muddy anastomosing channels in some lowland settings. Generalising and conceptualising are important for attaining a greater understanding of highly complex natural systems, yet when perceptions obtained from specific and restricted settings are converted in scale to interpret a much larger entity, serious misconceptions can result.

REFERENCES

Adamson, D.A. 1982. The integrated Nile. In M.A.J. Williams and D.A. Adamson (Eds), *A Land Between Two Niles*. Balkema, Rotterdam, 221–234.

Allan, R.J. 1985. *The Australasian Summer Monsoon, Teleconnections and Flooding in the Lake Eyre Basin*. Royal Geographical Society of Australasia, South Australian Branch, South Australian Geographical Papers 2.

Allan, R.J. 1988. El Niño Southern Oscillation influences in Australasia. *Progress in Physical Geography*, 12, 4–40.

Argue, J.R. and Salter, L.E.M. 1977. Waterhole development: a viable water resource option in the arid zone? *Hydrology Symposium 1977: The Hydrology of Northern Australia, National Conference Publication* No. 77/5. Institution of Engineers, Australia, Brisbane, Queensland, 35–39.

Beaumont, P. 1989. *Drylands: Environmental Management and Development*. Routledge, London and New York.

Ben-Zvi, A., Massoth, S. and Schick, A.P. 1991. Travel time of runoff crests in Israel. *Journal of Hydrology*, 122, 309–320.

Blackburn, W.H., Knight, R.W. and Schuster, J.L. 1982. Saltcedar influence on sedimentation in the Brazos River. *Journal of Soil and Water Conservation*, 37, 298–301.

Bourke, M.C. and Pickup, G. 1999. Fluvial form variability in arid central Australia. In A.J. Miller and A. Gupta (Eds), *Varieties of Fluvial Form*. John Wiley & Sons, Chichester, 249–271.

Budel, J. 1980. Climatic and climatomorphic geomorphology. *Zeitschrift für Geomorphologie Supplement Band*, 36, 1–8.

Bull, W.B. 1979. Threshold of critical power in streams. *Geological Society of America Bulletin*, 90, 453–464.

Bull, W.B. 1991. *Geomorphic Responses to Climatic Change*. Oxford University Press, New York.

Burkham, D.E. 1972. Channel changes of the Gila River in Safford Valley, Arizona 1846–1970. *United States Geological Survey Professional Paper*, 655-G.

Chorley, R.J. 1965. A re-evaluation of the geomorphic system of W.M. Davis. In R.J Chorley and P. Haggett (Eds), *Frontiers in Geographical Teaching*. Methuen, London, 21–38.

Chorley, R.J., Schumm, S.A. and Sugden, D.E. 1984. *Geomorphology*. Methuen, London and New York.

Clark, P.B. and Davies, S.M.A. 1988. The application of regime theory to wadi channels in desert conditions. In W.R. White (Ed.) *International Conference on River Regime*, 18–20 May 1988. Hydraulics Research Limited. John Wiley & Sons, Chichester, 67–82.

Clarke, M.L. and Rendell, H.M. 1998. Climate change impacts on sand supply and the formation of desert sand dunes in the southwest USA. *Journal of Arid Environments*, 39, 517–531.

Clements, T., Merriam, R.H., Eymann, J.L., Stone, R.O. and Reade, H.L. 1957. *A Study of Desert Surface Conditions*. US Army Environmental Protection Research Division Technical Report EP-53, Natck, MA.

Coleman, M., in prep. *Alluvial, aeolian and lacustrine evidence for climate and flow regime change over the past 250 ka, Cooper Creek near Innamincka, South Australia*. PhD thesis, University of Wollongong.

Cooke, R.U., Warren, A. and Goudie, A.S. 1993. *Desert Geomorphology*. University College London Press, London.

Cordery, I., Pilgrim, D.H. and Doran, D.G. 1983. Some hydrological characteristics of arid western New South Wales. In *Hydrology and Water Resources Symposium 1983*. Institution of Engineers, Australia, 287–292.

Dunkerley, D.L. 1992. Channel geometry, bed material and inferred flow conditions in ephemeral stream systems, Barrier Range, western N.S.W, Australia. *Hydrological Processes*, 6, 417–433.

Fagan, S.D. 2001. *Channel and floodplain characteristics of Cooper Creek, central Australia*. PhD thesis, University of Wollongong.

Farquharson, F.A.K., Meigh, J.R. and Sutcliffe, J.V. 1992. Regional flood frequency analysis in arid and semi-arid areas. *Journal of Hydrology*, 138, 487–501.

Fielding, C.R., Alexander, J. and Newman-Sutherland, E. 1997. Preservation of in situ, aborescent vegetation and fluvial bar construction in the Burdekin River of north Queensland, Australia. *Palaeogeography, Palaeoclimatology, Palaeoecology*, 135, 123–144.

Gibling, M.R., Nanson, G.C. and Maroulis, J.C. 1998. Anastomosing river sedimentation in the Channel Country of central Australia. *Sedimentology*, 45, 595–619.

Gilbert, G.K. 1877. *Report on the Geology of the Henry Mountains*. United States Geographical and Geological Survey of the Rocky Mountains Region. United States Government Printing Office, Washington, DC.

Goudie, A.S. 1987. Change and instability in the desert environment. In M.J. Clark, K.J. Gregory and A.M. Gurnell (Eds) *Horizons in Physical Geography*. Macmillan Education, London, 250–267.

Graeme, D. and Dunkerley, D.L. 1993. Hydraulic resistance by the river red gum, *Eucalyptus camaldulensis*, in ephemeral desert streams. *Australian Geographical Studies*, 31, 141–154.

Graf, W.L. 1978. Fluvial adjustments to the spread of tamarisk in the Colorado Plateau region. *Geological Society of America Bulletin*, 89, 1491–1501.

Graf, W.L. 1979. Fluvial adjustments to the spread of tamarisk in the Colorado Plateau region: reply. *Geological Society of America Bulletin*, 90, 1183–1184.

Graf, W.L. 1980. Riparian vegetation: a flood control perspective. *Journal of Soil and Water Conservation*, 35, 158–161.

Graf, W.L. 1981. Channel instability in a braided, sand bed river. *Water Resources Research*, 17, 1087–1094.

Graf, W.L. 1983. Downstream changes in stream power in the Henry Mountains, Utah. *Annals of the Association of American Geographers*, 73, 373–387.

Graf, W.L. 1988. *Fluvial Processes in Dryland Rivers*. Springer-Verlag, Berlin.

Gregory, K.J. and Gurnell, A.M. 1988. Vegetation and river channel form and process. In H.A. Viles (Ed.) *Biogeomorphology*. Basil Blackwell, Oxford, 11–42.

Griffin, G.F., Smith, D.M.S., Morton, S.R., Allan, G.E. and Masters, K.A. 1989. Status and implications of the invasion of Tamarisk (*Tamarix aphylla*) on the Finke River, Northern Territory, Australia. *Journal of Environmental Management*, 29, 297–315.

Gupta, A. 1993. The changing geomorphology of the humid tropics. *Geomorphology*, 7, 165–186.

Gupta, A. 1995. Magnitude, frequency, and special factors affecting channel form and processes in the seasonal tropics. In J.E. Costa, A.J. Miller, K.W. Potter and P.R. Wilcock (Eds) *Natural and Anthropogenic Influences in Fluvial Geomorphology*. American Geophysical Union, Washington DC, 125–136.

Gurnell, A.M. 1987. Suspended sediment. In A.M. Gurnell and M.J. Clark (Eds) *Glacio-Fluvial Sediment Transfer: An Alpine Perspective*. John Wiley & Sons, Chichester, 305–353.

Hadley, R.F. 1961. Influence of riparian vegetation on channel shape, northeastern Arizona. *United States Geological Survey Professional Paper*, 424-C, 30–31.

Hayward, J.A. 1980. *Hydrology and Stream Sediment from Torless Stream Catchment*. Tussock Grasslands and Mountain Lands Institute, Lincoln College, Special Publication 17.

Hereford, R. 1984. Climate and ephemeral-stream processes: twentieth-century geomorphology and alluvial stratigraphy of the Little Colorado River, Arizona. *Geological Society of America Bulletin*, 95, 654–668.

Hereford, R. 1986. Modern alluvial history of the Paria River drainage basin, southern Utah. *Quaternary Research*, 25, 293–311.

Heritage, G.L., van Niekerk, A.W. and Moon, B.P. 1999. Geomorphology of the Sabie River, South Africa: an incised bedrock-influenced channel. In A.J. Miller and A. Gupta (Eds) *Varieties of Fluvial Form*. John Wiley & Sons, Chichester, 53–79.

Hey, R.D. and Thorne, C.R. 1986. Stable channels with mobile gravel beds. *Journal of the Hydraulics Division, Proceedings of the ASCE*, 112, 671–689.

Hickin, E.J. 1984. Vegetation and river channel dynamics. *Canadian Geographer*, 28, 111–126.

Huang, H.Q. and Nanson, G.C. 1997. Vegetation and channel variation; a case study of four small streams in southeastern Australia. *Geomorphology*, 18, 237–249.

Huang, H.Q. and Nanson, G.C. 1998. The influence of bank strength on channel geometry: an integrated analysis of some observations. *Earth Surface Processes and Landforms*, 23, 865–876.

Hupp, C.R. 1990. Vegetation patterns in relation to basin hydrogeomorphology. In J.B. Thornes (Ed.) *Vegetation and Erosion: Processes and Environments*. John Wiley & Sons, Chichester, 217–237.

Hupp, C.R. and Osterkamp, W.R. 1985. Bottomland vegetation distribution along Passage Creek, Virginia, in relation to fluvial landforms. *Ecology*, 66, 670–681.

Hupp, C.R. and Osterkamp, W.R. 1996. Riparian vegetation and fluvial geomorphic processes. *Geomorphology*, 14, 277–295.

Knight, D.W. and Shiono, K. 1996. River channel and floodplain hydraulics. In M.G. Anderson, D.E. Walling and P.D. Bates (Eds) *Floodplain Processes*. John Wiley & Sons, Chichester, 139–181.

Knighton, A.D. and Nanson, G.C. 1994a. Waterholes and their significance in the anastomosing channel system of Cooper Creek, Australia. *Geomorphology*, 9, 311–324.

Knighton, A.D. and Nanson, G.C. 1994b. Flow transmission along an arid zone anastomosing river, Cooper Creek, Australia. *Hydrological Processes*, 8, 137–154.

Knighton, A.D. and Nanson, G.C. 1997. Distinctiveness, diversity and uniqueness in arid zone river systems. In D.S.G. Thomas (Ed) *Arid Zone Geomorphology: Process, Form and Change in Drylands* (2nd Edn). John Wiley & Sons, Chichester, 185–203.

Knighton, A.D. and Nanson, G.C. 2000. Waterhole form and process in the anastomosing channel system of Cooper Creek, Australia. *Geomorphology*, 35, 101–117.

Knighton A.D. and Nanson, G.C. in press a. An event based approach to the hydrology of arid zone rivers in the Channel Country of Australia. *Journal of Hydrology*.

Knighton A.D and Nanson, G.C. in press b. Inbank and overbank velocity conditions in an arid zone anastomosing river. *Hydrological Processes*.

Kondolf, G.M. and Curry, R.R. 1984. The role of vegetation in channel bank stability: Carmel River, California. In R.E. Warner and K.M. Hendrix (Eds) *California Riparian Systems*. University of California Press, Berkeley and Los Angeles, 124–133.

Kondolf, G.M. and Curry, R.R. 1986. Channel erosion along the Carmel River, Monterey County, California. *Earth Surface Processes and Landforms*, 11, 307–319.

Kotwicki, V. 1986. *Floods of Lake Eyre*. Engineeering and Water Supply Department, Adelaide.

Kotwicki, V. and Isdale, P. 1991. Hydrology of Lake Eyre, Australia: El Niño link. *Palaeogeography, Palaeoclimatology, Palaeoecology*, 84, 87–98.

Kotwicki, V. and Allan, R. 1998. La Niña de Australia – contemporary and paleohydrology of Lake Eyre. *Palaeogeography, Palaeoclimatology, Palaeoecology*, 144, 265–280.

Lancaster, N. 1997. Response of eolian geomorphic systems to minor climate change: examples from the southern Californian deserts. *Geomorphology*, 19, 333–347.

Lancaster, N. and Teller, J.T. 1988. Interdune deposits of the Namib Sand Sea. *Sedimentary Geology*, 55, 91–107.

Langbein, W.B. and Schumm, S.A. 1958. Yield of sediment in relation to mean annual precipitation. *Transactions of the American Geophysical Union*, 39, 1076–1084.

Laronne, J.B. and Reid, I. 1993. Very high rates of bedload sediment transport by ephemeral desert rivers. *Nature*, 366, 148–150.

Laronne, J.B., Reid, I., Yitshak, Y. and Frostick, L.E. 1994. The non-layering of gravel streambeds under ephemeral flood regimes. *Journal of Hydrology*, 159, 353–363.

Leopold, L.B. and Miller, J.P. 1956. Ephemeral streams – hydraulic factors and their relation to the drainage net. *United States Geological Survey Professional Paper*, 282-A.

Leopold, L.B., Wolman, M.G. and Miller, J.P. 1964. *Fluvial Processes in Geomorphology*. W.H. Freeman & Company, San Francisco.

Mabbutt, J.A. 1971. The Australian arid zone as a prehistoric environment. In D.J. Mulvanney and P.J. Golson (Eds) *Aboriginal Man and Environment in Australia*. ANU Press, Canberra, 66–79.

Mabbutt, J.A. 1977. *Desert Landforms*. Australian National University Press, Canberra.

Macklin, M.G., Lewin, J. and Woodward, J.C. 1995. Quaternary fluvial systems in the Mediterranean basin. In J. Lewin, M.G. Macklin and J.C. Woodward (Eds) *Mediterranean Quaternary River Environments*. Balkema, Rotterdam, 1–25.

Makaske, B. 1998. *Anastomosing Rivers: Forms, Processes and Sediments*. Faculteit Ruimtelijke Wetenschappen, Universiteit Utrecht, Utrecht.

Maroulis, J.C. and Nanson, G.C. 1996. Bedload transport of aggregated muddy alluvium from Cooper Creek, central Australia: a flume study. *Sedimentology*, 43, 771–790.

McCarthy, T.S. and Ellery, W.N. 1995. Sedimentation on the distal reaches of the Okavango fan, Botswana, and its bearing on calcrete and silcrete (ganister) formation. *Journal of Sedimentary Research*, A65, 77–90.

McEwen, L.J. and Matthews, J.A. 1998. Channel form, bed material and sediment sources of the Sprongdøla, southern Norway: evidence for a distinct periglacio-fluvial system. *Geografiska Annaler*, 80A, 17–36.

McMahon, T.A. 1979. Hydrological characteristics of arid zones. *Proceedings of the Canberra Symposium, The Hydrology of Areas of Low Precipitation*. IAHS-IASH Publication No. 128, 105–123.

McMahon, T.A., Finlayson, B.L., Haines, A.T. and Srikanthan, R. 1992. *Global Runoff – Continental Comparisons of Annual Flows and Peak Discharges*. Catena Paperback. Catena Verlag, Cremlingen-Destedt.

Middleton, N.J. and Thomas, D.S.G. 1997. *World Atlas of Desertification* (2nd Edn). UNEP/Edward Arnold, London.

Nanson, G.C. 1974. Bedload and suspended-load transport in a small, steep, mountain stream. *American Journal of Science*, 274, 471–486.

Nanson, G.C. and Beach, H.F. 1977. Forest succession and sedimentation on a meandering river floodplain, northeast British Columbia, Canada. *Journal of Biogeography*, 4, 229–251.

Nanson, G.C. and Huang, H.Q. 1999. Anabranching rivers: divided efficiency leading to fluvial diversity. In A.J. Miller and A. Gupta (Eds) *Varieties of Fluvial Form*. John Wiley & Sons, Chichester, 477–494.

Nanson, G.C. and Tooth, S. 1999. Arid-zone rivers as indicators of climate change. In A.K. Singhvi and E. Derbyshire (Eds) *Paleoenvironmental Reconstruction in Arid Lands*. A.A. Balkema, Rotterdam, 175–216.

Nanson, G.C. and Young, R.W. 1981. Downstream reduction in rural channel size with contrasting urban effects in small coastal streams of southeastern Australia. *Journal of Hydrology*, 52, 239–255.

Nanson, G.C., Chen, X.Y. and Price, D.M. 1995. Aeolian and fluvial evidence of changing climate and wind patterns during the past 100 ka in the western Simpson Desert, Australia. *Palaeogeography, Palaeoclimatology, Palaeoecology*, 113, 87–102.

Nanson, G.C., Price, D.M., Short, S.A., Young, R.W. and Jones, B.G. 1991. Comparative uranium-thorium and thermoluminescence dating of weathered Quaternary alluvium in the tropics of northern Australia. *Quaternary Research*, 35, 347–366.

Nanson, G.C., Rust, B.R. and Taylor, G. 1986. Coexistent mud braids and anastomosing channels in an arid-zone river: Cooper Creek, central Australia. *Geology*, 14, 175–178.

Nanson, G.C., Young, R.W., Price, D.M. and Rust, B.R. 1988. Stratigraphy, sedimentology and late-Quaternary chronology of the Channel Country of western Queensland. In R.F. Warner (Ed.) *Fluvial Geomorphology of Australia*. Academic Press, Sydney, 151–175.

Park, C.C. 1977. World-wide variations in hydraulic geometry exponents of stream channels: an analysis and some observations. *Journal of Hydrology*, 33, 133–146.

Peltier, L.C. 1975. The concept of climatic geomorphology. In W.N. Melhorn and R.C. Flemal (Eds) *Theories of Landform Development*. George Allen & Unwin, London, 129–143.

Petryk, S. and Bosmajian, G. 1975. Analysis of flow through vegetation. *Journal of the Hydraulics Division, Proceedings of the ASCE*, 101(HY7), 871–884.

Pickup, G. 1986. Fluvial landforms. In D.N. Jeans (Ed.) *Australia – A Geography. Volume One: The Natural Environment*. Sydney University Press, Sydney, 148–179.

Pickup, G., Allan, G. and Baker, V.R. 1988. History, palaeochannels and palaeofloods of the Finke River, central Australia. In R.F. Warner (Ed) *Fluvial Geomorphology of Australia*. Academic Press, Sydney, 105-127.

Pilgrim, D.H., Chapman, T.G. and Doran, D.G. 1988. Problems of rainfall–runoff modelling in arid and semiarid regions. *Hydrological Sciences Journal*, 33, 379–400.

Reid, I. 1994. River landforms and sediments: evidence of climatic change. In A.D. Abrahams and A.J. Parsons (Eds) *Geomorphology of Desert Environments*. Chapman & Hall, London, 571–592.

Reid, I. and Frostick, L.E. 1994. Fluvial sediment transport and deposition. In K. Pye (Ed.) *Sediment Transport and Depositional Processes*. Blackwell Scientific Publications, Oxford, 89–156.

Reid, I. and Frostick, L.E. 1997. Channel form, flows and sediments in deserts. In D.S.G. Thomas (Ed.) *Arid Zone Geomorphology: Process, Form and Change in Drylands* (2nd Edn). John Wiley & Sons, Chichester, 205–229.

Reid, I. and Laronne, J.B. 1995. Bed load sediment transport in an ephemeral stream and a comparison with seasonal and perennial counterparts. *Water Resources Research*, 31, 773–781.

Reid, I., Laronne, J.B. and Powell, D.M. 1998. Flash flood and bedload dynamics of desert gravel-bed streams. *Hydrological Processes*, 12, 543–557.

Rhoads, B.L. 1990. Hydrologic characteristics of a small desert mountain stream: implications for short-term magnitude and frequency of bedload transport. *Journal of Arid Environments*, 18, 151–163.

Richards, K.S. 1982. *Rivers: Form and Process in Alluvial Channels*. Methuen, London.

Robinson, T.W. 1965. Introduction, spread and aerial extent of saltcedar (*Tamarix*) in the western States. *United States Geological Survey Professional Paper*, 491-A.

Rodier, J.A. 1985. Aspects of arid zone hydrology. In J.C. Rodda (Ed.) *Facets of Hydrology Volume II*. John Wiley & Sons, Chichester, 205–247.

Rust, B.R. 1981. Sedimentation in an arid-zone anastomosing fluvial system: Cooper Creek, central Australia. *Journal of Sedimentary Petrology*, 51, 745–755.

Rust, B.R. and Nanson, G.C. 1986. Contemporary and palaeochannel patterns and the late Quaternary stratigraphy of Cooper Creek, southwest Queensland, Australia. *Earth Surface Processes and Landforms*, 11, 581–590.

Rust, B.R. and Nanson, G.C. 1989. Bedload transport of mud as pedogenic aggregates in modern and ancient rivers. *Sedimentology*, 36, 291–306.

Schick, A.P. 1970. Desert floods. *Symposium on the Results of Research on Representative Experimental Basins*. International Association Scientific Hydrologists/Unesco, 478–493.

Schick, A.P. 1974. Formation and obliteration of desert stream terraces – a conceptual analysis. *Zeitschrift für Geomorphologie Supplement Band*, 21, 88–105.

Schumann, R.R. 1989. Morphology of Red Creek, Wyoming, an arid-region anastomosing channel system. *Earth Surface Processes and Landforms*, 14, 277–288.

Schumm, S.A. 1965. Quaternary paleohydrology. In H.E. Wright and D.G. Fry (Eds) *The Quaternary History of the United States*. Princeton University Press, New Jersey, 783–794.

Schumm, S.A. 1973. Geomorphic thresholds and complex response of drainage systems. In M.E. Morisawa (Ed.) *Fluvial Geomorphology*. Binghampton Publication in Geomorphology, New York, 299–310.

Schumm, S.A. 1977. *The Fluvial System*. John Wiley & Sons, New York.

Schumm, S.A. and Lichty, R.W. 1963. Channel widening and flood-plain construction along Cimarron River in southwestern Kansas. *United States Geological Survey Professional Paper* 352-D, 71–88.

Sharma, K.D. and Murthy, J.S.R. 1994. Modelling sediment transport in stream channels in the arid zone of India. *Hydrological Processes*, 8, 567–572.

Sharma, K.D. and Murthy, J.S.R. 1998. A practical approach to rainfall–runoff modeling in arid zone drainage basins. *Hydrological Sciences*, 43, 331–348.

Sharma, K.D., Murthy, J.S.R. and Dhir, R.P. 1994. Streamflow routing in the Indian arid zone. *Hydrological Processes*, 8, 27–43.

Shaw, P.A. and Nash, D.J. 1998. Dual mechanisms for the formation of fluvial silcretes in the distal reaches of the Okavango Delta fan, Botswana. *Earth Surface Processes and Landforms*, 23, 705–714.

Slatyer, R.O. and Mabbutt, J.A. 1964. Hydrology of arid and semiarid regions. In V.T. Chow (Ed.) *Handbook of Applied Hydrology*. McGraw-Hill, New York, 24–1 to 24–46.

Smith, D.G. 1976. Effect of vegetation on lateral migration of anastomosed channels of a glacial meltwater river. *Geological Society of America Bulletin*, 86, 857–860.

Sullivan, M.E. 1976. *Drainage Disorganisation in Arid Australia and its Measurement*. MSc. Thesis (unpubl.), University of New South Wales.

Summerfield, M.A. 1991. *Global Geomorphology*. Longman, London.

Taylor, C.F.H. 1999. The role of overbank flow in governing the form of an anabranching river: the Fitzroy River, northwestern Australia. In N.D. Smith and J. Rogers (Eds) *Fluvial Sedimentology VI*.

International Association of Sedimentologists, Special Publication 28. Blackwell Scientific Publications, Oxford, 77–91.

Thomas, D.S.G., O'Connor, P.W., Bateman, M.D., Shaw, P.A., Stokes, S. and Nash, D.J. 2000. Dune activity as a record of late Quaternary aridity in the Northern Kalahari: new evidence from northern Namibia interpreted in the context of regional arid and humid chronologies. *Palaeogeography, Palaeoclimatology, Palaeoecology*, 156, 243–259.

Thornes, J.B. 1976a. Semi-arid erosional systems: case studies from Spain. *London School of Economics Geographical Papers*, No. 7.

Thornes, J.B. 1976b. Autogeometry of semi-arid channel systems. In *Rivers 76. Symposium on Inland Waterways for Navigation, Flood Control and Water Diversions. Volume 2*. American Society of Civil Engineers, Fort Collins, Colorado, 1715–1725.

Thornes, J.B. 1977. Channel changes in ephemeral streams: observations, problems and models. In K.G. Gregory (Ed.) *River Channel Changes*. John Wiley & Sons, Chichester, 317–335.

Thornes, J.B. 1994a. Catchment and channel hydrology. In A.D. Abrahams and A.J. Parsons (Eds) *Geomorphology of Desert Environments*. Chapman & Hall, London, 257–287.

Thornes, J.B. 1994b. Channel processes, evolution and history. In A.D. Abrahams and A.J. Parsons (Eds) *Geomorphology of Desert Environments*. Chapman & Hall, London, 288–317.

Tooth, S. 1997. *The morphology, dynamics and late Quaternary sedimentary history of ephemeral drainage systems on the Northern Plains of central Australia*. PhD. thesis (unpubl.), University of Wollongong.

Tooth, S. 1999a. Floodouts in central Australia. In A.J. Miller and A. Gupta (Eds) *Varieties of Fluvial Form*. John Wiley & Sons, Chichester, 219–247.

Tooth, S. 1999b. Downstream changes in floodplain character on the Northern Plains of arid central Australia. In N.D. Smith and J. Rogers (Eds) *Fluvial Sedimentology VI*. International Association of Sedimentologists, Special Publication 28. Blackwell Scientific Publications, Oxford, 93–112.

Tooth, S. 2000a. Process, form and change in dryland rivers: a review of recent research. *Earth Science Reviews*, 51, 67–107.

Tooth, S. 2000b. Downstream changes in dryland river channels: the Northern Plains of arid central Australia. *Geomorphology*, 34, 33–54.

Tooth, S. and McCarthy, T.S., submitted. Anabranching in mixed alluvial-bedrock rivers: the example of the Orange River above Augrabies Falls, Northern Cape Province, South Africa. *Geomorphology*.

Tooth, S. and Nanson, G.C. 1995. The geomorphology of Australia's fluvial systems: retrospect, perspect and prospect. *Progress in Physical Geography*, 19, 35–60.

Tooth, S. and Nanson, G.C. 1999. Anabranching rivers on the Northern Plains of arid central Australia. *Geomorphology*, 29, 211–233.

Tooth, S. and Nanson, G.C. 2000a. The role of vegetation in the formation of anabranching channels in an ephemeral river, Northern Plains, arid central Australia. *Hydrological Processes*, 14, 3099–3117.

Tooth, S. and Nanson, G.C. 2000b. Equilibrium and nonequilibrium conditions in dryland rivers. *Physical Geography*, 21, 183–211.

Tooth, S., McCarthy, T.S., Brandt, D., Hancox, P.J. and Morris, R., submitted. Geological controls on the formation of alluvial meanders and floodplain wetlands: the example of the Klip River, eastern Free State, South Africa. *Earth Surface Processes and Landforms*.

Tooth, S., McCarthy, T.S., Hancox, P.J., Brandt, D., Buckley, K., Nortje, E. and McQuade, S., in press. The geomorphology of the Nyl River and floodplain in the semi-arid Northern Province, South Africa. *South African Geographical Journal*.

Tricart, J. and Cailleux, A. 1972. *Introduction to Climatic Geomorphology* (translated by C.J.K. de Jonge). Longman, London.

Tyson, P.D. and Preston-Whyte, R.A. 2000. *The Weather and Climate of Southern Africa* (2nd Edn). Oxford University Press Southern Africa, Cape Town.

UNEP 1992. *World Atlas of Desertification*. Edward Arnold, Sevenoaks, UK.

Walling, D.E. and Kleo, A.H.A. 1979. Sediment yields of rivers in areas of low precipitation. In *The Hydrology of Areas of Low Precipitation, Proceedings of the Canberra Symposium*, IAHS-AISH Publication No. 128, 479–493.

Walling, D.E. and Webb, B.W. 1983. Patterns of sediment yield. In K.G. Gregory (Ed.) *Background to Palaeohydrology: A Perspective*. John Wiley & Sons, Chichester, 69–100.

Ward, J.D. and von Brunn, V. 1985. Sand dynamics along the lower Kuiseb River. In B.J. Huntley (Ed.) *The Kuiseb Environment: The Development of a Monitoring Baseline*. South African National Scientific Programmes Report No. 106. Foundation for Research Development/Council for Scientific and Industrial Research, Pretoria, 51–72.

Warren, A. 1985. Arid geomorphology. *Progress in Physical Geography*, 9, 434–441.

Wasson, R.J. 1983. The Cainozoic history of the Strzelecki and Simpson dunefields (Australia), and the origin of the desert dunes. *Zeitschrift für Geomorphologie Supplement Band*, 45, 85–115.

Watts, J.F. and Watts, G.D. 1990. Seasonal change in aquatic vegetation and its effect on river channel flow. In J.B. Thornes (Ed.) *Vegetation and Erosion: Processes and Environments*. John Wiley & Sons, Chichester, 257–267.

Wende, R. and Nanson, G.C. 1998. Anabranching rivers: ridge-form alluvial channels in tropical northern Australia. *Geomorphology*, 22, 205–224.

Wolman, M.G. and Gerson, R. 1978. Relative scales of time and effectiveness of climate in watershed geomorphology. *Earth Surface Processes and Landforms*, 3, 189–208.

Woodfull, J., Rutherfurd, I and Bishop, P. 1996. Downstream increasing flood frequency on Australian floodplains. *Proceedings of the First National Conference on Stream Management in Australia*, February 1996, 81–86.

Woodyer, K.D., Taylor, G. and Crook, K.A.W. 1979. Depositional processes along a very low-gradient, suspended-load stream: the Barwon River, New South Wales. *Sedimentary Geology*, 22, 97–120.

Part II Processes in Dryland Catchments

3 Runoff Generation in Semi-arid Areas

KEITH BEVEN

Institute of Environmental and Natural Sciences, Lancaster University, UK

3.1 A PERCEPTUAL MODEL OF RUNOFF GENERATION IN SEMI-ARID AREAS

A perceptual model is a term coined to represent a summary of our perceptions of the processes controlling how a catchment responds to rainfall under different conditions (Beven, 1989b, 2000b). An appreciation of the perceptual model for a particular catchment is important because it must be remembered that any quantitative model of the processes will inevitably be a simplification of the perceptual model, in many cases a gross simplification, but perhaps still sufficient to provide adequate and useful predictions. There are good reasons for this necessary simplification and approximation. The perceptual model is not constrained by any mathematical theory. It exists primarily in the head of each hydrologist and need not even be written down in words or equations. We can perceive complexities of the flow processes in a purely qualitative way that may be very difficult indeed to describe in the language of mathematics. However, a mathematical description (or conceptual model) is, traditionally, the first stage in the formulation of a model that will make quantitative predictions.

There are many outlines of the processes of catchment response available in the literature. Most general hydrological texts deal, in greater or less detail, with the processes of catchment response. The volumes edited by Kirkby (1978) and Anderson and Burt (1990) are of particular interest in that the different chapters reflect the experience of a number of different hydrologists. Indeed, the starting point for any perceptual model of runoff processes should still always be the chapter by Dunne (1978) in the Kirkby volume. Hydrological systems are sufficiently complex that each hydrologist will have his or her own impression or perceptual model of what is most important in the rainfall–runoff process. Different hydrologists might not necessarily agree about what are the most important processes or the best way of describing them. There are sure to be general themes in common, as reflected in hydrological texts, but our understanding of hydrological processes is still evolving and the details will depend on experience, in particular the type of hydrological environments that a hydrologist has experienced. Different processes may be dominant in different environments and in catchments with different characteristics of topography, soil, vegetation and bedrock.

One of the problems involved in having a complete understanding of hydrological systems is that most of the water flows take place underground in the soil or bedrock, and our ability to measure and assess subsurface flow processes is generally very limited. The measurement techniques available reflect conditions only in the immediate area of the probe. When the characteristics of the flow domain vary rapidly in space and also, during the wetting up of semi-arid catchments, rapidly in time, the small-scale nature of such measurements can give

Dryland Rivers: Hydrology and Geomorphology of Semi-arid Channels. Edited by L.J. Bull and M.J. Kirkby.
© 2002 John Wiley & Sons, Ltd.

only a very partial picture of the nature of the flow. Thus, there is much that remains unknown about subsurface flow processes, and is indeed unknowable given the limitations of current measurement techniques. It is necessary to make inferences about the processes from the available measurements. Such inferences add information to the perceptual model of hydrological response, but they are only inferences.

Many studies of runoff generation in semi-arid catchments have demonstrated the complexity of the controlling processes and a strong non-linear dependency on antecedent wetness. Figure 3.1 from Nicolau et al. (1996) is a typical example, demonstrating that there is generally a threshold of storm rainfall below which no runoff is measured, while above that threshold, a storm with a particular rainfall volume may yield a very wide range of runoff coefficients. Even storms greater than 100 mm may produce only a few millimetres of runoff. This complex variability is a function of the antecedent wetness of the catchment, storm duration, and the pattern of rainfall intensities. Antecedent wetness controls the infiltration capacity of the soil surface and the connectivity of surface and subsurface runoff pathways. When the catchment is dry, or in storms of short duration, runoff may be generated locally but may never reach a stream channel. The result is apparent as a strong non-linearity in the hydrograph responses (e.g. Figure 3.2) and a decline in the effective runoff coefficients both at the hillslope plot scale (Figure 3.3(a)) and catchment scale (Figure 3.3(b)) as a result of both limited storm extent in both space and time and, on larger catchments, channel transmission losses due to infiltration of stream discharge into the channel bed (Figure 3.4). The complex nature of the processes involved is revealed in data such as that shown in Figure 3.5, where there is a general trend for times to initiate runoff to fall as antecedent rainfall increases, but with a very high scatter. Such complexity is a feature of this hydrologist's perceptual model of the response of semi-arid catchments and will recur throughout this chapter.

One way of gaining further understanding is to examine a part of the system in much greater detail. Many studies have been made of the flow processes on particular hillslopes or plots, or columns of undisturbed soil brought back to the laboratory. It has generally been found in such studies that more detailed investigation reveals greater complexity and variability in the flow pathways. The same has generally been true of adding different types of information, such as

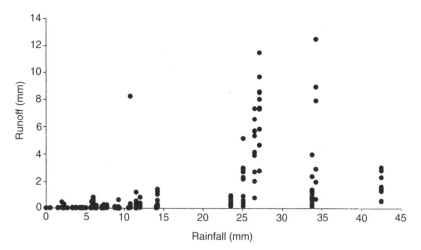

Figure 3.1   Event rainfall–runoff relation at the semi-arid Rambla Honda field plots (MAP 300–350 mm), Almeria, Spain. (After Puigdefàbregas et al., 1996)

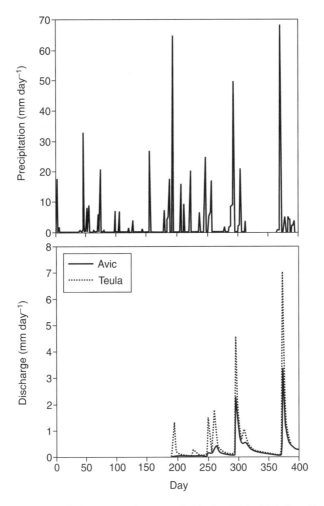

Figure 3.2 Rainfall and runoff during a wetting-up period in the L'Avic (52 ha) and La Teula (39 ha) catchments (MAP 547 mm), Prades, Spain. (After Piñol et al., 1997)

the use of artificial or environmental tracers. Such complexity can be made part of the perceptual model. As noted above, it is not necessary that the perceptual model be anything more than a set of qualitative impressions, but complexity inevitably creates difficulty in the choice of assumptions in moving from the perceptual model to a set of equations defining a conceptual model. Choices must be made at this point to simplify the description and, as we will see, such choices have not always had a good foundation in hydrological reality.

3.1.1 A Perceptual Model of Point Runoff Generation

The traditional perception of runoff generation in semi-arid areas is that the primary generation mechanism is the result of rainfall falling at an intensity greater than the local infiltration

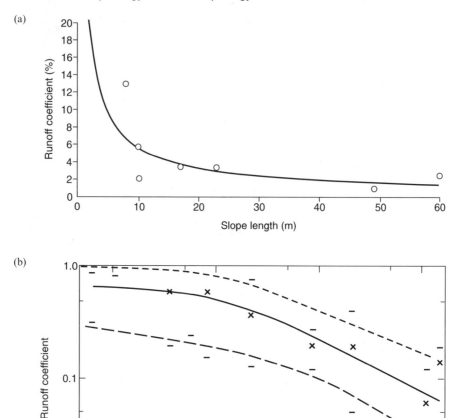

Figure 3.3 (a) Decline in runoff coefficients at the hillslope scale at the Rambla Honda and Solar Power Plant field sites, Almeria. (After Puigdefàbregas and Sánchez, 1996). (b) Decline in runoff coefficients at the catchment scale, southwestern US (after Dunne, 1978)

capacity of the soil (e.g. Horton, 1933; Dunne, 1978; Yair and Lavee, 1985). The resulting surface runoff is called Hortonian or infiltration excess overland flow. This model is named after Robert E. Horton (1875–1945), the celebrated American hydrologist, who worked as both hydrological scientist and consultant. Horton (1933) suggested that the soil acts as 'a separating surface'

> Infiltration divides rainfall into two parts, which thereafter pursue different courses through the hydrological cycle. One part goes via overland flow and stream channels to the sea as surface runoff; the other goes initially into the soil and thence to the groundwater flow again to the stream or else is returned to the air by evaporative processes.

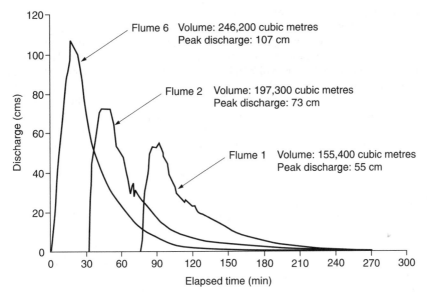

Figure 3.4   The effect of channel transmission losses on storm hydrographs, Walnut Gulch (MAP 324 mm), Arizona. (After Goodrich et al., 1997)

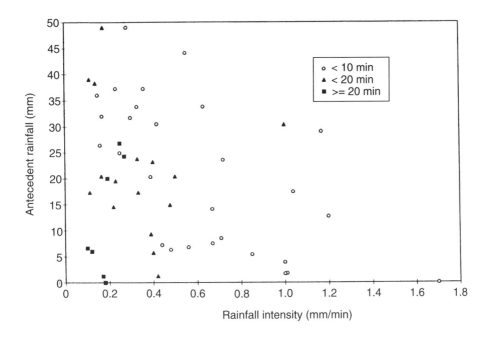

Figure 3.5   Times to initiate runoff on field plots (MAP 496 mm), Shanxi province, China. (After Zhu et al., 1997)

While the term Hortonian overland flow will be widely found in hydrological texts (e.g. Hornberger et al., 1998), I am not sure that he would have totally approved of the widespread use of the infiltration excess concept. Although he frequently used the infiltration excess concept as a way of calculating the volume of runoff production from a particular rainfall (e.g. Horton, 1933), he also had a hydrological laboratory in his wooded back garden in Voorheesville, New York State (Horton, 1936) where he would surely not have observed infiltration excess overland flow very often. Horton was an insightful scientist who published papers on a wide variety of hydrological and meteorological phenomena. His perceptual model surely involved a much wider range of processes than the model that now bears his name (including, for example, the 'concealed surface runoff' that he used to describe flow through cracks in the soil (see Horton, 1942).

It is certainly true that recorded rainfall intensities in semi-arid areas can be very high. Wainwright (1996a and references quoted therein) notes the 264 mm in 10 hours recorded in the Nîmes, southern France, flood event (and up 420 mm to the northwest of Nîmes itself); 556 mm in 48 hours in 1992 in the Spanish Pyrenees; 650 mm in 18 hours in Valencia in 1982; and 866 mm in 72 hours in 1940 in Catalonia. Peak intensities in the Vaison-la-Romaine, southern France, flood event in 1992 were of the order of 200 mm h^{-1} over 6-minute periods (Figure 3.6), while the catastrophic event of 1996 at Biescas in the Arás basin in the Spanish Pyrenees had a maximum recorded intensity of 153 mm h^{-1} over a 10-minute period (White et al., 1997; Gutiérrez et al., 1998). An analysis of rainfall intensity–duration–frequency relationships relative to measured infiltration rates suggests that the infiltration excess runoff generation in semi-arid areas must be a more important source of runoff than in the humid-temperate zones (see, for example, Kirkby, 1969). Surface runoff, or indications that flow has occurred over the surface after an event, are commonly observed in semi-arid catchments. However, we will consider here a perceptual model that is more complex than simple Hortonian runoff generation, and recognises that we still have much to learn about runoff generation in the semi-arid environment.

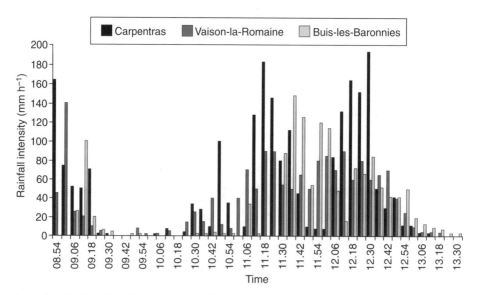

Figure 3.6   Rainfall intensities recorded for 6-minute periods during the Vaison-la-Romaine flood event, expressed as mm h^{-1}. (After Wainwright, 1996a)

All studies of the variability of infiltration rates in semi-arid areas have revealed great spatial heterogeneity and temporal variability in infiltration rates. Indeed, it will be commonly the case that not all the surface is producing runoff such that, even where some infiltration excess overland flow is generated, it may infiltrate (as *'run-on'*) further downslope and may not reach the stream to contribute directly to the hydrograph. This may be particularly true where there is local enhancement of infiltration due to the presence of vegetation and soil cracking (e.g. Dunne and Dietrich, 1980a). Some field studies have demonstrated an apparently random surface runoff generation at the plot scale (Hjelmfelt and Burwell, 1984). Infiltration into the stream bed can also be an important process controlling runoff volumes and peak discharges at larger scales in this type of environment. The role of antecedent moisture conditions is always important in controlling the non-linear responses of runoff to rainfalls, but may have particular importance in the context of the seasonality of semi-arid environments in the Mediterranean and elsewhere.

Rainfalls are not spatially uniform, but can show rapid changes in intensity and volume over relatively short distances, particularly in convective events (e.g. Figure 3.7). Extreme intensities at ground level, after the pattern of intensities has been affected by the vegetation canopy, may be even greater. Some of the rainfall will fall directly to the ground as direct throughfall. Some of the rainfall will be intercepted and evaporated from the canopy back to the atmosphere. Some evaporation of intercepted water may occur even during events, especially from rough canopies, under windy conditions, and when the air close to the canopy may not be saturated with vapour. In general, it is expected that the proportion of a storm lost to interception will decrease as storm volume increases (e.g. Llorens et al., 1997; Bergkamp, 1998b; Figure 3.8). The remaining rainfall will drip from the vegetation canopy as throughfall or run down the branches, trunks and stems as stemflow. The latter process may be important since stemflow may result in local concentrations of water at the soil surface at much higher intensities than the incident rainfall. Some plants that have evolved under semi-arid conditions, such as maize, shrubs and grasses, have a structure designed to channel water to their roots in this way. Slatyer (1965) and Gonzalez-Hidalgo (quoted in Thornes, 1994), for example, have reported stemflow of up to 40% of incident rainfalls.

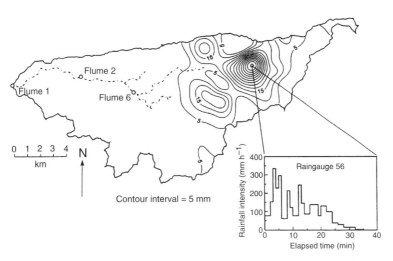

Figure 3.7 Pattern of rainfall volumes recorded during an event on the Walnut Gulch catchment, Arizona. (After Goodrich et al., 1997)

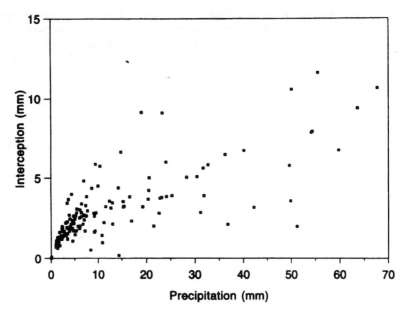

Figure 3.8 Interception losses at the event scale, Ramon Poch field plot, Vallcebre (MAP 850 mm), Catalonia, Spain. (After Gallart et al., 1997)

Once the water has reached the ground it will start to infiltrate the soil surface, except on impermeable areas of bare rock or some human-made surfaces where surface runoff will start almost immediately. The rate and amount of infiltration will be limited by the rainfall intensity and the infiltration capacity of the soil. Soils tend to be locally heterogeneous in their characteristics, so that infiltration capacities might vary greatly from place to place (e.g. Musgrave and Holtan, 1964; Scoging, 1982; Berndtsson and Larson, 1987; Loague and Gander, 1990). In many places, particularly on vegetated surfaces, rainfalls will only very rarely exceed the infiltration capacity of the soil unless the soil is already completely saturated. When infiltration capacities have been exceeded, this will tend to occur in areas where soil permeabilities are lowest and, since infiltration capacities tend to decrease with increased wetting, runoff will gradually expand to areas with higher permeability. Bare soil areas will be particularly vulnerable to such an infiltration of excess runoff since the energy of the raindrops can rearrange the soil particles at the surface and form a surface crust, effectively sealing the larger pores, or an armour layer of stones. A vegetation or litter layer will tend to protect the surface and also create root channels that may act as pathways for infiltrating water.

It has also been suggested that overland flow may occur in semi-arid areas as a saturation excess mechanism (e.g. Scoging and Thornes, 1979; Gallart et al., 1994; Fitzjohn et al., 1998; Martinez-Mena et al., 1998; Taha et al., 1997; Gresillon and Taha, 1998). This suggestion has been made on the basis of inference from field data which show that runoff generation may be more related to antecedent conditions and rainfall volumes than rainfall intensities. In such environments, a saturation excess mechanism does not imply that the soil profile is completely saturated, only that the primary control on infiltration into the soil profile may not be at the surface (a 'true' Hortonian infiltration excess mechanism) but at some (perhaps shallow) depth into the soil, perhaps associated with a tillage layer.

As in more humid environments, a saturation excess mechanism requires that rainfall volumes will be greater than the effective storage deficit in the soil above the controlling layer. Areas of saturated soil will therefore tend to occur first where the antecedent soil moisture deficit is smallest prior to an event, but significant saturation may only occur when there has been antecedent wetting (Gresillon and Taha, 1998). This may be more related to variations in soil characteristics than to the topographic controls that are important in humid areas (e.g. Beven and Kirkby, 1979). The area of saturated soil will tend to expand with increased wetting during a storm, and reduce again after rainfall stops. Thus the effective contributing area for both types of surface runoff generation should be expected to be dynamic during the storm (e.g. the modelling study of Coles et al., 1997, discussed in Section 3.2.4 below).

It is clear that it may be very difficult to distinguish these forms of runoff generation in the field without detailed and careful investigations, particularly if the controlling layer is close to the surface. Indeed, it is quite possible that both mechanisms might occur on the same slope in the same storm and that the threshold rainfall volume for surface runoff generation might be highly variable; on the one hand because of variations in infiltration capacities and on the other due to variations in the effective storage deficit above any controlling layer. When overland flow, by whatever mechanism, is generated, some surface depression storage may need to be satisfied before there is a consistent downslope flow. Even then, surface flow will tend to follow discrete pathways and rills rather than occurring as a sheet flow over the whole surface (Emmett, 1978).

Another feature of rainstorms in semi-arid areas is that the periods of maximum intensities over a point can be short. This may mean that, during an event, runoff generated during a period of high intensity may infiltrate into the soil, perhaps some distance downslope, during a period of lower intensity or immediately after the cessation of rainfall. Bergkamp (1998a), for example, suggests that for point and plot scale sprinkling experiments, in an event with intensities of $70 \, mm \, h^{-1}$, maximum travel distances for overland flow were less than 1 m.

3.1.2 Factors Affecting Infiltration Rates

Infiltration rates are affected by soil texture, soil structure, vegetation cover and land management, raindrop impact and surface crusting, chemical effects controlling disaggregation and clay dispersion, hydrophobicity due to drought or fire, surface stoniness, air pressure effects and the pattern of rainfall intensities at the soil surface in time and space. In general, during a rainstorm that is sufficiently intense to saturate the soil surface, infiltration rates will decrease with time (Figure 3.9). This decrease is predicted by Darcian flow theory as a result of the decline in the effective capillary pressure gradient as the wetting front moves into the soil. In the field, the decline may also be due to the effects of crusting at the surface and the filling or blocking of preferential flow pathways by the redistribution of fine sediment due to rainsplash and overland flow.

Infiltration into the soil surface can be significantly affected by such crusts which, by blocking the larger pore spaces at the soil surface, greatly reduce the permeability of the surface layer (see, for example, the review by Römkens et al., 1990 and more recent papers by Valentin and Bresson, 1992; Vandervaere et al., 1997; Zhu et al., 1997; Smith et al., 1999), although recovery can be rapid (see Figure 3.10). Bare surfaces of dispersive soil materials are particularly prone to crusting and crusts, once formed, will persist between storms unless broken up by vegetation growth, freeze–thaw action, soil faunal activity, cultivation or erosion. Studies of crusted soils have shown that, in some cases, infiltration rates after ponding might increase over time in a

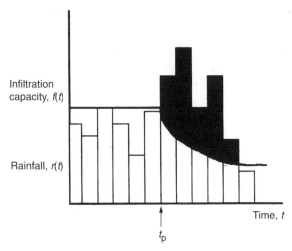

Figure 3.9 Infiltration decrease with time

rainstorm more than would be expected as a result of the depth of ponding (e.g. Fox et al., 1998). This was thought to be due to the breakdown or erosion of the crust.

It has long been speculated that during widespread surface ponding there could be a significant effect of air entrapment and pressure build-up within the soil on infiltration rates. This has been shown to be the case in the laboratory (e.g. Wang et al., 1998) and, in a smaller number of studies in the field (e.g. Dixon and Linden, 1972). It has also been suggested that air

Figure 3.10   Re-establishment of macropores by ants immediately after a high-intensity rainfall event

pressure effects can cause a response in local water tables (e.g. Linden and Dixon, 1973) and that the lift associated with the escape of air at the surface might be a cause of initiation of motion of surface soil particles. The containment of air will be increased by the presence of a surface crust of fine material, but significant air-pressure effects would appear to require ponding over extensive areas of a relatively smooth surface. In the field, surface irregularities (such as vegetation mounds) and the presence of macropores might be expected to reduce the build-up of entrapped air by allowing local pathways for the escape of air to the surface.

In the absence of a surface crust, the underlying soil structure, and particularly the macroporosity of the soil, will be an important control on infiltration rates. Since discharge of a laminar flow in a cylindrical channel varies with the fourth power of the radius (see, for example, Childs, 1969), larger pores and cracks may be important in controlling infiltration rates. However, soil cracks and some other macropores, such as earthworm channels and ant burrows, may only extend to limited depths so that their effect on infiltration may be limited by storage capacity and infiltration into the surrounding matrix rather than potential flow rates. Some root channels, earthworm and ant burrows can, of course, extend to depths of metres below the surface (for reviews of the effects of preferential flows in such large pores, see Beven and Germann, 1982). Even on bare soil and rock surfaces, however, some rainfall may enter cracks and fractures, perhaps reappearing at the surface further downslope (as seen, for example, in the plot scale artificial rainfall experiments of Hodges and Bryan, 1982, and Oostwoud Wijdenes and Ergenzinger, 1998).

There is an important positive feedback between infiltration and vegetation in semi-arid areas where water stress is such an important control on vegetation growth and competition. In an environment where annual potential evapotranspiration rates may be a factor of 2 or more greater than annual rainfall rates, infiltration and subsurface storage will make more water available to the vegetation. Growth of vegetation will tend to protect the soil surface from rainsplash and crust formation as well as improving the soil structure and macroporosity, thereby enhancing infiltration rates. Many studies, using both artificial and natural rainfall, have concluded that vegetation is a dominant control on runoff generation in semi-arid areas (see, for example, Yair and Lavee, 1985; Wilcox et al., 1988; Lavee et al., 1991; Böhm and Gerold, 1995; Snelder and Bryan, 1995; Sorriso-Valvo et al., 1995; Nicolau et al., 1996; Wainwright, 1996b; Kosmas et al., 1997; Solé-Benet et al., 1997). Bergkamp (1998a) suggests that vegetation patterns both reflect and control the water movement on at least parts of Mediterranean hillslopes.

Factors affecting the vegetation cover will therefore also have an impact on infiltration rates. The density of vegetation increases with available water, and this will tend to offset the possibility of increased runoff generation with increasing rainfall. Kosmas et al. (1997) show that for a variety of shrubland sites around the Mediterranean, runoff coefficients are greatest when the annual rainfall is in the range 200–300 mm. Higher rainfalls result in increased vegetation, decreasing runoff coefficients. Lower rainfalls will not be able to support vegetation but will necessarily have less runoff, even if runoff coefficients are higher. They also show that different cultivations have different tendencies to produce runoff in the Mediterranean, with the smallest runoff coefficients being associated with olive trees grown with a semi-natural understorey and the highest (up to 32%) with perennial crops subject to weed control and mechanical tillage. A similar pattern of increasing runoff and erosion with decreasing annual rainfall has been reported from Spain by Imeson et al. (1998).

Another important factor affecting vegetation in semi-arid areas is fire and there have been many studies of the effects of fire on runoff generation. The general effect is that fire reduces infiltration rates and increases runoff coefficients, at least in the period immediately after a fire (e.g. Cerda, 1998b; Prosser and Williams, 1998), especially after high severity burns (Robichaud and Waldrup, 1994; Martin and Lavabre, 1997). At Mount Carmel in Israel, in the first year after the fire of 1989, runoff coefficients on steep burned plots were up to 500 times greater than on unburned plots (Inbar et al., 1998). Fire reduces protection of the surface resulting from the loss in vegetation cover and will tend to increase the water repellency or *hydrophobicity* of the soil surface (Savage, 1974; Soto and Díaz-Fierros, 1998).

Hydrophobicity does not necessarily result in the generation of surface runoff (Burch et al., 1989; Imeson et al., 1992; Kutiel et al., 1995). Doerr et al. (1996), for example, report plot scale studies of infiltration and runoff generation in Portugal where the soil had high hydrophobicity but low runoff generation. They show that at their sites hydrophobicity is mostly associated with the finer particle fractions of the soil and does not decrease with depth into the soil. Hydrophobicity tends to decrease with wetting and fire has a relatively small effect. Other studies have also shown that one effect of hydrophobicity may be to enhance the deep infiltration of water along macropores and lines of preferential flow (e.g. Cerda, 1998a; Cerda et al., 1998).

Stone cover may also have both positive and negative impacts on infiltration rates and erosion (e.g. Brakensiek and Rawls, 1994). Stone fragments at the soil surface intercept rainwater and reduce the surface area available for infiltration, especially if sufficient fines are removed to form a gravel lag (Abrahams and Parsons, 1991). Stones will also, however, protect the surface from rainsplash and crusting, with a tendency to increase infiltration rates, particularly if the fragments are lying on the surface rather than partially buried (Poesen and Ingelmo-Sanchez, 1992; Poesen and Bunte, 1996). Small rock fragments will be more effective in enhancing infiltration than large fragments which will tend to produce more concentrated flows of water to the surrounding soil surface (Poesen and Lavee, 1991). Buried rock fragments will tend to reduce the permeability of the soil matrix but, for a given quantity of infiltrated water, may also result in deeper penetration of the wetting front into the soil.

A further important factor on infiltration in semi-arid regions that cannot be neglected is the effect of humans. It is a common perception that the effect of humans has led to land degradation and, in extreme or marginal areas, desertification. Certainly the effects of tillage practices on topsoil structure can have significant effects on infiltration and runoff generation and sediment production, and in some studies appear to be the dominant factor (e.g. Peugeot et al., 1997; Leonard and Andrieux, 1998).

3.1.3   Infiltration and Runoff at the Plot Scale

There have been many plot-scale experiments to study runoff generation under both natural and artificial rainfalls in semi-arid areas. In most cases, the perception that runoff is dominated by a Hortonian infiltration excess mechanism has meant that no other process of runoff generation has even been considered. However, many such studies have shown that runoff coefficients tend to decrease as plot sizes increase (Yair and Lavee, 1985; Yair, 1992; Hawkins, 1982). In a recent study, Bergkamp (1998a) used both sprinkler experiments on large plots and field plots under natural rain in Central Spain and concluded that there may be apparently little connection between local fine-scale runoff generation and the surface runoff seen at the hillslope scale as a result of the complexity of local soil and vegetation characteristics and the reinfiltration of runoff generated on upslope areas into soil of higher infiltration capacity further downslope. Vegetation patterns in many semi-arid regions are a mixture of bare soil and vegetation clumps or bands. Studies have shown that the vegetated areas may be dependent on runoff generated on the adjacent bare soil as a source of additional moisture (e.g. Puigdefàbregas and

Sánchez, 1996; Thiéry et al., 1995; Valentin and d'Herbès, 1999). The same suggestion has also been made for cultivated plots in the Sahel region in Niger (Rockström et al., 1998).

There have been a much smaller number of plot-scale experiments of runoff generation involving multiple replicate plots. The study of Hjelmfelt and Burwell (1984) involved the measurement of runoff from 40 adjacent cultivated plots (each 27.4 by 3.2 m) at the Claypan Experimental Watershed, near Kingdom City, Missouri. The results were surprising. Measurements suggested that the soil on the plots was relatively uniform. Total runoff volume was measured in tanks for 26 events over a six-month period in 1981, giving data for 984 plot runoff volumes (95% of the possible total). Mean storm runoff over all plots ranged from 0.22 to 78.56 mm. Distributions of runoff volumes in any storm were broadly Gaussian with coefficients of variation ranging from 7.1 to 109%. The mean total runoff over all storms measured was 457 mm with a coefficient of variation of 10%. In fact only two of the individual storms had coefficients of variation less than 10%, suggesting that the plots were not consistent in their runoff production relative to the measured rainfall. The differences could not be attributed to carryover effects from previous cultivations, quantifiable surface variations or to spatial variation in soil texture.

In Australia, Bonell and Williams (1986) and Williams and Bonell (1988) used unbounded runoff troughs on a eucalypt woodland hillslope (MAP 552 mm) placed at different distances downslope to examine the differences in runoff and infiltration parameters at different scales. They showed that point infiltration measurements showed much greater spatial variability than the plots. Spatial variability in runoff for the plots was relatively small but temporal variability in the effective infiltration parameters was significant. Runoff was mostly generated on bare soil areas beneath the trees and then redistributed downslope. Some areas acted as net source areas for runoff; other areas gained more overland flow than they exported. Overall, the net runoff coefficient was only 2%.

3.1.4 The Role of Subsurface Flow

Downslope subsurface flows are important in controlling runoff generation in humid environments but, as noted above in the context of the design of runoff plot-scale experiments, are often ignored in hydrological assessments of semi-arid areas, despite the fact that storm hydrographs can continue for several hours after rainfall has stopped (albeit with a tendency for the recession limb to be steep). It is certainly true that semi-arid areas will normally have, at least seasonally, a cumulative potential evapotranspiration that is considerably greater than the volume of water that infiltrates the soil surface. Thus, it is commonly assumed that the infiltrated water is a 'loss', eventually evaporating or transpiring back to the atmosphere. Field studies have shown that in some areas soil moisture levels may actually decrease downslope (e.g. Yair and Danin, 1980), suggesting that downslope flows might be considered negligible.

However, during short periods when significant rainfall does infiltrate into the soil, there may still be flow to depth and it is of interest to pose the question whether subsurface flow can be an important control on runoff generation in semi-arid areas. The perception is that the runoff coefficients for subsurface flow will generally be low in semi-arid areas because of the possibility of storage in a dry soil. It is worth remembering that a 50 cm depth of soil, with an average porosity of 0.4 has a storage capacity of 200 mm. Thus, if the infiltration capacity of the soil is not exceeded, a large 100-mm rainstorm could, in principle, be totally absorbed by that soil layer, even if the antecedent storage deficit is only half the porosity (and it will often be less, prior to a storm, in a semi-arid soil). When saturation does start to build up at the base of the soil over a relatively impermeable bedrock, it will start to flow downslope. The connectivity of

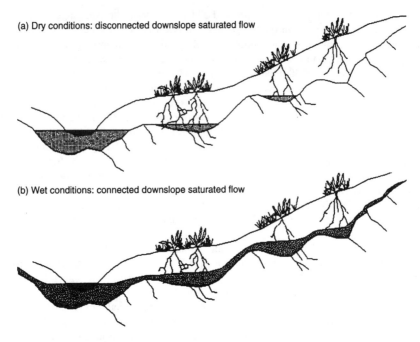

Figure 3.11 Depression storage/bedrock storage and connectivity

saturation in the subsurface will, however, initially be important. It may be necessary to satisfy some initial bedrock depression storage before there is a consistent flow downslope (Figure 3.11). Effective upslope subsurface contributing areas will be limited where the soil is at least seasonally dry (Barling et al., 1994; Western et al., 1999). The dominant flow pathways may be localised, at least initially, related to variations in the form of the bedrock surface (e.g. McDonnell et al., 1996) or to preferential flows in bedrock cracks and fractures. In semi-arid areas, such subsurface flows may be important in maintaining levels of water storage in valley bottoms over long periods of time, but in general it is expected that the soil will act to store rainfalls and reduce runoff coefficients.

Soil pipes and pipe erosion do serve as additional evidence that subsurface flows can be generated in semi-arid areas (e.g. Harvey, 1982; López-Bermúdez and Romero-Díaz, 1989). Piping requires significant subsurface flow velocities and poorly cohesive soil materials. Such soils often have low infiltration capacities, but there may be sufficient spatial heterogeneity to allow the generation of subsurface runoff with locally steep hydraulic gradients. Abandoned cultivation terraces are particularly favourable locations for the initiation of piping of this sort. Torri et al. (1994), as a result of sprinkling experiments on plots in Tuscany, Italy, have even suggested that cracks and piping may be more important in moving water downslope than surface rills (perhaps a true 'concealed surface runoff' in the sense of Horton).

Even where piping is not an important process, however, subsurface water storage may have important long-term feedback controls on erosion and vegetation development. Bryan et al. (1998) have shown in flume experiments that erosivity of the soil surface can be increased by increased topsoil moisture contents due to a decrease in soil strength. Higher soil moisture content might therefore lead to greater erosion, at least on bare slopes (Gabbard et al., 1998).

On the other hand, Bergkamp (1998b) has shown that vegetation recovery after disturbance is related to access by the vegetation to subsurface water as well as other factors such as slope and aspect. More rapid vegetation recovery will mean more protection for the soil surface, less likelihood of crusting and consequently less surface runoff generation and erosion. A further level of complexity is added by consideration of the downslope effects of erosion. Not all storm runoff and surface material that is mobilised will reach a stream channel. The water will often infiltrate further downslope and the particles being transported will be deposited, leading to greater storage capacity. It is often vegetation that serves to trap both runoff and sediment (e.g. Bergkamp, 1998a), either locally on the hillslope or in the valley bottom. Certainly valley bottoms are often wetter and support better vegetation cover and greater plant productivity. Thus, at least in some circumstances, subsurface flow will have a role in the factors, particularly vegetation cover, that control surface infiltration. If a water table is near to the surface, subsurface flow may also have a role in runoff generation. In particular, valley bottoms may change from acting as storage zones for runoff generated on hillslopes and trails and become runoff generation areas under wet conditions (e.g. Ceballos and Schnabel, 1998; Croke et al., 1999).

Direct evidence of the importance of subsurface flows in semi-arid areas has been reported by Wilcox et al. (1997) from a study of runoff generation in a small forested catchment in New Mexico (MAP 500 mm; RC 3–11%). Measured infiltration rates varied by two orders of magnitude on the hillslopes studied. Surface runoff was generated by an infiltration excess mechanism during occasional summer thunderstorms – by prolonged frontal storm periods lasting several days and producing larger volumes of surface runoff – and also during snowmelt where the soil had been frozen prior to the development of the snowpack. Subsurface runoff was the result of the development of a perched water table during wet periods (see Figure 3.12). The rapid subsurface responses and low hydraulic conductivity of the soil suggested that the bulk of the subsurface flow may have been taking place through macropores. Differences in surface runoff between north- and south-facing slopes were observed. It was suggested that the more connected patchwork of bare surfaces on south-facing slopes were more efficient in allowing surface runoff to reach the base of the slope as overland flow.

There is also some indirect evidence that subsurface flows may contribute to storm runoff in semi-arid areas. As in humid temperate areas, the contributions of subsurface flows can be inferred from analyses of measured hydrochemical characteristics of catchment discharges. The idea is to represent the discharge as a mixture of different components or *end members*, each having its own distinctive hydrochemical signature (e.g. Sklash, 1990; Hooper et al., 1990). Such an analysis is limited by some important assumptions, in particular that each end member has a distinct and identifiable signature that does not change rapidly during an event. End member mixing analysis has been mostly used in humid temperate catchments, but there have been just a small number of studies in semi-arid catchments. Turner et al. (1991) showed that for pasture areas in Australia, event rainfalls made up some 65–75% of the storm hydrograph. The remainder must have been related to subsurface responses, as displaced soil water, but they suggested that there was no evidence for contributions from deep groundwaters. Neal et al. (1992) suggested that the bulk of the storm hydrograph for the Montseny and Prades catchments in Catalonia, Spain, was made up of pre-event water, but that because there were at least two types of source waters the isotope data could not easily be used for hydrograph separation.

Sandström (1996), in a study of forested and degraded catchments in semi-arid Tanzania, did detect contributions from a saturated zone in the soil, making up some 22% of the total discharge, and between 0 and 62% of the increase in discharge during storms in the forested catchment (Figure 3.13). There was an inverse relationship between the proportion of sub-surface stormflow and rainfall intensities in a storm. He suggests that in this catchment

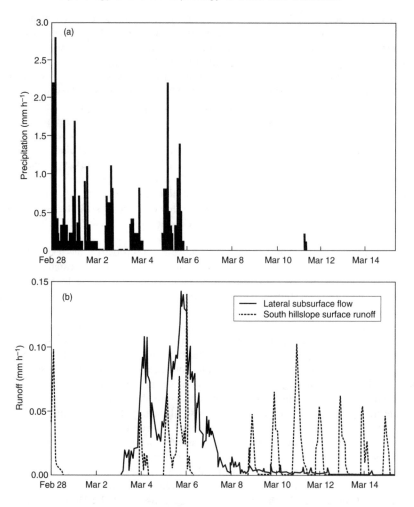

Figure 3.12   Subsurface flow at the hillslope scale, Pajarito plateau (MAP 500 mm) Los Alamos National Laboratory, New Mexico. (Wilcox et al., 1997)

macropores play a critical role in the delivery of throughflow to the stream channel by enhancing deep percolation to the saturated zone. In degraded areas, this effect may be lost, leading to reduced percolation, increased flashiness and increased surface erosion. Throughflow trough experiments in the seasonally dry Panola catchment in Georgia, USA, also consistently produced both macropore flows and downslope seepage in rainstorms with a volume greater than about 100 mm (Freer et al., 1997).

Isotope measurements from the Maurets catchment in the south of France suggested that up to 80% of the storm hydrograph was derived from pre-event subsurface water (Taha et al., 1997). Here, it was suggested that the infiltration capacity of the soil surface was generally high, but that a perched water table developed within the top 20 cm of the soil profile. This soil layer had a high hydraulic conductivity and allowed subsurface runoff to reach the stream within the time-scale of the storm hydrograph, given wet antecedent conditions (Gresillon and Taha,

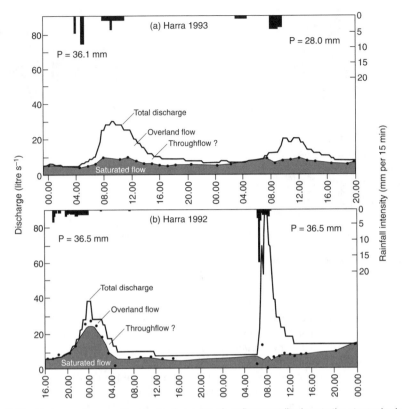

Figure 3.13 Hydrograph separations indicating subsurface flow contributions to the stream hydrograph, North-central Tanzania (MAP 800 mm) (after Sandström, 1996)

1998). The evidence that, at the catchment scale, the stream runoff was mostly pre-event water implies that sufficient water was already stored in the soil and was displaced by the incoming rainwater. This interpretation was tested using a two-layer Darcian model of a representative hillslope. Nicolau et al. (1996) suggest that the compaction of soil following abandonment of cultivation on an alluvial fan may be a factor in the generation of this type of subsurface runoff.

As another example from personal experience, in the summer of 1981 the USDA organised some artificial rainfall experiments on reclaimed mine spoil sloping plots (25 m by 5 m) near to Steamboat Springs, Colorado (Figure 3.14). At the base of each plot was a runoff collector and H-flume to measure the rate of surface runoff from the plot. Each collector discharged into a shallow ditch to take the runoff to the nearest channel. Surface runoff was generated on each plot during the rainfall, but with highly heterogeneous flow depths and sediment transport. At the end of the rainfall, surface runoff continued to be collected for about 15 minutes. It was then apparent that there was a larger discharge as a shallow subsurface flow coming from beneath the collector. This flow (which was not being measured) continued for some 90 minutes after the rainfall (Figure 3.14(c)).

This latter experiment was, in many ways, a special case. Because the plots were on reclaimed mine spoil, the 'soil' had been reworked by heavy machinery. It is quite possible that this had created a compacted layer at a shallow depth before the final 30 cm of topsoil had been added to

(a)

(b)

(c)

Figure 3.14 Field runoff experiments under artificial rainfalls at the plot scale, reclaimed mine spoil sites, Steamboat Springs, Colorado. (a) How many hydrologists does it take to measure an infiltration rate? (b) 'Deltaic' sedimentary deposition upslope of the surface runoff collector immediately after the cessation of sprinkling. (c) Subsurface runoff from beneath the collector flume continuing after the cessation of surface runoff

the plots. It is, however, perhaps not so different from having a shallow layer of permeable soil over bedrock, which is a common situation in many semi-arid areas. Could it be that the role of subsurface flow in runoff generation has been neglected in semi-arid areas only because it has been perceived as being unimportant and, therefore, has not been measured? Such flows may be the explanation of the extended recession limb sometimes found in the discharge hydrographs from semi-arid catchments. If such a flow is to be modelled as a surface runoff alone, an artificially low infiltration capacity would be required to predict the runoff volume and an unrealistically high effective roughness coefficient would be required to reproduce the recession limb of the hydrograph.

3.1.5 Integration to the Catchment Scale

Runoff generation at the hillslope and catchment scale is the end result of all the spatial and temporal complexity of processes recognised in the perceptual model of point runoff generation. Locally at the point or plot scale, great heterogeneity of inputs, surface and subsurface characteristics, and antecedent conditions may result in a complexity in runoff generation that is almost impossible to either measure or predict in detail. Some of this variability will, however, be integrated out as the interest moves to larger scales, as a result of a complex, non-linear averaging process.

This integration is complex because it involves a balance of concentrating and dispersive effects. The concentrating effects arise as a result of downslope flow of surface runoff leading to deeper flow depths and a tendency for self-organisation into channelled flows in rills and gulleys. The dispersive effects arise partly as a result of the dispersive nature of dynamic waves in shallow free surface flows, but also because surface runoff that is generated upslope may later infiltrate into unsaturated soil or cracks as run-on further downslope or at the edges of rills and gulleys (Dunne and Dietrich, 1980a, 1980b). In cracking clay soils, deep cracks can absorb significant quantities of surface water and the duration of rainstorms may be too short to allow the cracks to seal, or for runoff to flow to a channel without infiltrating. The result is often that long slopes may produce less runoff per unit area than short slopes and that there may be a significant decrease in runoff coefficient with increasing catchment area (see Figure 3.3). Indeed, it has been suggested that, except in the most extreme events, runoff production on semi-arid hillslopes may be effectively disconnected from the stream channel response (Yair and Lavee, 1985; de Boer, 1992; Yair, 1992; Bergkamp, 1998a). This implies that the main source area for stream runoff at the catchment scale may be subsurface contributions from valley bottom deposits, perhaps with some displacement resulting from inputs of water from the hillslopes. The effective contributing area for this mechanism will increase with increasing wetness or rainstorm volume, until for the larger events subsurface and surface runoff from the hillslopes may also contribute directly to the stream.

The valley bottom can also act as a sink for runoff due to infiltration through the channel bed. This *transmission loss* can have an important effect on both the shape and magnitude of hydrographs, particularly in ephemeral stream channels and where the local valley bottom water table is at depth below the stream channel (e.g Osterkamp et al., 1994; Goodrich et al., 1997). Such losses may be very difficult to estimate, since they may be controlled by the hydraulic conductivity of thin layers of the channel bed, but in many semi-arid areas they are a very important source of groundwater recharge to valley bottom aquifers (e.g. Osterkamp et al., 1994).

It seems that it may be impossible to describe the end result at the catchment scale by a simple aggregation of the small-scale processes, even if it were feasible to obtain some statistical characterisation of the small-scale properties. Characterising the system by means of effective Darcian parameter values would also not seem to be a viable approach for coupled surface/subsurface flow systems (Binley et al., 1989). How, then, should the complex hydrological responses in semi-arid catchments be modelled?

3.2 MODELLING RUNOFF GENERATION IN SEMI-ARID AREAS

The previous sections have reviewed, briefly, current perceptual understanding of the processes of runoff generation arising from a variety of studies in different semi-arid environments. The perceptual model that results is complex and recognises effects such as macroporosity, hydrophobicity, and the effects of vegetation, surface rock fragments and crusting that may be difficult to quantify. Spatial heterogeneity of soil and vegetation characteristics is recognised as an important control on the (highly non-linear) runoff generation process that will not only be difficult to describe mathematically but will be difficult to assess in the field for any particular application.

There is, however, a need for quantitative predictions of runoff generation in semi-arid areas for flow forecasting and water resources management under current conditions and to allow the impacts of different management strategies to be assessed. This will then require the definition of a conceptual model in the form of mathematical equations that can be solved to allow quantitative predictions. It is important to recognise at this point that any conceptual model, however complex the mathematical description on which it is based, will necessarily be a simplification of the qualitative perceptual understanding reviewed above (Beven, 1989a, 1993, 2000b). The trick is always to find a form of mathematical description that mimics the dominant process controls in a way that results in acceptable predictions, even if the model is not correct in all details.

Pilgrim et al. (1988) reviewed the problem of rainfall–runoff modelling in arid and semi-arid regions, in the light of experience in humid regions, and summed up the differences and difficulties succinctly. They noted in particular that the whole nature of the vegetation and hydrology in semi-arid regions can change as a result of a prolonged wet or dry spell; the variability of rainfalls in time and space; the importance of channel transmission losses; and the complex controls on the soil surface properties that affect infiltration rates. A major conclusion of their review was that there was a lack of data for catchment responses in semi-arid and arid regions that could be used for the calibration and evaluation of modelling approaches. One significant development in the decade since then has been the development of projects combining both modelling and field data collection programmes, such as the European Union-funded MEDALUS project (Brandt and Thornes, 1996).

3.2.1 Point-Scale Runoff

At the point scale it might appear that producing an appropriate conceptual model is relatively straightforward, at least where the dominant process is infiltration excess runoff generation. What is required is an input time sequence of rainfall intensities, and a description of the way in which the infiltration capacity of the soil changes over time. The excess of rainfall over infiltration will then be available as surface runoff at that point. This is, in effect, the classic Horton model:

$$r(t) = p(t) - f(t); \qquad p(t) > f(t) \tag{3.1}$$

where $r(t)$ is the runoff generated at time t, $p(t)$ is the precipitation rate and $f(t)$ is the infiltration rate. For rainfall rates that are initially less than the infiltration capacity of the soil, this equation will only apply after the *time to ponding*.

This apparently simple equation turns out to be extremely difficult to apply for three reasons. One is the problem of knowing exactly what the precipitation (or throughfall plus stemflow) input intensities are at a point; the second is knowing what the soil characteristics that control infiltration are at a point; the third is the difficulty of solving the soil water flow equations for particular soil and rainfall characteristics. Most of the research in this area has concentrated on the third problem, since the first two are essentially problems of data availability. As we shall see, however, they are also a major constraint on our capabilities to predict runoff generation.

Obtaining a solution of the soil water equations presupposes that we can define a descriptive equation for the flow processes. Most of the available descriptions are based on Darcy's law, which for the case of partially saturated flows, leads to a non-linear partial differential equation first proposed by Richards (1931). The details can be found in any soil physics text. For now, we should note that the Darcy and Richards equations assume that flow velocity is proportional to a definable hydraulic gradient in the soil matrix, where the constant of proportionality (the hydraulic conductivity) changes in a non-linear way, and over orders of magnitude, as the soil wets and dries. The Richards equation does not explicitly take account of flows in macropores.

For the infiltration case, the Richards equation can only be solved analytically for some very specific soil characteristic curves and boundary conditions. A wide variety of solutions have been published, the majority of which are based on the assumption that the rainfall intensity is constant in time. Time variable rainfalls are then introduced by a further approximation by integrating equation (3.1) over time such that:

$$R(t) = P(t) - F(t); \qquad P(t) > F(t) \tag{3.2}$$

where $R(t)$ is the cumulative runoff generated at intervals up to time t, $P(t)$ is the cumulative rainfall volume and $F(t)$ is the cumulative infiltrated volume. Again, for rainfall rates that are initially less than the infiltration capacity of the soil, this equation will only apply after the *time to ponding.* For the case of time variable rainfalls, estimating the time to ponding by the time at which cumulative rainfall is equal to cumulative infiltration predicted by one of the equations below, is called the *time compression approximation* (Parlange et al., 2000).

The simplest solution to the Richards equation for infiltration actually pre-dates it and is based on assuming that infiltrated water enters the soil as a sharp front between the initially dry soil ahead of the front and saturated soil behind it. With these assumptions, Green and Ampt (1911) used a simple form of Darcy's law to represent the infiltration such that the infiltration rate f is calculated as:

$$f(t) = K_\mathrm{s} \left(\frac{h_0 + \psi_\mathrm{f}}{z_\mathrm{f}} + 1 \right) \tag{3.3}$$

where K_s is the hydraulic conductivity of the soil at field saturation, h_0 is the depth of ponded water on the soil surface, ψ_f is the initial capillary potential of the dry soil and z_f is the depth of penetration of the wetting front at time t. The first term in the brackets in this equation is due to the capillary potential gradient, here estimated from the difference in capillary potential across the wetting front averaged over the depth of penetration. The magnitude of this term gets less as the wetting front goes deeper, leading to the decline in infiltration capacity of the soil (Figure 3.9). The second term in the gradient component of (3.3) is the gravitational term. It stays at unity, regardless of the depth of the wetting front, and when multiplied by the effective saturated hydraulic conductivity gives the final infiltration capacity.

The original Green and Ampt infiltration equation assumes constant soil characteristics with depth. An analysis of a wide variety of soil moisture characteristics data by Rawls and Brakensiek (1989) has led to a classification of the Green and Ampt parameters by soil texture. These types of relationship should be used with care, however, since they are based on mea-surements of small samples brought back to the laboratory, not field measurements at the plot scale. Beven (1995) has produced a solution with equivalent assumptions for the case where hydraulic conductivity declines exponentially with depth – often a useful approximation of real soil characteristics.

Morel-Seytoux and Khanji (1974) have shown that a parameter called the *capillary drive*, with units of length and defined as $C_\mathrm{D} = \psi_\mathrm{f} \Delta\theta$, is a relatively constant parameter for a range of initial moisture conditions defined as $\Delta\theta = (\theta_\mathrm{S} - \theta_\mathrm{i})$ which is the change in moisture content between the initial state of the soil, θ_i, and field saturation, θ_S. The Green–Ampt equation is then better applied in the form:

$$f(t) = \frac{K_\mathrm{s}}{B} \left(\frac{h_0 \Delta\theta + C_\mathrm{D}}{z_\mathrm{f} \Delta\theta} + 1 \right) \tag{3.4}$$

where B, an additional parameter proposed by Morel-Seytoux and Khanji (1974) to allow for air pressure effects, is called the viscous resistance correction factor ($1 < B < 1.7$).

A different set of assumptions about the soil was used by Philip (1957) to derive an infiltra-tion equation as an approximate solution of the Richards equation, of the form:

$$f(t) = 0.5St^{-0.5} + A \tag{3.5}$$

where S, the *sorptivity* of the soil, is calculated from knowledge of the moisture characteristics of the soil, and A is a final infiltration capacity equivalent to the f_c of the Horton equation (equation (3.7) below) or K_s of the Green–Ampt equation. The effects of the sorptivity term gradually reduce with increasing time, eventually leaving the final infiltration capacity as a function of the effective saturated hydraulic conductivity of the soil. Williams and Bonell (1988) have fitted the Philip equation to large plot-scale infiltration and runoff measurements, as well as at the point scale.

Under assumptions of a diffusivity that changes exponentially with moisture content, Smith and Parlange (1978) derived another widely used infiltration equation that takes the form:

$$f(t) = K_s \left(\frac{\exp F(t)/C_D}{\exp F(t)/C_D - 1} \right) \qquad (3.6)$$

where $F(t)$ is the cumulative infiltration and C_D is the capillary drive as above. A variety of other approximate solutions of the infiltration case, based on different assumptions, have been summarised by Smith (1981), Parlange and Haverkamp (1989) and Corradini et al. (1997).

It is worth noting that in Horton's original work, he used an empirical equation to describe infiltration into the soil based on a large number of field infiltration measurements, rather than a solution of the theoretical Richards equation. Horton (1933) used an exponential equation to describe the decline of infiltration rate towards a final capacity, f_c, of the form:

$$f(t) = (f_0 - f_c) \exp\{-kt\} + f_c \qquad (3.7)$$

where $f(t)$ is infiltration rate at time t, f_0 is an initial infiltration rate, f_c is a final infiltration capacity and k is an empirical constant. Eagleson (1970) has shown that this equation can also be derived from the Richards equation under simplifying assumptions. All infiltration experiments tend to show that, at the start of rainfall, infiltration rates are high and decline gradually over time. The initially high rates are due to the effects of capillary potential drawing water into the initially dry soil in addition to the effects of gravity. The final infiltration capacity, f_c, will be close to the hydraulic conductivity of the soil at field saturation. Horton noted that the parameters of his infiltration equation could be time dependent – in particular, varying seasonally with the activities of soil fauna (Horton, 1940).

Another widely used empirical infiltration model, originally derived from runoff experiments on field plots and small catchments, is the USDA Soil Conservation Service (SCS) model (McCuen, 1982). The SCS method can be applied by specifying a single parameter called the curve number, CN, and the popularity of the method arises from the tabulation of CN values by the USDA for a wide variety of soil types and conditions. In the SCS method it is postulated that the cumulative volume of infiltration $F(t)$ can be predicted as:

$$F(t) = S_{max} \left(\frac{R(t)}{P(t) - I} \right) \qquad (3.8)$$

where $P(t)$ is the cumulative rainfall volume, I is an initial abstraction, S_{max} is a potential maximum retention, and $R(t)$ is the volume of rainfall excess over retention. From empirical studies it was assumed that $I \approx 0.2S_{max}$ so that it can be shown that

$$R(t) = \frac{(P(t) - 0.2S_{max})^2}{P + 0.8S_{max}} \qquad (3.9)$$

The value of S_{max} (mm) for a given soil is related to the curve number as

$$S_{max} = 245 \left(\frac{100}{CN} - 1 \right) \qquad (3.10)$$

The curve number ranges in value from 0 to 100 and varies with soil type, land use and antecedent conditions. Tables are provided for the estimation of the curve number for different circumstances. For areas of complex land use, it is suggested that a linear combination of the component curve numbers, weighted by the area to which they apply, should be used to determine an effective curve number for the area. The curve number approach to predicting runoff generation has been the subject of a number of critical reviews (e.g. Rallison and Miller,

1982; Hjelmfelt et al., 1982; Bales and Betson, 1982) but it has seen somewhat of a recent rehabilitation due to an interpretation of the equation in terms of variable contributing areas (Steenhuis et al., 1995; Yu, 1998; Beven, 2000b). Mishra and Singh (1999) have shown how the form of the equation can be derived from simple assumptions about the water balance of a catchment. It does also have the advantage that the original formulation of the method was based on data collected at scales larger than point infiltration measurements (and may also have reflected subsurface as well as surface contributions). However, the continuing popularity of the method is primarily due to the fact that a GIS database of soils and vegetation information can be easily related to tabulated curve number values whether those values are appropriate in a particular location or not.

Treating infiltration capacity as a function of infiltrated volume can also be used to treat the case where overland flow is produced as a result of the topsoil layer becoming saturated owing to a limitation on vertical flow at some depth within the soil. This can occur where either a thin soil overlies an impermeable bedrock or where there is some horizon of lower permeability at some depth into the soil profile (e.g. Taha et al., 1997). In these circumstances, infiltration rates might be controlled more by a saturation excess than by a surface infiltration excess process. This situation was addressed by Kirkby (1975; see also Scoging and Thornes, 1979) using an infiltration equation of the form:

$$f(t) = \mathrm{B}/H + A \qquad (3.11)$$

where H is the current depth of storage, A is the long-term infiltration rate (which may now be controlled at depth) and B is a constant. For steady rainfall inputs this results in infiltration capacity being an inverse function of time rather than the square root of time as in the Philip (1957) equation.

The search for simple analytical solutions of the Richards equation was necessary in the past because computer limitations meant that the full non-linear partial differential equation could not easily be solved in a reasonable time. These limitations have now largely disappeared and most of the distributed modelling systems that are now available in hydrology use direct numerical solutions of the Richards equation to calculate surface runoff generation due to an infiltration excess mechanism. One example of such a modelling system is the SHE model (Système Hydrologique Européen; see, for example, Abbott and Refsgaard, 1996). In the SHE model, a catchment is represented as a large number of grid elements in space. A one-dimensional (vertical) finite difference solution of the Richards equation is carried out for every grid element at every time step, effectively treating each grid element as a 'point'. This approach allows arbitrary variations in rainfall intensities over time and arbitrary variations in soil characteristics with depth to be incorporated into the solution, at the expense of introducing many more parameter values that must be specified to characterise the soil. In general, such models assume that the soil characteristics stay constant over the period of a simulation. It is worth noting that these models remain only approximate numerical solutions to a purely Darcian description of the vertical flow processes, and that care is needed in specifying solution algorithms and time steps to obtain a solution that is consistent with a true solution of the Richards equation.

The perceptual model also recognises other processes that might be important in runoff generation at a point. These include hydrophobicity, crusting and flow in soil cracks and other macropores due to roots or soil fauna. All of these processes have also been the subject of modelling studies (see, for example, Römkens et al., 1990; Germann, 1990; Smith et al., 1999) and are dynamic and difficult to describe in mathematical form such that, unlike Darcy's law for flow in a soil matrix, there are as yet no commonly agreed conceptual models. Thus, in many models they are simply ignored or, at best, assumed to be reflected implicitly in the effective parameter values for any description of the infiltration process. Empirical descriptions based on field measurements might, in fact, have some advantage over a purely Darcian description in this respect. The SCS curve number approach, for example, does not imply any particular process, only a relationship between total runoff and input precipitation. The difficulty then is in knowing whether any particular application conforms to the conditions under which the relationships between curve numbers and other soil, vegetation and antecedent condition variables were established. Recent approaches to this problem have started to reflect the fact that the parameters of such descriptions can only be specified in an imprecise or fuzzy way.

Given some field observations on infiltration rates, it is generally not too difficult to fit one of the infiltration models described above to the data since only one or two parameters are involved. If data are available on the changes in the soil water profile during infiltration, then calibration of the parameters of a Richards equation model, which may have multiple parameters for several different soil layers, might be more difficult. These more theoretically acceptable solutions are almost always over-parameterised with respect to the data available for calibration. This would not be a problem if the necessary parameters could be estimated independently on the basis of soil texture or other characteristics. There are methods available to do so for basic soil hydraulic characteristics (for example the *pedotransfer functions* suggested by Rawls and Brakensiek, 1989, Vereecken et al., 1992, and others), but it is not yet clear whether such methods can give reliable estimates of effective parameter values since they are themselves based on calibration to data collected at specific sites and at specific scales (see discussion in Beven, 1996). Nearly all are based on regression relationships associated with large standard errors that should be taken into account in making predictions (though this is rarely done). There are, as yet, no such methods that attempt to account for the effects of macroporosity or crusting on infiltration rates, although Brakensiek and Rawls (1994) have addressed the effect of stoniness on the soil hydraulic characteristics.

A number of models have been proposed for predicting infiltration into crusted soils. A crust can be introduced as an additional layer into a numerical solution of the Richards equation for water flow in unsaturated soil. This will require some estimate of the soil hydraulic characteristics of the crust (as well as the soil beneath) and, to resolve the effects of the crust on infiltration rates, small depth and time increments in the solution. While this is computationally feasible, at least where only a small number of solutions are required, simplified methods are still computationally attractive (see Section 3.2.2 below). Smith et al. (1999) review previous simplified models for crusted soils, including variants on the Green–Ampt equation and propose a new model. There are also a number of models available that aim to simulate the evolution and characteristics of surface crusts.

Thus, even at the point scale, there is much that remains to be learned about how best to represent the dynamic characteristics of infiltration and surface runoff generation due to infiltration excess (subsurface runoff generation will be considered below). The prediction problem becomes even more difficult at the hillslope and catchment scales when the heterogeneity of soil and vegetation characteristics will start to play an important role.

3.2.2   Integration to Hillslope and Catchment Scale Models

The type of estimation of excess rainfall at a point, based on descriptions of the infiltration process, serves the functional requirement of having a loss function that is non-linear with respect to total rainfall. All these models have been widely used to estimate storm runoff generation at catchment scales, regardless of whether the runoff generation process is actually due to an infiltration excess mechanism. They all have the right sort of functional form to calculate the 'losses' to infiltration, despite the fact that they were originally derived from point-scale theory or measurements (or in the case of the SCS method from plot scale and small catchment data). The fact that they have been successful at larger scales is largely due to the fact that when such models are applied the parameters are generally calibrated against some dis-charge variables. This leaves plenty of scope for getting reasonable answers in calibration with (possibly) quite the wrong process description or parameter set.

The major problems in extending the point description to the hillslope and catchment scale are: (a) taking account of the heterogeneity of precipitation inputs and soil and surface char-acteristics; routeing any surface runoff downslope over an irregular surface; and characterising the subsurface flow domain on the hillslopes (where subsurface flow is a significant runoff source or has an impact on the antecedent moisture controls on infiltration capacities). The perceptual model suggests that soil heterogeneity is important in both inducing variable amounts of surface runoff generation at different points and in the run-on process where some runoff may infiltrate having flowed over the surface for some distance downslope. Downslope connectivity of both surface and subsurface flows is a crucial aspect of the runoff hydrology of semi-arid regions that are generally wetting-up during any individual event.

This creates a difficult modelling situation. Even in the most detailed field studies it is difficult to obtain sufficient measurements to characterise heterogeneity in the moisture, infiltration and surface characteristics in plan (see discussion of the R-5 Chickasha catchment case study below). Variations in the third vertical dimension and changes in the controls on infiltration over time will be even more difficult to characterise, while models of these processes tend to introduce more and more parameters, many of which might be expected to be time and space variable. Thus, as in the case of point infiltration above, it may be necessary to accept that any model at the hillslope and catchment scale will be very much an approximation of the full physical (biological and chemical) controls on the infiltration and runoff processes. If it is necessary to also consider the contribution of subsurface flows to runoff then the same will be true, especially since it will be even more difficult to characterise the properties of the subsurface flow system. It should therefore also be expected that any model will be associated with some uncertainty in the predictions that should be evaluated.

What then is a promising approach to model surface runoff at the hillslope and catchment scale? It is certainly not going to be possible to produce a simulation model that characterises all the point-to-point heterogeneity in infiltration, microtopography, surface roughness character-istics, and surface and subsurface connectivity (and their changes over time) explicitly. The number of grid points required in such a model would not only be computationally infeasible but, more importantly, there would be no way of measuring or estimating all the different parameters required at every grid point. Again, the same will be true of a subsurface flow description. If a distributed model is to be used there will then be a choice as to what scale of calculation element to use in a model, from the finest scale that is computationally feasible up to treating the catchment as a lumped unit.

A search through the hydrological literature will reveal little guidance as to what scale is most appropriate. There are many studies that recognise that there is an interaction between the scale chosen and the effective parameter values required in a particular model structure (e.g. Lane and Woolhiser, 1977; Willgoose and Kuczera, 1995; Parsons et al., 1997; Saulnier et al., 1997; Tayfur and Kavvas, 1998) but there are many other studies that have taken the grid scale of the digital terrain map available for a catchment (or something larger if computational times are too long at the finest resolution) and then assumed that there is not a problem in specifying parameter values. The majority of distributed models simply assume that the equations and parameters that have been defined on the basis of small-scale studies can be used at larger scales (see discussions in Beven, 1989a, 1996, 2000b), but this lacks any theory in hydrology that suggests how to move from a representation at one scale to another. There are many reasons to believe that larger scale process equations should be different from smaller-scale descriptions, especially given the small-scale heterogeneity of the controls on runoff generation in semi-arid regions. The scale problem is essentially too difficult (as yet), and it has been argued that it may be an impossible problem, in which case *scale-dependent* models may be all that is possible (Beven, 1995, 2000a).

With these considerations in mind, let us look at a possible distributed model of the surface runoff process on a hillslope that allows some of these problems to be taken into account. Such a model needs to reflect the complex controls on infiltration, runoff generation, run-on and surface flow hydraulics at the grid element scale, together with the linkages between elements in the downslope direction, without requiring massive computational resources or unnecessarily detailed parameterisations. The key here is in the subgrid element parameterisations of those processes at the element scale of the model.

The issues involved can be illustrated by the example of a simple hillslope scale distributed hydrological model. The kinematic wave (KW) description of surface flow has been used widely in models of runoff at hillslope and catchment scales (Singh, 1996; Beven, 2000b). The earliest recorded application was by Merrill Bernard in 1937 (see Hjelmfelt and Amerman, 1980). The KW equation arises from the combination of a mass balance or continuity equation expressed in terms of storage and flows, and a functional relationship between storage and flow that may be non-linear but is single-valued – that is to say, there is a single value of flow at a point corresponding to any value of storage at that point. Consider a one-dimensional downslope overland flow on a slope of constant width. Let x be the distance along the slope, h the depth of flow (which acts as the storage variable), and q the mean downslope velocity at any x (which is the flow variable). The mass balance equation can then be expressed as a differential equation:

$$w_x \frac{\partial h}{\partial t} = -\frac{\partial w_x q}{\partial x} + w_x r \qquad (3.12)$$

where w_x is the width of the slope segment at distance x downslope, h is the storage per unit area, q is the discharge per unit slope width, r is a rate of addition or loss per unit area, and t is time.

The functional relation between h and q may take many forms (see, for example, Beven, 1979) but a common assumption is the power law:

$$q = bh^a \qquad (3.13)$$

Thus, assuming that the coefficients a and b are constant, and combining the two equations to yield a single equation, in the storage variable h:

$$w_x \frac{\partial h}{\partial t} = -abh^{(a-1)} \frac{\partial w_x h}{\partial x} + w_x r \qquad (3.14)$$

or

$$w_x \frac{\partial h}{\partial t} = -c \frac{\partial w_x h}{\partial x} + w_x r \qquad (3.15)$$

where c is the kinematic wave velocity or celerity. The celerity is the speed with which any disturbance to the system will be propagated downslope. It is worth noting here that kinematic wave equations can only propagate the effects of disturbances in a downslope or downstream direction. They cannot therefore predict any backwater effects in rivers or drawdown effects due to a channel for subsurface hillslope flows. They do have the advantage, however, that for some simple cases, such as the case of constant slope width, constant input rates, analytical solutions exist (see, for example, Eagleson, 1970, or Singh, 1996). For more complex cases, such as variable width slopes and arbitrary patterns of inputs, it may be necessary to resort to a numerical solution but a finite difference approximation is very easy to formulate and include in a hydrological model.

For surface runoff, both overland flow and channel flow, the kinematic wave approach will be a good approximation to the full dynamic equations as the roughness of the surface or channel gets greater and as the bedslope gets steeper. A number of studies have examined the theoretical limits of acceptability of the kinematic wave approximation relative to the more complete solution of the full St Venant equations for free surface flow (e.g. Ponce et al., 1978; Morris and Woolhiser, 1980; DaLuz Vieira, 1983) but in most circumstances the errors in routeing will be small relative to estimating the amount of runoff to route.

The kinematic wave approach can also be adapted for the case of saturated downslope subsurface flow, in which h now represents a depth of saturation above an impermeable bed and account must be taken of the effective storage deficit in the unsaturated zone above the water table, which will affect the rise and fall of the water table. For saturated subsurface runoff on a hillslope, Beven (1981) showed that the kinematic wave description was a good approximation to a more complete Dupuit–Forchheimer equation description if the value of a non-dimensional parameter $\lambda = 4K \sin \beta / r$ was greater than about 0.75. In this case, r is the effective rate of storm recharge to the slope, K is the saturated hydraulic conductivity of the soil and β is the slope angle. When this condition is met, any drawdown of the water table at the lower end of the slope due to an incised channel, is unlikely to have a great effect on the predicted discharges and only a small effect on the predicted water table shape close to the downstream boundary (although this theoretical condition takes no account of heterogeneities of soil or effective rainfalls).

The kinematic wave approach has been widely employed in distributed hydrological models that have been used for the prediction of runoff for semi-arid hillslopes and catchments. Most of these models are based on coupling an infiltration model with a kinematic wave routeing algorithm on one-dimensional hillslope planes. Predictions at the catchment scale are made by combining a number of hillslope and channel elements in a dendritic network. They can therefore only predict infiltration excess runoff generation, although most will take account of the infiltration of run-on water further downslope. One of the first was the model developed by the Agricultural Research Service of the USDA (Smith and Woolhiser, 1971) that later developed into the KINEROS model that predicts both runoff and sediment transport (Woolhiser and Goodrich, 1988; Smith et al., 1995). The Corps of Engineers HEC-1 modelling package is also based on infiltration excess concepts and kinematic wave routeing with a variety of infiltration options (Feldman et al., 1982; Feldman, 1995). In Europe, the MEDALUS (Thornes et al., 1996; Kirkby et al., 1996) and EUROSEM (Morgan et al., 1998) models are both based on the kinematic wave assumptions and Field and Williams (1987) have used a similar approach in Australia. The model of Corradini et al. (1998) has combined kinematic wave routeing with a spatially heterogeneous representation of infiltration, demonstrating the importance of run-on effects at the hillslope scale.

Models such as the Institute of Hydrology Distributed Model Version 4 (Beven et al., 1987; Calver and Wood, 1995) use a kinematic wave description for surface runoff routeing, but use a fully two-dimensional (vertical slice) solution for partially saturated subsurface flow on the hillslope. This model is therefore capable of predicting perched saturated zones and surface runoff due to saturation of the soil. The model of Wigmosta et al. (1994) uses a two-dimensional kinematic wave solution in plan for predicting subsurface flow (see warning below). Recently models based on GIS grids have been developed (although the Bernard model of 1937 and Huggins and Monke model of 1968 were also essentially gridded kinematic wave models). Such models can use GIS overlays to estimate the characteristics of individual grid elements in a catchment, such as infiltration parameters. Runoff generation predicted on each grid can then be routed downslope using a kinematic wave algorithm. Most such models route the runoff downslope from a grid element only in the direction of steepest descent, and this may mean that flow pathways on the hillslopes are poorly represented even if fine grids are used. Examples of such models are given by Huggins and Monke (1968) and Parsons et al. (1996) and this is one of the possible models that can be implemented in the USGS Modular Modelling System (Leavesley and Stannard, 1995). Goodrich et al. (1991) have used a model based on a triangular irregular network (TIN) that can potentially improve the representation of flow pathways (if the triangles are sufficiently small).

Kinematic wave models are relatively simple and easy to implement numerically. They have to be used with some care, however, because of the problem of kinematic shocks. During the rising limb of a hydrograph there are many situations where waves with fast speeds overtake waves with slow speeds, this can happen, for example, on concave slopes, where the downslope decrease in slope angle may reduce the wave speed, resulting in a steepening of any wave fronts until a discontinuity develops: a kinematic shock (e.g. Kibler and Woolhiser, 1972; Borah et al., 1980; Woolhiser et al., 1996). This can also happen where two hillslopes of different characteristics meet; each hillslope will have a different wave speed, creating a shock at the meeting point. Two-dimensional descriptions based on kinematic wave equations should therefore be avoided.

It is evident that kinematic shocks are essentially a mathematical problem, resulting from the nature of the simplified description used to describe the flow processes. Shock following solutions, based on the method of characteristics, have been developed (e.g. Woolhiser et al., 1996) but in reality, for many reasons, any tendency towards such a shock will be reduced by dispersive processes in the flow. In fact, the problem of shocks in kinematic wave solutions can also be mitigated by the effects of numerical dispersion in the approximate finite difference solutions. This seems to be true in two dimensions (see, for example, the model of Wigmosta et al., 1994, which uses a two-dimensional kinematic wave solution for subsurface flows), although personal experience suggests that such shocks can produce numerical instabilities in convergent surface flows. Thus, users of kinematic wave-based models should be aware that the problem of kinematic shocks in a particular application might be offset by (uncontrolled) numerical dispersion. This has two implications: first, in some extreme circumstances the solution may become unstable; second, the approximate solution will not be a true solution of the original kinematic wave equation.

Application of the kinematic wave model requires the specification of the parameters a and b and the time and space variable net addition or loss r for every point on the catchment. For the prediction of surface runoff the law parameters governing the flow are often assumed to be constant over the slope (or catchment), while the addition or loss is calculated on the basis of a net rainfall in excess of local infiltration rates. The power law form of equation (3.11) is very flexible. Under the assumption that the surface runoff is a sheet flow, the storage/discharge relationship is often represented in terms of one of the uniform flow relationships with the hydraulic radius set equal to the flow depth. For the Manning equation, for example, the power b can then be set equal to 1.667 and the value of a is proportional to the inverse of the Manning n; for the Chezy equation, to 1.5. Engman (1986) and Gilley et al. (1992) have summarised values of the Manning n for surface runoff.

It is important to remember, however, that the KW routing model is not dependent on the form of equation (3.11). It only requires a specified storage/discharge relationship. Lawrence (1997), for example, has proposed a modified equation to provide a better fit to measured data. In addition, the Manning equation is an equation for turbulent flow conditions. Other studies have suggested that thin sheet flows are more likely to be in the laminar range (Emmett, 1978;

Dunne and Dietrich, 1980b; Wainwright, 1996b). Making the analogy of a laminar flow equation with the power law (3.11) implies a larger value of the power b. Both vegetation and tillage can have very important effects on the microtopography, effective roughness coefficients and overland flow velocities (e.g. Römkens and Wang, 1986; Woolhiser et al., 1996). Emmett (1978) also reports field plot experiments where apparent roughness increases with depth of flow and Reynolds number. Abrahams et al. (1994) have derived relationships between the Darcy–Weisbach roughness coefficient with vegetation cover, litter cover and Reynolds number, implying not only an effect of vegetation on flow velocities but also a roughness that varies with discharge. Willgoose and Kuczera (1995) have shown that in fitting a and b, significant interaction may be found between these parameters, which may also be scale dependent as the value of b was found to depart significantly from 1.667.

A calculation of local infiltration is also required for every point in the catchment, in principle taking account of the temporal and spatial heterogeneity in soil and rainfall characteristics. In practice, such information is rarely available in a distributed form and even where significant effort is expended in trying to characterise spatial variability, the improvements in model performance are not significant (see the study by Loague and Kyriakidis, 1997, discussed in Section 3.2.3 below). This problem far outweighs any loss of accuracy arising from the use of a kinematic wave model to route the resulting runoff.

These arguments would suggest that there is perhaps still scope for lumped conceptual modelling at the catchment scale even though it may be particularly difficult to capture the strong degree of non-linearity in surface and subsurface runoff generation in semi-arid catchments. A number of comparisons of rainfall–runoff models have been reported for semi-arid areas, including catchments in Australia (Pilgrim et al., 1988; Ye et al., 1997), USA (Michaud and Sorooshian, 1994), Zimbabwe (Refsgaard and Knudsen, 1996) and South Africa (Hughes, 1994). The results of Ye et al. (1997) are representative. They found that in tests on three different catchments in Western Australia (0.82 to 517 km^2; RC 2–12%) a simpler model (IHACRES, with six parameters to be calibrated) was no less successful in discharge prediction than a more complex model (LASCAM, with 22 parameters). Both could achieve calibration modelling efficiencies of the order of 90% over a five-year daily time step simulation, and over 75% in simulation of a second five-year validation period (but see also the discussion of the study by Michaud and Sorooshian in the next section).

In the next three sections some case studies will be introduced. These are in no way a complete review of the applications of hydrological models to semi-arid catchments but have been selected to introduce some considerations that will be relevant to all model applications in this kind of environment, in terms of process representations, scale issues and parameter estimation.

3.2.3 Runoff Modelling Case Studies in the USA

One of the most important study areas in the USA for the investigation of surface runoff processes has been the R-5 catchment (0.1 km^2) at Chickasha, Oklahoma. This small catchment area was the subject of an intensive set of infiltration measurements by Sharma et al. (1980). Loague and Freeze (1985) attempted to use these measurements in a physically based (infiltration model plus kinematic wave routeing) distributed model of surface runoff with only limited success. Loague and Gander (1990) later added to the density of infiltration measurements, but without great improvements in the performance of the model (Loague, 1990). A total of 247 measured infiltration capacity values are now available. In the most recent paper arising from this work, Loague and Kyriakidis (1997) have used kriging interpolation to map the permeability derived from the infiltration measurements to all points in the catchment. They have then applied the same distributed model to predict runoff for a sample of storms, testing the sensitivity to the spatial discretisation of the catchment into between 22 and 959 overland flow plane segments and to a variety of stochastically generated permeability fields, all consistent with the observed infiltration rates. The new model, however, is still limited in its success in predicting the observed runoff. They conclude that, despite the detailed data available for this small catchment, the model structure is not adequate to represent the runoff processes and that it is 'now time to go beyond the Horton model for the R-5 catchment' to include both surface and subsurface runoff generation processes. The problem is, of course, that this will then require detailed information about the heterogeneity of the soil and subsoil characteristics to predict the transient, spatial patterns of the water table (and possibly preferential flows).

One of the continuing challenges of semi-arid hydrology must be to model the extensive dataset collected by the USDA Agricultural Research Service on the very well-known Walnut Gulch catchment (149 km^2; MAP 324 mm) in Arizona (e.g. the hydrographs shown in Figure 3.4). This has been the subject of numerous experimental and modelling studies. Michaud and Sorooshian (1994) compared three different models at the scale of the whole catchment, including a lumped SCS curve number model, a simple distributed SCS curve number model and the more complex distributed KINEROS model. The modelled events were 24 severe thunderstorms (mean RC 11%), with a raingauge density of 1 per 20 km^2. Their results suggested that none of the models could adequately predict peak discharges and runoff volumes, but that the distributed models did somewhat better in predicting time to runoff initiation and time to peak. The lumped model was, in this case, the least successful.

At the hillslope runoff plot scale in Walnut Gulch, Scoging et al. (1992) and Parsons et al. (1997) have compared observed and predicted discharges, together with flow depths and velocities at several cross-sections. The model used the simplified infiltration model of equation (3.9) with two-dimensional KW routeing downslope, assuming a power law storage–discharge relationship with parameters that varied with the percentage cover of desert pavement in each grid cell of the model. In the first application to the shrubland site, the model under-predicted the runoff generation but was relatively successful in predicting the shape of the experimental hydrograph. The second application to the grassland plot reported in Parsons et al. (1997) was less successful, despite a number of modifications to the model, including the introduction of stochastic parameter values.

At a somewhat larger scale, Faurès et al. (1995) have applied KINEROS to the 4.4-ha Lucky Hills subcatchment to examine the importance of wind direction and velocity and rainfall pattern on the predicted discharges. They showed that taking account of wind direction and velocity has a relatively small effect, but in this semi-arid environment it is very important to have an adequate representation of the variability in the rainfall. Within the 4.4-ha area they had access to observations from five raingauges. Checking the model predictions for different combinations of different numbers of raingauges showed that combinations of four gauges (that is a density of 1 per hectare) gave a variation in predicted discharges that spanned the observed discharges and had a similar coefficient of variation to that estimated for the discharge measurements.

More recently, Goodrich et al. (1997) have used the data from all the 29 nested subcatchments within Walnut Gulch, with drainage areas ranging from 0.2 to 13 100 ha, to investigate the effects of storm area and catchment scale on runoff coefficients. They conclude that, unlike humid areas, there is a tendency for runoff responses to become *more* non-linear with increasing catchment scale in this type of semi-arid catchment as a result of the loss of water into the bed of ephemeral channels (see Ponce et al., 1999, for another case study) and the decreasing relative size of rainstorm coverage with catchment area for any individual event. These effects also lead to an apparent decrease in the curve numbers of the SCS infiltration equation with increase in catchment scale, when runoff and infiltration are treated as lumped catchment scale quantities (Simanton et al., 1996). Goodrich et al. (1997) reported greater success in modelling the discharges from the full catchment and 11 different subcatchments using the model of Lane (1982), which uses the SCS curve number to compute runoff volumes and an empirical channel-routeing algorithm that implicitly accounts for channel transmission losses. Three of the subcatchments (ranging in size from 0.0034 to $6.3\,km^2$) were also modelled using KINEROS (Smith et al., 1995). This performed very well for the two smaller catchments, but less well for both calibration and validation datasets on the larger catchment.

3.2.4 Runoff Modelling Case Studies in Australia

An interesting study by Coles et al. (1997) goes beyond a process representation that is based only on infiltration excess runoff generation. They report on modelling experiments for a number of small experimental agricultural catchments in the eastern wheatbelt of Western Australia (0.15–$0.41\,km^2$, MAP 300-450 mm). This is an interesting environment where the conversion of land to agriculture has generally led to a rise in water tables and a consequent increase in the saturation and salinisation of valley bottoms subject to return flow.

Coles et al. (1997) used a version of TOPMODEL, the rainfall–runoff model which is based on a simplified mechanistic description that uses an analysis of the topography of a catchment as the basis for the prediction of patterns of storage deficits in a catchment and, in particular, for the prediction of dynamic contributing areas of saturated soil where the deficit is zero. In this version, it includes components to predict infiltration excess overland flow, saturation excess overland flow and subsurface contributions to stream discharge (Sivapalan et al., 1987; Beven et al., 1995). Saturated areas are related to the distribution of a topographic index derived from a 10-m grid of elevations. Thus, the predictions can be mapped back into space (Figure 3.15). The simulations were found to be very sensitive to antecedent conditions and soil surface properties including crusting and cracking. Some storms occurring under conditions of high antecedent moisture conditions could not be simulated successfully. Winter storms were simulated more successfully in a way that suggested that saturation excess runoff generation was an important mechanism in these catchments, although the predicted patterns of runoff generation were not verified in the field. One limitation of the model was that infiltration excess runoff generation was routed too quickly to the stream channel. It was suggested that a delay due to the filling of depression storage should be incorporated to allow for the lack of connectivity of surface flow pathways in the initial stages of runoff production, although the authors were reluctant to add additional complexity and parameters to their model.

3.2.5 Mediterranean Case Studies

There have been a number of hydrological modelling studies of Mediterranean catchments, many of which have been based on either purely conceptual model representations (e.g the ARNO model of Todini, 1996) or an assumption of infiltration excess-dominated runoff generation. One example of the latter is the MEDALUS slope catena model (Kirkby et al., 1996), which includes model components for vegetation growth and development, soil profile development, surface armouring and crusting and soil erosion as well as runoff generation. The only runoff generation mechanism considered in the slope catena model is infiltration excess.

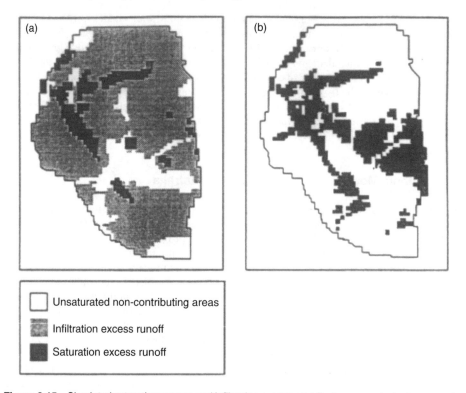

Figure 3.15 Simulated saturation excess and infiltration excess contributing areas, based on a version of TOPMODEL, for the Moolanooka catchment (24 ha, MAP 350 mm), Western Australia. (After Coles et al., 1997)

Infiltration is predicted as a solution of the Richards equation but in a way that takes account of the variable depths of surface water across an irregularly sloped surface. The predicted distribution of depths of flow can be used to represent the effects of rilling on the surface and is also used in routeing the runoff downslope. The model has been tested on single storm events for runoff plots near Murcia in Spain (MAP 250 mm; Thornes et al., 1996). Hydraulic conductivity, roughness and detachment coefficients were calibrated, but the model tended to over-predict runoff generation, with levels of explained variance of only 40%, although it did better for soil erosion. Prediction errors were less for plots covered with organic residues. The authors suggest that the predictive capabilities of the model may have been limited by the input data, since rainfall inputs were only available at hourly time increments.

The SHE mechanistic model (Abbott et al., 1986) does not include vegetation growth and soil development components although some versions can be used to predict erosion (e.g. Bathurst et al., 1995, 1996). The SHETRAN version has been used by Bathurst et al. (1996) in applications to a number of catchments in Portugal and Spain at scales ranging from soil plots to the 701 km^2 Cobres basin. The grid scale used by the model varies from representing a runoff plot as a single grid element; to 50-m squares in simulations of the 32-ha Santa Clara do Louredo subcatchment; 1-km squares in the 159-km^2 Mula basin near Murcia, Spain (MAP 300 mm); and 2-km squares in representing the Cobres basin in Portugal (MAP 700 mm).

The parameters of the model (particularly vertical saturated hydraulic conductivities for the unsaturated zones, lateral saturated hydraulic conductivity for the saturated zone, and overland flow roughness) were calibrated by a trial and error procedure. They note the difficulty of calibrating model parameters by comparing observed and predicted discharges when the runoff may represent only a very small proportion of the rainfall input. The results for the Cobres basin showed that the calibrated values of hydraulic conductivity were constant over the different scales while overland flow roughness tended to show a slight increase at larger grid scales. On the basis of these results, Bathurst et al. (1996) suggest that plot-scale calibration of the parameters of such a mechanistic model might be feasible where overland flow is the dominant runoff mechanism and is unaffected by lateral subsurface flows.

It was suggested, however, in the discussion of a perceptual model for runoff generation in semi-arid areas, that lateral subsurface flows may not always be a negligible contribution to the storm hydrograph and there have been a number of applications of models based on subsurface controls of runoff generation, such as TOPMODEL, to semi-arid basins. Applications of versions of TOPMODEL in the Mediterranean have been reported by Durand et al. (1992) at Mount Lozère, France (MAP 1900 mm); by Obled et al. (1994) in the Réal Collobrier, Maures Massif, France (MAP 1000–1200 mm); by Saulnier et al. (1997) in the Maurets subcatchment of the Réal Collobrier; by Franchini et al. (1996) in the Réal Collobrier and Sieve basin in Italy; and by Piñol et al. (1997) in the Prades catchments Catalonia, Spain (MAP 547 mm). Colosimo and Mendicino (1996) have combined infiltration excess and TOPMODEL saturation excess calculations in an application to the Turbolo basin in Italy. Obled et al. (1994) studied the impact of knowledge of rainfall patterns on TOPMODEL simulations at the scale of the 70-km^2 Réal Collobrier catchment. As in the Faurès et al. (1995) study at Walnut Gulch discussed above, they found that knowledge of rainfall patterns is important, but in this case it is more important to get the volume of inputs correct. The pattern itself was a second-order effect, which may be the result of the larger scale of the catchment and the greater significance of subsurface flows.

TOPMODEL should not be expected to be highly effective in this type of environment, particularly during dry and wetting-up periods. The topographic analysis on which the model predictions are based depends on an assumption of a water table that is shallow, parallel to the soil surface, and in equilibrium at every time step with a spatially averaged rate of recharge. This implies that there is always some downslope subsurface flow everywhere on the hillslope so that effective subsurface contributing areas extend to the divide. Barling et al. (1994), for example, showed that better predictions of storage deficits could be attained by relaxing this assumption, so that the effective upslope contributing area is smaller under drier conditions.

However, the results of these applications do suggest that good simulations of the responses of Mediterranean catchments can be obtained using TOPMODEL, at least with some calibration of parameter values after comparing observed and predicted discharges. This, however, may not be for the right reasons as it seems likely that the high effective calibrated downslope transmissivity values in such applications is a result of a compensating effect for overestimating the effective contributing areas in the topographic analysis and index calculation. Franchini et al. (1996) and Saulnier et al. (1997) have also demonstrated in these applications that there is an interaction between the grid size used in the topographic analysis and the calibrated transmissivity value that may be due to this compensating effect (although high transmissivity values may also reflect preferential flow pathways – see discussion in Beven, 1997). These applications suggest, however, that the subsurface component of TOPMODEL can, to some extent, reflect the controls of the subsurface on runoff production, even during a wetting-up period (Figure 3.16). In the Maurets catchment, the importance of subsurface flow has been reinforced by the alternative modelling studies of Taha et al. (1997) based on a finite element solution of the Darcian flow equations for a single hillslope.

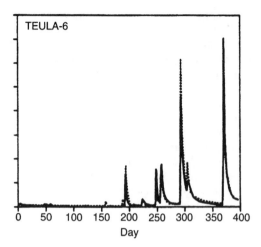

Figure 3.16 Observed discharges and TOPMODEL predictions for a wetting-up period in La Teula catchment, Prades, Catalonia, Spain. Predictions are based on best-fit parameter values for 55 000 Monte Carlo realisations in the most complex of six model structures tried. Results were better in this catchment than in the adjacent l'Avic catchment. (After Piñol et al., 1997)

Given all the difficulties of direct modelling of runoff generation, whether by distributed or lumped models, the recent study of Meirovich et al. (1998) is of some interest. They propose a stochastic model of runoff at the catchment scale for the Negev region of Israel that reproduces the arrival time, event discharge and event volume distributions as well as regional relationships between catchment area and maximum discharges. They show that, as a result of the limited area extent of rainfall events and the effects of channel transmission losses, only small catchments (up to $250–300\,km^2$) show a consistent increase in runoff volumes with area. At larger catchment areas, discharge volumes may decrease with increasing area. The turning point for this relationship varies between catchments. Such relationships will be useful in the estimation of the frequencies of events of different magnitudes; they do not necessarily help to predict the runoff generation in any particular event given some knowledge of the event rainfall. Another recent stochastic model for daily rainfalls and runoff based on Markov transition matrices has been applied to the $44\text{-}km^2$ Algecíras catchment in Spain (MAP 331–437 mm) by Conesa-García and Alonso-Sarría (1997). Such models make no assumptions about the nature of the processes of runoff generation but rely on available observations of rainfalls and runoff to establish the stochastic relationships assumed within the model. This may therefore make the resulting models quite basin specific.

3.3　UNIQUENESS OF PLACE, SCALE AND PREDICTIVE UNCERTAINTY IN MODELLING RUNOFF GENERATION IN SEMI-ARID AREAS

Models such as SHE, the MEDALUS model, EUROSED, KINEROS, and other distributed models that have been applied to predict runoff generation in semi-arid catchments rely on the distributed nature of the model to take account of the spatial complexities of the real world processes. They all, however, are limited in the scale of heterogeneity that they can take account of by the model grid scale. All need some way of parameterising the effects of processes operating at the subgrid scales. In some cases this is made explicit. The MEDALUS model, and the hillslope models of Dunne et al. (1991), Ruan and Illangasekare (1998) and Tayfur and Kavvas (1998), all try to parameterise the effects of variations in flow depth across the hillslope on flow velocities and infiltration rates. In the case of models such as SHE and KINEROS, each element is given parameters for the infiltration and surface flow components that are assumed to be effective values, implicitly taking account of the effects of any subgrid scale heterogeneity.

This raises something of a dilemma in applications to specific catchments in semi-arid regions. The literature reviewed above has revealed the importance of heterogeneity on small scales of runoff generation, even down to scales of individual plants, terracettes, patches of surface crust and rock fragments. How then can this variability be reproduced by a model that assumes effective parameters for grid scale infiltration and surface flow roughness parameters, especially when every grid model element will essentially be unique in its characteristics? The difficulties of characterising unique sites, even at the runoff plot scale, have been well illustrated by the study of Parsons et al. (1997) discussed above. They concluded that their results 'are not encouraging for the general use of distributed dynamic modelling of interrill overland flow as a basis for process based erosion models' (p. 1858) since each site would require a huge data input and possible model modifications for only limited predictive success.

The simple answer, of course, is that effective parameter values cannot and do not properly represent the effects of such variability in anything other than an approximate way (see, for example, Binley et al., 1989). The more complex answer is that if the parameters are calibrated to allow the volume of runoff and the timing of that runoff in an event to be reproduced reasonably well, then the model will be considered useful, even if the mechanisms of the runoff generation are not being reproduced correctly.

The calibration of model parameters is only possible if some observations are available against which to compare observed and predicted variables. Normally, it is discharge predictions that are compared and many models have been shown to reproduce discharges reasonably after calibration. In fact, the problem in calibration is not that it is difficult to get acceptable fits to the observations by adjusting parameter values but rather that there are so many sets of parameter values that will give acceptable fits to the data (the *equifinality problem* of Beven, 1993). Many parameter sets may be acceptable, but there is no way of knowing which one, if any, is correct. In fact, because of the limitations of the model descriptions (and in some cases of the input data on rainfalls and the observed discharges), *none* can be considered correct, only adequate for the purposes of prediction.

Much more rarely, models are applied without calibration using *a priori* estimates of parameter values. The problem then is that there are no techniques available for either estimating or measuring the values of the effective parameters required at the grid element or catchment scale. The values needed will therefore be subject to some uncertainty. A methodology for the 'blind testing' of models in this way has been outlined by Ewen and Parkin (1996) with an application using the SHE model on the 1-km^2 Rimbaud catchment in southern France reported by Parkin et al. (1996). Their approach involves estimating ranges for the parameters required and making runs of the model for different combinations of parameters within those ranges. The computational requirements of the SHE model are such that, even on this small catchment, only a limited number of runs could be made. Choosing to run the extremes of the ranges requires only a small number of runs, but it is then difficult to know the significance of the simulated outputs if all the parameters are considered to be at the edges of the feasible ranges. To counter this, some runs are rejected by the modellers as unrealistic before the range of predictions is presented. Rejection at this stage appears to be made purely on the basis of expert judgement by the modellers.

At this point, no use has been made of any observed data for the predicted variables. The catchment has been treated as if it is ungauged. For the Rimbaud catchment, however, observed discharges are available and for the blind test, the authors had set a number of criteria for success, for the prediction of continuous discharge, peak discharges, monthly discharges and total discharge in the simulation period, *before* making comparisons with the observations. For continuous discharge, the observations were within the simulation bounds for 78% of the time steps (Figure 3.17; criterion for success, 90%); for discharge peaks, 13 out of 32 storms were within the bounds (criterion of success, 28); for monthly discharges, 10 out of 13 months were within the bounds (criterion of success, 11), and the total discharge was satisfactorily predicted. The model therefore failed three of the four tests in this case.

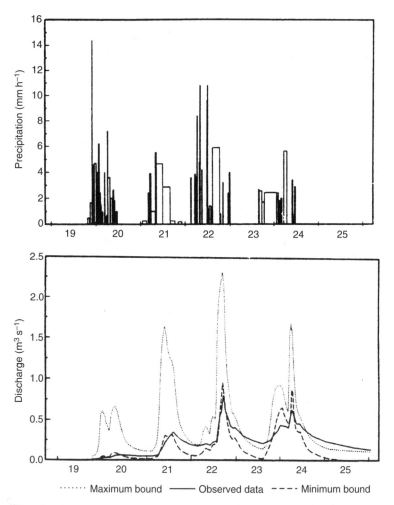

Figure 3.17 Observed discharges and SHE prediction bounds for the 1 km² Rimbaud catchment, Maures Massif, south France. Prediction bounds are based on *a priori* estimates of parameter ranges without model calibration but with subjective elimination of predictions from some parameter sets. (After Parkin et al., 1996)

The reasons for this may arise from any of the sources of error noted above. Parkin et al. (1996) note themselves that the simulations do show systematic errors, in particular, that the simulated recession curves were in general too steep for all models (perhaps a result of assuming a uniform hydraulic conductivity with depth into the soil profile). The model predicts widespread overland flow only in one storm (in September 1968); all other simulated storm peaks were dominated by subsurface runoff. The version of the model used did not have any capability of predicting shallow throughflow (though it is not clear how this might have helped to improve the simulations without introducing additional parameters).

With models that are less computationally demanding, this type of methodology can be extended to a more rigorous exploration of the model space using many different parameter sets after making some assumptions about the possible distributions of parameter values. The difficulty, still, is knowing what are the appropriate distributions of effective parameter values at the scale required by the model *and* all their covariation. Such information will not normally be available, even to expert modellers.

One way around this problem has been suggested by Beven (2000a) as an extension of the Generalised Likelihood Uncertainty Estimation (GLUE) methodology of Beven and Binley (1992; see also Freer et al., 1996; Beven, 2000b; Beven et al., 2001). In GLUE, randomly chosen parameter sets are evaluated by comparing simulated variables with any available observations, either quantitative or qualitative, on the catchment under study. Many parameter sets will be rejected at this stage as *non-behavioural*, but many others will be retained as providing simulations that are considered to be acceptable according to the evaluation criteria. Where quantitative evaluation is possible, a weight or likelihood can be associated with each parameter set. Those that have performed well in the past will have the highest weight; those that have only just avoided rejection will be given low weight. The weights can then be used in prediction, as a way of combining the predictions from all the retained parameter sets into a cumulative weighted distribution of any selected predicted variable. From this distribution, prediction quantiles for the selected variable can be easily determined.

This approach has been applied in modelling the same small Rimbaud catchment in southern France in order to follow changes in the distributions of effective parameters of a variant of TOPMODEL as determined using the GLUE approach, before and after a fire that affected the major part of the catchment in August 1989 (Lavabre et al., 1993; Martin and Lavabre, 1997). The aim was to determine whether, given the uncertainties of modelling any individual rainfall–runoff dataset, the impact of a major change (in this instance, the effects of fire) could be identified. Some problems in doing so were immediately apparent in a close look at the data available. While this was generally of good quality, there were some periods of data for which there were flow events with no recorded rainfall, periods with very low runoff coefficients that were very difficult to model correctly, and one storm period where the apparent runoff coefficient was greater than 100% – a challenge for any rainfall–runoff model that does not include a rainfall adjustment parameter! Eventually six periods were modelled, two prior to the fire and four after.

The version of the model used had eight parameters, of which four were held constant at their initial estimates and four were sampled uniformly across specified ranges in the Monte Carlo sampling. These were an interception loss parameter (RI), a root zone storage capacity ($SR_{\max}$), the transmissivity at soil saturation (T_0) and a parameter controlling the spread of the contributing area (f). These parameters were all assumed to be effective values at the scale of the 1 km^2 catchment. No additional infiltration excess component was added in order to reduce the dimensionality of the parameter space.

Fifteen thousand runs of the model were made and the simulated discharges were compared with the observed discharges for each of the six simulation periods. In GLUE a likelihood value is associated with each of the 15 000 runs according to how well the model fits the available data. Many of the models were rejected at this stage as *non-behavioural* and given a likelihood value of zero. The resulting cumulative likelihood distributions for each of the four varied parameters over all of the behavioural parameter sets are compared for the different periods in Figure 3.18. All the plots show that the behavioural parameter sets span a significant proportion of the parameter ranges (and in some cases some of the best-fitting models were found right up to the edges of the ranges considered feasible). This reflects the difficulty of finding an optimal parameter set for each period. Note that these are the marginal distributions integrated over all the behavioural parameter sets. Non-behavioural simulations are also found everywhere across the parameter ranges. Essentially it is the *set* of parameter values that is important in determining the goodness of fit of the model.

However, it is also clear from these plots that the effects of the fire do have an impact on the distributions of effective parameter values. The effect on the transmissivity parameter is small. The behavioural values of the root zone storage parameter are much smaller in the two periods after the fire and then recover to values higher than the pre-fire distribution. Similarly the f parameter becomes dramatically smaller after the fire, indicating that the contributing area expands much more rapidly during storms (this may be trying to simulate the effects of the fire

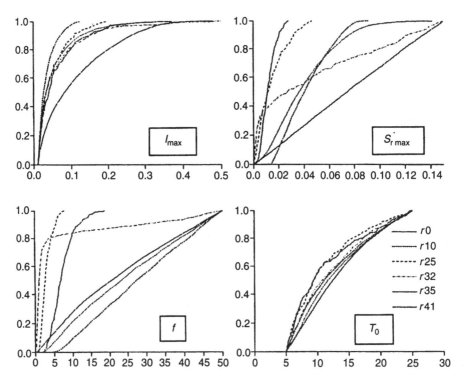

Figure 3.18 Cumulative distribution functions for different TOPMODEL parameters conditioned by comparisons with observed discharges for periods before and after the August 1989 fire, and during vegetation recovery, in the 1 km² Rimbaud catchment, Maures Massif, south France. I_{max} is an interception parameter, S_{rmax} is the storage capacity of the root zone, T_0 is the transmissivity of the soil at saturation, and f controls the rate of decline of an exponential transmissivity function as the water table falls. The periods r0 and r10 are from before the fire; r25, r32, r35, and r41 are during recovery

on the surface infiltration characteristics of the soil, rather than reflecting a flashier saturation excess mechanism in the catchment after the fire). For the interception parameter, most of the periods have similar distributions of values, including those just after the fire, but with the exception of the last period when the vegetation regrowth had become well developed. These changing distributions can therefore be interpreted in a way that is consistent with a perceptual model of the effects of fire on a semi-arid catchment, including an appreciation of deficiencies in the model structure.

There remains, of course, a problem of predicting changes in distributions of sets of effective parameter values ahead of such a change for making decisions about the impacts of land use management strategies on runoff generation. It is clear from the type of work described above that comparing single parameter sets before and after a change is not an adequate strategy, and yet predicting changes in distributions of sets of effective parameter values is undoubtedly more difficult. Even estimating changes in the marginal distributions may not be very helpful since choosing combinations of parameter values from those distributions will not necessarily result in a behavioural simulation; the overlap with the distributions of non-behavioural simulations was, in this case at least, much too great.

However, the GLUE methodology can also be considered as a form of mapping of the landscape into the model space (Beven, 2000a, 2000b). The weights that arise from the model evaluations are essentially a form of non-linear transformation by which the landscape or catchment of interest, with all its unique characteristics, is represented in the hyperdimensional space of the model parameters. Because there may be many parameter sets that are acceptable representations of the catchment (or part of it), this transformation will not be a single, crisp, mapping but will rather be imprecise. This type of transformation has been demonstrated in the study of land surface to atmosphere flux modelling by Franks et al. (1997).

When considering the application of a distributed model to a semi-arid catchment within this framework, the model represents the catchment as a number of elements. There is known to be strong heterogeneity in the surface and subsurface characteristics in each element that has an important effect on the local runoff generation (in this case, the item of interest is the surface or subsurface discharge that reaches a downslope element or stream channel). The model represents this complex of processes in terms of some subgrid scale parameterisation involving a number of parameters. It is likely that this parameterisation has been tested against plot experiment data collected elsewhere but no direct experimental evidence may be available for the catchment under study. In addition, generally only a classification of soils and vegetation types will be available with no additional information available on stand densities, soil depths, porosity and hydraulic conductivity profiles, or surface roughnesses (or their spatial and temporal variability).

Thus, the relevant effective parameters must be estimated by a combination of past experience and literature search. This can be done (as in the case of the application of the SHE model to the Rimbaud catchment described above) but the estimations will inevitably be imprecise or fuzzy. This represents the first stage of mapping the catchment into the model space by identifying the fuzzy feasible regions of the model space to represent each grid element. Regions for different elements (and even for different types of elements) will almost inevitably overlap within the model space. In the same way as in the GLUE procedure, simulations made with different sample parameter sets taken from the feasible regions can be evaluated against any information about the behaviour of different elements within the catchment with a view to rejecting some of the models and refining the mapping.

It is interesting, however, to turn this process around. Within the model space, every simulation arising from a parameter set and particular sequence of input data is still considered to be deterministic. Within this framework, any uncertainty in the predictions comes from the fuzziness of the mapping of any particular element into the model space. This means that within the model space, in principle, all outcomes can be determined beforehand. In particular, regions of the model space that give very similar simulations can be determined, and this will purely reflect the functioning of the model with the different parameter sets. Of particular interest here are parameter sets that give very similar functional behaviour and could be considered to be different functional types (see Beven, 1995; Beven and Franks, 1999). The fuzzy mapping of the elements of the landscape that is the catchment can then be considered as a mapping of the landscape into these functional types. It may be necessary to have only a small number of functional types (see Franks and Beven, 1997), but note that the mapping may still not be unique in the sense that every part of the landscape maps to a single functional type. Given our current state of knowledge, the mapping will generally remain fuzzy, i.e. it may only be possible to evaluate the *possibility* of any part of the landscape belonging to a functional type.

This focus on functional types within the model space is of interest because it raises the question of the type of data that are required to classify the landscape into different functional types. This has traditionally been the role of the model itself, but can it be done more empirically, in a way that does not depend on a particular model structure and parameters? This may be the key to developing a new modelling strategy for catchment hydrology that is based on closer links between modelling and fieldwork as advocated by Dunne (1983). This may require the use of novel types of measurement, aimed at determining a hillslope-scale functional type rather than the small-scale process parameters of the past.

The modelling of runoff generation in semi-arid catchments is known to be particularly difficult. So much of the surface and subsurface heterogeneity and connectivity is unknown or unknowable. It seems unlikely that the current generation of models based on small-scale process representations will ultimately be successful. Some new ideas are needed, but will probably be dependent on the development of novel measurement techniques that give a more direct indication of functional behaviour at the hillslope element scale.

ACKNOWLEDGEMENTS

It is always somewhat presumptuous for a predominantly humid temperate environment hydrologist to expound on the processes of runoff production in semi-arid regions on the basis of limited experience, albeit that the field experience extends back to 1972 in Murcia, Spain. I am therefore grateful to those colleagues and friends who have given me the chance to think about the processes involved in the field, especially Dave Woolhiser in Colorado, Charles Obled, Jacques Lavabre and Jean-Michel Gresillon in France, and Ferran Rodà and Pep Piñol in Cataluña. I am also grateful to Pilar Montesinos-Barrios for her helpful comments. The calculations on which Figure 3.18 are based were made by James Fisher at Lancaster as part of the EU-funded DM2E project led by Jacques Lavabre of CEMAGREF, Aix-en-Provence, France, who also made the hydrological data available.

REFERENCES

Abbott, M.B., Bathurst, J.C., Cunge, J.A., O'Connell, P.E. and Rasmussen, J. 1986. An introduction to the European Hydrological System – Système Hydrologique Européen. 2. Structure of a physically-based distributed modelling system. *Journal of Hydrology*, 87, 61–77.
Abbott, M.B. and Refsgaard, J.-C. 1996. *Distributed Hydrological Modelling*. Kluwer, Dordrecht.

Abrahams, A.D. and Parsons, A.J. 1991. Relation between infiltration and stone cover on a semi-arid hillslope, S. Arizona, *Journal of Hydrology*, 122, 49–59.

Abrahams, A.D, Parsons, A.J. and Wainwright, J. 1994. Resistance to flow on semi-arid grassland and shrubland hillslopes, Walnut Gulch, southern Arizona. *Journal of Hydrology*, 156, 431–446.

Anderson, M.G. and Burt, T.P. 1990. *Process Studies in Hillslope Hydrology*. Wiley, Chichester.

Bales, J. and Betson, R.P. 1982. The curve number as a hydrologic index. In V.P. Singh (Ed.) *Rainfall–Runoff Relationships*. Water Resource Publications, Littleton, CO, 371–386.

Barling, R.D., Moore, I.D. and Grayson, R.B. 1994. A quasi-dynamic wetness index for characterising the spatial distribution of zones of surface saturation and soil water content, *Water Resources Research*, 30, 1029–1044.

Bathurst, J.C., Wicks, J.M. and O'Connell, P.E. 1995. The SHE/SHESED Basin Scale Water Flow and Sediment Transport Modelling System. In V.P. Singh (Ed.) *Computer Models of Watershed Hydrology*. Water Resource Publications, Highlands Ranch, CO., 563–594.

Bathurst, J.C., Kilsby, C. and White, S. 1996. Modelling the impacts of climate and land-use change on basin hydrology and soil erosion in Mediterranean Europe. In C.J. Brandt and J.B. Thornes (Eds) *Mediterranean Desertification and Land Use*. Wiley, Chichester, 355–387.

Bergkamp, G. 1998a. A hierarchical view of the interactions of runoff and infiltration with vegetation and microtopography in semiarid shrublands. *Catena*, 33, 201–220.

Bergkamp, G. 1998b, Hydrological influences on the resilience of *Quercus* spp. dominated geoecosystems in central Spain. *Geomorphology*, 23, 101–126.

Berndtsson, R. and Larson, M. 1987. Spatial variability of infiltration in a semi-arid environment, *Journal of Hydrology*, 90, 117–133.

Beven, K.J. 1979. On the generalised kinematic routing method. *Water Resources Research*, 15(5), 1238–1242.

Beven, K.J. 1981. Kinematic subsurface stormflow. *Water Resources Research*, 17(5), 1419–1424.

Beven, K.J. 1989a. Changing ideas in hydrology: the case of physically-based models. *Journal of Hydrology*, 105, 157–172.

Beven, K.J. 1989b. Interflow. In H.J. Morel-Seytoux (Ed.) *Unsaturated Flow in Hydrologic Modelling, Proc. NATO ARW*. Reidel, Dordrecht, 191–219.

Beven, K.J. 1993. Prophecy, reality and uncertainty in distributed hydrological modelling. *Advances in Water Resources*, 16, 41–51.

Beven, K.J. 1995. Linking parameters across scales: subgrid parameterisations and scale dependent hydrological models. In J.D. Kalma and M. Sivapalan (Eds) *Scale Issues in Hydrological Modelling*. Wiley, Chichester, 263–281.

Beven, K.J. 1996. A discussion of distributed hydrological modelling. In M.B. Abbott and J.C. Refsgaard (Eds) *Distributed Hydrological Modelling*. Kluwer, Dordrecht, 255–278.

Beven, K.J. 1997. TOPMODEL: a critique, *Hydrological Processes*, 11(3), 1069–1086.

Beven, K.J. 2000a. On uniqueness of place and process representations in hydrology. *Hydrology and Earth Systems Science*, 4(2), 203–213.

Beven, K.J. 2000b. *Rainfall–Runoff Modelling – The Primer*. Wiley, Chichester.

Beven, K.J. and Binley, A.M. 1992. The future of distributed models: model calibration and uncertainty prediction. *Hydrological Processes*, 6, 279–298.

Beven, K.J., Calver, A. and Morris, E.M. 1987. *The Institute of Hydrology Distributed Model. Institute of Hydrology Report No. 98*, Wallingford, UK.

Beven, K.J. and Franks, S.W. 1999. Functional similarity in landscape scale SVAT modelling. *Hydrology and Earth Systems Science*, 3(1), 85–94.

Beven, K.J. and Germann, P.F. 1982. Macropores and water flow in soils. *Water Resources Research*, 18, 1311–1325.

Beven, K.J. and Kirkby, M.J. 1979. A physically-based variable contributing area model of basin hydrology. *Hydrological Sciences Bulletin*, 24(1), 43–69.

Beven, K.J., Lamb, R., Quinn, P.F., Romanowicz, R. and Freer, J. 1995. TOPMODEL. In V.P. Singh (Ed.) *Computer Models of Watershed Hydrology*. Water Resources Publications, Highlands Ranch, CO., 627–668.

Beven, K.J., Freer, J., Hankin, B. and Schulz, K. 2001. The use of generalised likelihood measures for uncertainty estimation in high order models of environmental systems. In W. Fitzgerald, R.L. Smith, A.T. Walden and P.C. Young (Eds) *Proc. Newton Institute on Nonlinear and Nonstationary Signal Processing*. Cambridge University Press, Cambridge (in press).

Binley, A.M., Beven, K.J. and Elgy, J. 1989. A physically-based model of heterogeneous hillslopes. II. Effective hydraulic conductivities. *Water Resources Research*, 25(6), 1227–1233.

Böhm, P. and Gerold, G. 1995. Pedo-hydrological and sediment responses to simulated rainfall on soils of the Konya Uplands (Turkey). *Catena*, 25, 63–76.

Bonell, M. and Williams, J. 1986. The generation and redistribution of overland flow on a massive oxic soil in a eucalyptus woodland within the semi-arid tropics of north Australia. *Hydrological Processes*, 1, 31–46.

Borah, D.K., Prasad, S.N. and Alonso, C.V. 1980. Kinematic wave routing incorporating shock fitting, *Water Resources Research*, 16, 529–541.

Brakensiek, D.L. and Rawls, W.J. 1994. Soil containing rock fragments: effects on infiltration. *Catena*, 23, 99–110.

Brandt, C.J. and Thornes, J.B. (Eds) 1996. *Mediterranean Desertification and Land Use*. Wiley, Chichester.

Bryan, R.B., Hawke, R.M. and Rockwell, D.L. 1998. The influence of subsurface moisture on rill system evolution. *Earth Surface Processes and Landforms*, 23, 773–789.

Burch, G.J., Moore, I.D. and Burns, J. 1989. Soil hydrophobic effects on infiltration and catchment runoff. *Hydrological Processes*, 3, 211–222.

Calver, A. and Wood, W.L. 1995. The Institute of Hydrology Distributed Model. In V.P. Singh (Ed.) *Computer Models of Watershed Hydrology*. Water Resources Publications, Littleton, CO, 595–626.

Ceballos, A. and Schnabel, S. 1998. Hydrological behaviour of a small catchment in the *dehesa* landuse system (Extremadura, SW Spain). *Journal of Hydrology*, 210, 146–160.

Cerda, A. 1998a. The influence of geomorphological position and vegetation cover on the erosional and hydrological processes on a Mediterranean hillslope. *Hydrological Processes*. 12, 661–672.

Cerda, A. 1998b. Changes in overland flow and infiltration after a rangeland fire in a Mediterranean scrubland. *Hydrological Processes*. 12, 1031–1042.

Cerda, A., Schnabel, S., Ceballos, A. and Gomez-Amelia, D. 1998. Soil hydrological response under simulated rainfall in the Dehesa land system (Extremadura, SW Spain) under drought conditions. *Earth Surface Processes and Landforms*, 23, 195–209.

Childs, E.C. 1969. *An Introduction to the Physical Basis of Soil Water Phenomena*. Wiley, London.

Coles, N.A., Sivapalan, M., Larsen, J.E., Linnet, P.E. and Fahrner, C.K. 1997. Modelling runoff generation on small agricultural catchments: can real world runoff responses be captured. In K.J. Beven (Ed.) *Distributed Hydrological Modelling: Applications of the TOPMODEL Concept*. Wiley, Chichester, 289–314.

Colosimo, C. and Mendicino, G. 1996. GIS for distributed rainfall–runoff modelling. In V.P. Singh and M. Fiorentino (Eds) *Geographical Information Systems in Hydrology*. Kluwer, Dordrecht, 185–235.

Conesa-García, C. and Alonso-Sarría, F. 1997. Stochastic matrices applied to the probabilistic analysis of runoff events in a semi-arid stream. *Hydrological Processes*, 11, 297–310.

Corradini, C., Melone, F. and Smith, R.E. 1997. A unified model for infiltration and redistribution during complex rainfall patterns. *Journal of Hydrology*, 192, 104–124.

Corradini, C., Morbidelli, R. and Melone, F. 1998. On the interaction between infiltration and Hortonian runoff. *Journal of Hydrology*, 204, 52–67.

Croke, J., Hairsine, P. and Fogarty, P. 1999. Runoff generation and re-distribution in logged eucalyptus forests, south-eastern Australia. *Journal of Hydrology.*, 216, 56–77.

DaLuz Vieira, J.H. 1983. Conditions governing the use of approximations for the Saint-Vénant equations for shallow surface water flow. *Journal of Hydrology*, 60, 43–58.

De Boer, D.H. 1992. Constraints on spatial transference of rainfall–runoff relationships in semi-arid basins drained by ephemeral streams. *Hydrological Science Journal*, 37, 491–504.

Dixon, R.M. and Linden, D.R. 1972. Soil air pressure and water infiltration under border irrigation. *Proc. of Soil Science Society of America*, 36, 948–953.

Doerr, S.H., Shakesby, R.A. and Walsh, R.P.D. 1996. Soil hydrophobicity variations with depth and particle size fraction in burned and unburned *Eucalytpus globulus* and *Pinus pinaster* forest terrain in the Agueda Basin, Portugal. *Catena*, 27, 25–47.

Dunne, T. 1978. Field studies of hillslope flow processes. In M.J. Kirkby and R.J. Chorley (Eds) *Hillslope Hydrology*. Wiley, Chichester, 227–293.

Dunne, T. 1983. Relation of field studies and modelling in the prediction of storm runoff. *Journal of Hydrology*, 65, 25–48.

Dunne, T. and Dietrich, W.E. 1980a. Experimental study of Horton overland flow on tropical hillslopes. 1. Soil conditions, infiltration and frequency of runoff. *Zeitschrift für Geomorphologie Supplement Band*, 35, 40–59.

Dunne, T. and Dietrich, W.E. 1980b. Experimental study of Horton overland flow on tropical hillslopes. 2. Hydraulic characteristics and hillslope hydrographs. *Zeitschrift für Geomorphologie Supplement Band*, 35, 60–80.

Dunne, T., Zhang, W. and Aubrey, B.F. 1991. Effects of rainfall, vegetation and microtopography on infiltration and runoff. *Water Resources Research*, 27, 2271–2285.

Durand, P., Robson, A. and Neal, C. 1992. Modelling the hydrology of submediterranean montane catchments (Mont Lozere, France), using TOPMODEL: initial results. *Journal of Hydrology*, 139, 1–14.

Eagleson, P. 1970. *Dynamic Hydrology*. McGraw-Hill, New York.

Emmett, W.W. 1978. Overland flow. In M.J. Kirkby and R.J. Chorley (Eds) *Hillslope Hydrology*. Wiley, Chichester, 145–176.

Engman, E.T. 1986. Roughness coefficients for routing of surface runoff. *Journal of Irrigation and Drainage Engineers, ASCE*, 112, 39–53.

Ewen, J. and Parkin, G. 1996. Validation of catchment models for predicting land-use and climate change impacts. 1. Method. *Journal of Hydrology*, 175, 583–594.

Faurès, J.-M., Goodrich, D.C., Woolhiser, D.A. and Sorooshian, S. 1995. Impact of small-scale rainfall variability on runoff modelling. *Journal of Hydrology*, 173, 309–326.

Feldman, A.D., Ely, P.B. and Goldman, D.M. 1982. The new HEC-1 flood hydrograph package. In V.P. Singh (Ed.) *Applied Modelling in Catchment Hydrology*. Water Resource Publications, Littleton, CO, 121–144.

Feldman, A.D. 1995. HEC-1 Flood Hydrograph Package. In V.P. Singh (Ed.) *Computer Models of Watershed Hydrology*. Water Resources Publications, Littleton, CO, 119–150.

Field, W.G. and Williams, B.J. 1987. A generalised kinematic catchment model. *Water Resources Research*, 23(8), 1693–1696.

Fitzjohn, C., Ternan, J.L. and Williams, A.G. 1998. Soil moisture variability in a semi-arid gully catchment: implications for runoff and erosion. *Catena*, 32, 55–70.

Fox, D.M., Le Bissonnais, Y. and Bruand, A. 1998. The effect of ponding depth on infiltration in a crusted surface depression. *Catena*, 32, 87–100.

Franchini, M., Wendling, J., Obled, C. and Todini, E. (1996). Physical interpretation and sensitivity analysis of the TOPMODEL. *Journal of Hydrology*, 175, 293–338.

Franks, S. and Beven, K.J. 1997. Estimation of evapotranspiration at the landscape scale: a fuzzy disaggregation approach. *Water Resources Research*, 33(12), 2929–2938.

Franks, S.W., Beven, K.J., Quinn, P.F. and Wright, I.R. 1997. On the sensitivity of soil–vegetation–atmosphere transfer (SVAT) schemes: equifinality and the problem of robust calibration. *Agricuture Forestry and Meteorology*, 86, 63–75.

Freer, J., Beven, K.J. and Ambroise, B. 1996. Bayesian estimation of uncertainty in runoff prediction and the value of data: an application of the GLUE approach. *Water Resources Research*, 32(7), 2161–2173.

Freer, J., McDonnell, J., Beven, K.J., Brammer, D., Burns, D., Hooper, R. and Kendal, C. 1997. Topographic controls on subsurface stormflow at the hillslope scale for two hydrologically distinct small catchments. *Hydrological Processes*, 11(a), 1347–1352.

Gabbard, D.S, Huang, C., Norton, L.D. and Steinhardt, G.C. 1998. Landscape position, surface hydraulic gradients and erosion processes. *Earth Surface Processes and Landforms*, 23, 83–93.

Gallart, F., Llorens, P. and Latron, J. 1994. Studying the role of old agricultural terraces on runoff generation in a small Mediterranean mountainous basin. *Journal of Hydrology*, 159, 291–304.

Gallart, F., Latron, J., Llorens, P. and Rabada, D. 1997. Hydrological functioning of Mediterranean mountain basins in Vallcebre, Catalonia: some challenges for hydrological modelling. In K.J. Beven (Ed.) *Distributed Hydrological Modelling: Applications of the TOPMODEL Concept*. Wiley, Chichester, 182–190.

Germann, P.F. 1990. Macropores and hydrologic hillslope processes. In M.G. Anderson and T.P. Burt (Eds) *Process Studies in Hillslope Hydrology*. Wiley, Chichester, 327–364.

Gilley, J.E., Flanagan, D.C., Kottwitz, E.R. and Weltz, M.A. 1992. Darcy-Weisbach roughness coefficients for overland flow. In A.J. Parsons and A.D. Abrahams (Eds) *Overland Flow: Hydraulics and Erosion Mechanics*, 25–52.

Goodrich, D.C., Woolhiser, D.A. and Keefer, T.O. 1991. Kinematic routing using finite elements on a triangular irregular network. *Water Resources Research*, 27(6), 995–1003.

Goodrich, D.C., Lane, L.J., Shillito, R.M., Miller, S.N., Syed, K.H. and Woolhiser, D.A. 1997. Linearity of basin response as a function of scale in a semi-arid watershed. *Water Resources Research*, 33(12), 2951–2965.

Green, W.H. and Ampt, G. 1911. Studies of soil physics. Part 1. The flow of air and water through soil. *J. Agric. Soc.*, 4, 1–24.

Gresillon, J.-M. and Taha, A. 1998. Les zones saturées contributives en climat méditerranéen: condition d'apparition et influence sur les crues. *Hydrological Science Journal*, 43, 267.

Gutiérrez, F., Gutiérrez, M. and Sancho, C. 1998. Geomorphological and sedimentological analysis of a catastrophic flash flood in the Arás drainage basin (Central Pyrenees, Spain). *Geomorphology*, 22, 265–283.

Harvey, A. 1982. The role of piping in the development of badlands and gully systems in south-east Spain. In R.B. Bryan and A. Yair (Eds) *Badland Geomorphology and Piping*. GeoBooks, Norwich, 317–335.

Hawkins, R.H. 1982. Interpretations of source area variability in rainfall–runoff relations. In V.P. Singh (Ed.) *Rainfall–Runoff Relationships*. Water Resource Publications, Littleton, CO, 303–324.

Hjelmfelt, A.T. and Amerman, C.R. 1980. The mathematical basin model of Merrill Bernard. *IAHS Publication*, 130, 343–349.

Hjelmfelt, A.T., Kramer, L.A. and Burwell, R.E. 1982. Curve numbers as random variables. In V.P. Singh (Ed.) *Rainfall–Runoff Relationships*. Water Resource Publications, Littleton, CO, 365–370.

Hjelmfelt, A.T. and Burwell, R.E. 1984. Spatial variability of runoff. *Journal of Irrigation and Drainage Engineering ASCE*, 110, 46–54.

Hodges, W.K. and Bryan, R.B. 1982. The influence of material behavior on runoff initiation in the Dinosaur Badlands, Canada. In A. Yair and R. Bryan (Eds) *Badland Geomorphology*. GeoBooks, CUP, Cambridge, 13–47.

Hooper, R.P., Christophersen, N. and Peters, N.E. 1990. Modelling streamwater chemistry as a mixture of soilwater end-members: an application to the Panola Mountain Catchment, Georgia, USA. *Journal of Hydrology*, 116, 321–343.

Hornberger, G.M., Raffensberger, J.P., Wiberg, P.L. and Eshleman, K.N. 1998. *Elements of Physical Hydrology*. The Johns Hopkins University Press, Baltimore, 302pp.

Horton, R.E. 1933. The role of infiltration in the hydrological cycle. *Transactions of the American Geophysical Union*, 14, 446–460.

Horton, R.E. 1936. Maximum ground-water levels. *Transactions of the American Geophysical Union*, 17, 344–357.

Horton, R.E. 1940. An approach to the physical interpretation of infiltration capacity. *Soil Science Society of America, Proceedings*, 5, 399–417.

Horton, R.E. 1942. Remarks on hydrologic terminology. *Transactions of the American Geophysical Union*, 23, 479–482.

Huggins, L.F. and Monke, E.J. 1968. A mathematical model for simulating the hydrological response of a watershed. *Water Resources Research*, 4(3), 529–539.

Hughes, D.A. 1994. Soil moisture and runoff simulations using four catchment rainfall-runoff models. *Journal of Hydrology*, 158, 381–404.

Imeson, A.C., Verstraeten, J.M., Van Mulligan, E.J. and Sevink, J. 1992. The effects of fire and water repellency on infiltration and runoff under Mediterranean type forest. *Catena*, 19, 345–361.

Imeson, A.C., Lavee, H., Calvo, A. and Cerda, A. 1998. The erosional response of calcareous soils along a climatological gradient in Southeast Spain. *Geomorphology*, 24, 3–16.

Inbar, M., Tamir, M. and Wittenberg, L. 1998. Runoff and erosion processes after a forest fire in Mount Carmel, a Mediterranean area. *Geomorphology*, 24, 17–33.

Kibler, D.F. and Woolhiser, D.A. 1972. Mathematical properties of the kinematic cascade. *Journal of Hydrology*, 15, 131–147.

Kirkby, M.J. 1969. Infiltration, throughflow and overland flow. In R.J. Chorley (Ed.) *Water, Earth and Man*. Methuen, 215–228.

Kirkby, M.J. 1975. Hydrograph modelling strategies. In R.F. Peel, M.D. Chisholm and P. Haggett (Eds) *Progress in Physical and Human Geography*. Heinemann, London, 69–90.

Kirkby, M.J. (Ed.) 1978. *Hillslope Hydrology*. Wiley, Chichester.

Kirkby, M.J., Baird, A.J., Diamond, S.M., Lockwood, J.G., McMahon, M.L., Mitchell, P.L., Shao, J., Sheehy, J.E., Thornes, J.B. and Woodward, F.I. 1996. The MEDALUS slope catena model: a physically based process model for hydrology, ecology and land degradation interactions. In C.J. Brandt and J.B. Thornes (Eds) *Mediterranean Desertification and Land Use*. Wiley, Chichester, 303–354.

Kosmas, C., Danalatos, N., Cammeraat, L.H., Chabart, M., Diamantopoulos, J., Farand, R., Gutiérrez, L., Jacob, A., Marques, H., Martínez-Fernandez, J., Miozara, A., Moustakas, N., Nicolau, J.M., Oliveros, C., Pinna, G., Puddu, R., Puigdefàbregas, J., Roxo, M., Simao, A., Stamou, G., Tomais, N., Usai, D., and Vacca, A. 1997. The effect of land use on runoff and soil erosion rates under Mediterranean conditions. *Catena*, 29, 45–59.

Kutiel, P., Lavee, H., Segev, M. and Benyamini, Y. 1995. The effect of fire-induced surface heterogeneity on rainfall-runoff-erosion relationships in an eastern Mediterranean ecosystem, Israel. *Catena*, 25, 77–87.

Lane, L.J. 1982. Distributed model for small semi-arid watersheds. *Journal of Hydraulic Division, ASCE*, 108 (HY10), 1114–1131.

Lane, L.J. and Woolhiser, D.A. 1977. Simplifications of watershed geometry affecting simulation of surface runoff. *Journal of Hydrology*, 35, 173–190.

Lavabre, J., Sempere-Torres, D. and Cernesson, F. 1993. Changes in the hydrological response of a small Mediterranean basin a year after a wildfire. *Journal of Hydrology*, 142, 273–299.

Lavee, H., Imeson, A.C., Pariente, S. and Benyamini, Y. 1991. The response of soils to simulated rainfall along a climatological gradient in an arid and semi-arid region. *Catena Supplement*, 19, 19–37.

Lawrence, D.S.L. 1997. Macroscale surface roughness and frictional resistance in overland flow. *Earth Surface Processes and Landforms*, 22, 365–382.

Leavesley, G.H. and Stannard, L.G. 1995. The Precipitation-Runoff Modelling System – PRMS. In V.P. Singh (Ed.) *Computer Models of Watershed Hydrology*. Water Resource Publications, Highlands Ranch, CO., 281–310.

Leonard, J. and Andrieux, P. 1998. Infiltration characteristics of soils in Mediterranean vineyards in S. France. *Catena*, 32, 209–223.

Linden, D.R. and Dixon, R.M. 1973, Infiltration and water table effects of soil air pressure under border irrigation. *Soil Science Society of America, Proceedings*, 37, 95–98.

Llorens, P., Poch, R., Latron, J. and Gallart, F. 1997. Rainfall interception by a *Pinus sylvestris* forest patch overgrown in a Mediterranean mountainous abandoned area. 1. Monitoring design and results down to the event scale. *Journal of Hydrology*, 199, 331–345.

Loague, K.M. 1990. R-5 revisited. 2. Reevaluation of a quasi-physically-based rainfall-runoff model with supplemental information. *Water Resources Research*, 26, 973–987.

Loague, K.M. and Freeze, R.A. 1985. A comparison of rainfall-runoff modelling techniques on small upland catchments. *Water Resources Research*, 21, 229–248.

Loague, K.M. and Gander, G.A. 1990. R-5 revisited. 1. Spatial variability of infiltration on a small rangeland watershed. *Water Resources Research*, 26, 957–971.

Loague, K.M. and Kyriakidis, P.C. 1997. Spatial and temporal variability in the R-5 infiltration data set: déjà vu and rainfall–runoff simulations. *Water Resources Research*, 33(12), 2883–2895.

López-Bermúdez, F. and Romero-Diaz, M.A. 1989. Piping erosion and badland development in south-east Spain, *Catena Supplement*, 14, 59–73.

Martin, C. and Lavabre, J. 1997. Estimation de la part du ruisellement sur les versants dans les crues du ruisseau du Rimbaud (massif des Maures, Var, France) après l'incendie de forêt d'août 1990. *Hydrological Science Journal*, 42, 893–907.

Martinez-Mena, M., Albaladejo, J. and Castillo, V.M. 1998. Factors influencing surface runoff generation in a Mediterranean semi-arid environment, Chicamo watershed, SE Spain. *Hydrological Processes*, 12, 741–754.

McCuen, R.H. 1982. *A Guide to Hydrologic Analysis Using SCS Methods*. Prentice-Hall, Englewood Cliffs, NJ.

McDonnell, J.J., Freer, J., Hooper, R., Kendall, C., Burns, D., Beven, K. and Peters, J. 1996. New method developed for studying flow on hillslopes. *EOS, Transactions AGU*, 77(47), 465/472.

Meirovich, L., Ben-Zvi, A., Shentsis, I. and Yanovich, E. 1998. Frequency and magnitude of runoff events in the arid Negev of Israel. *Journal of Hydrology*, 207, 204–219.

Michaud, J. and Sorooshian, S. 1994. Comparison of simple versus complex distributed runoff models on a midsized semi-arid watershed. *Water Resources Research*, 30(3), 593–606.

Mishra, S.K. and Singh, V.P. 1999. Another Look at the SCS-CN Method. *Journal of Hydrological Engineering ASCE*, 4, 257–264.

Morel-Seytoux, H.J. and Khanji, J. 1974. Derivation of an Equation of Infiltration. *Water Resources Research*, 10, 795–800.

Morgan, R.P.C., Quniton, J.N., Smith, R.E., Govers, G., Poesen, J.W.A., Auerswald, K., Chisci, G., Torri, D. and Styczen, M.E. 1998. The European Soil Erosion Model (EUROSEM): a dynamic approach for predicting sediment transport from fields and small catchments. *Earth Surface Processes and Landforms*, 23, 527–544.

Morris, E.M. and Woolhiser, D.A. 1980. Unsteady one-dimensional flow over a plane: partial equilibrium and recession hydrographs. *Water Resources Research*, 16, 355–360.

Musgrave, G.W. and Holtan, H.N. 1964. Infiltration. In V.T. Chow (Ed.) *Handbook of Applied Hydrology*. McGraw-Hill, New York, section 12.

Neal, C., Neal, M., Warrington, A., Avila, A., Piñol, J. and Rodà, F. 1992. Stable hydrogen and oxygen isotope studies of rainfall and streamwaters for two contrasting holm oak areas of Catalonia, northeastern Spain. *Journal of Hydrology*, 140, 163–178.

Nicolau, J.M., Solé-Benet, A., Puigdefàbregas, J. and Gutiérrez, L. 1996. Effects of soil and vegetation on runoff along a catena in semi-arid Spain. *Geomorphology*, 14, 297–309.

Obled, Ch, Wendling, J. and Beven, K.J. 1994. The role of spatially variable rainfalls in modelling catchment response: an evaluation using observed data. *Journal of Hydrology*, 159, 305–333.

Oostwoud Wijdenes, D.J. and Ergenzinger, P. 1998. Erosion and sediment transport on steep marly hillslopes, Draix, Haute-Provence, France: an experimental field study. *Catena*, 179–200.

Osterkamp, W.R., Lane, L.J., and Savard, C.S. 1994. Recharge estimates using a geomorphic/distributed-parameter simulation approach, Armagosa river basin. *Water Resource Bulletin*, 30, 493.

Parkin, G., O'Donnell, G., Ewen, J., Bathurst, J.C., O'Connell, P.E. and Lavabre, J. 1996. Validation of catchment models for predicting land-use and climate change impacts. 2. Case study for a Mediterranean catchment. *Journal of Hydrology*, 175, 595–613.

Parlange, J.-Y. and Haverkamp, R. 1989. Infiltration and ponding time. In H.J. Morel-Seytoux (Ed.) *Unsaturated Flow in Hydrologic Modeling*, Kluwer Academic, Dordrecht, 105–126.

Parlange, J.-Y., Hogarth, W., Ross, P., Parlange, M.B., Sivapalan, M., Sander, G.C. and Liu, M.C. 2000. A note on the error analysis of time compression approximations. *Water Resources Research*, 36, 2401–2406.

Parsons, A.J., Wainwright, J. and Abrahams, A.D. 1996. Runoff and erosion on semi-arid hillslopes. In M.G. Anderson and S.M. Brooks (Eds) *Advances in Hillslope Processes*. Wiley, Chichester, 1061–1078.

Parsons, A.J., Wainwright, J., Abrahams, A.D. and Simanton, J.R. 1997. Distributed dynamic modelling of interrill overland flow. *Hydrological Processes*, 11, 1833–1859.

Peugeot, C., Esteves, M., Galle, S., Rajot, J.L. and Vandercaere, J.-P. 1997. Runoff generation processes: results and analysis of field data collected at the East Central Supersite of the HAPEX-Sahel experiment, *Journal of Hydrology*, 188–189, 179–202.

Philip, J.R. 1957. The theory of infiltration. 4. Sorptivity and algebraic infiltration equations. *Soil Science*, 84, 257–264.

Pilgrim, D.H., Chapman, T.G. and Doran, D.G. 1988. Problems of rainfall-runoff modelling in arid and semi-arid regions. *Hydrological Science Journal*, 33, 379–400.

Piñol, J., Beven, K.J. and Freer, J. 1997. Modelling the hydrological response of mediterranean catchments, Prades, Catalonia – the use of distributed models as aids to hypothesis formulation. *Hydrological Processes*, 11(9), 1287–1306.

Poesen, J. and Lavee, H. 1991. Effects of size and incorporation of synthetic mulch on runoff and sediment yield from interills in a laboratory study with simulated rainfall. *Soil and Tillage Research*, 21, 209–223.

Poesen, J. and Ingelmo-Sánchez, F. 1992. Interrill runoff and sediment yield from topsoils with different structure as affected by rock fragment cover and position. *Catena*, 19, 451–474.

Poesen, J. and Bunte, K. 1996. The effects of rock fragments on desertification processes in Mediterranean environments. In C.J. Brandt and J.B. Thornes (Eds) *Mediterranean Desertification and Land Use*. Wiley, Chichester, 247–269.

Ponce, V.M., Li, R.M. and Simons, D.B. 1978. Applicability of kinematic and diffusion models. *Journal of the Hydrology Division ASCE*, 104, 353–360.

Ponce, V.M., Pandey, R.P. and Kumar, S. 1999. Groundwater recharge by channel infiltration in El Barbon basin, Baja California, Mexico. *Journal of Hydrology*, 214, 1–7.

Prosser, I.P. and Williams, L. 1998. The effects of wildfire on runoff and erosion in native Eucalyptus forest. *Hydrological Processes*, 12, 251–265.

Puigdefàbregas, J. and Sánchez, G. 1996. Geomorphological implications of vegetation patchiness on semi-arid slopes. In M.G. Anderson and S.M. Brooks (Eds) *Advances in Hillslope Processes*. Wiley, Chichester, 1027–1060.

Puigdefàbregas, J., Alonso, J.M., Delgado, L., Domingo, F., Cueto, M., Gutiérrez, L., Lazaro, R., Nicolau, J.M., Sánchez, G., Solé, A. and Vidal, S. 1996. The Rambla Honda field site: interactions of soil and vegetation along a catena in semi-arid southeast Spain. In C.J. Brandt and J.B. Thornes (Eds) *Mediterranean Desertification and Land Use*. Wiley, Chichester, 137–168.

Rallison. R.E. and Miller, M. 1982. Past, present and future SCS runoff procedure. In V.P. Singh (Ed.) *Rainfall–Runoff Relationships*. Water Resource Publications, Littleton, CO, 353–364.

Rawls, W.S. and Brakensiek, D.L. 1989. Estimation of soil water retention and hydraulic properties. In H.J. Morel-Seytoux (Ed.) *Unsaturated Flow in Hydrologic Modeling: Theory and Practice*. Kluwer Academic, Dordrecht, 275–300.

Refsgaard, J.-C. and Knudsen, J. 1996. Operational validation and intercomparison of different types of hydrological models. *Water Resources Research*, 32, 2189–2202.

Richards, L.A. 1931. Capillary conduction of liquids through porous mediums. *Physics*, 1, 318–333.

Robichaud, P.R. and Waldrup, T.A. 1994. A comparison of surface runoff and sediment yields from low- and high-severity preparation burns. *Water Resource Bulletin*, 30, 27–34.

Rockström, J., Jansson, P.-E. and Barron, J. 1998. Seasonal rainfall partitioning under run-on and runoff conditions on sandy soil in Niger. On-farm measurements and water balance modelling. *Journal of Hydrology*, 210, 68–92.

Römkens, M.J.M. and Wang, J.Y. 1986. Effects of tillage on surface roughness. *Transactions of the American Society of Agricultural Engineers*, 29, 429–433.

Römkens, M.J.M., Prasad, S.N. and Whisler, F.D. 1990. Surface sealing and infiltration. In M.G. Anderson and T.P. Burt (Eds) *Process Studies in Hillslope Hydrology*. Wiley, Chichester, 127–172.

Ruan, H. and Illangasekare, T.H. 1998. A model to couple overland flow and infiltration into macroporous vadose zone. *Journal of Hydrology*, 210, 116–127.

Sandström, K. 1996. Hydrochemical deciphering of streamflow generation in semi-arid East Africa. *Hydrological Processes*, 10, 703–720.

Saulnier, G.-M., Obled, Ch., and Beven, K.J. 1997. Analytical compensation between DTM grid resolution and effective values of saturated hydraulic conductivity within the TOPMODEL framework. *Hydrological Processes*, 11, 1331–1346.

Savage, S.M. 1974. Mechanisms of fire induced water repellency in soils. *Soil Science Society of America Proceedings*, 38, 652–657.

Scoging, H.M. 1982. Spatial variations in infiltration, runoff and erosion on hillslopes in semi-arid Spain. In R.B. Bryan and A. Yair (Eds) *Badland Geomorphology and Piping*. GeoBooks, Norwich, 89–112.

Scoging, H.M. and Thornes, J.B. 1979. Infiltration characteristics in a semi-arid environment. *IAHS Publication No.* 128, 159–168.

Scoging, H.M., Parsons, A.J. and Abrahams, A.D. 1992. Application of a dynamic overland flow hydraulic model to a semi-arid hillslope, Walnut Gulch, Arizona. In A.J. Parsons and A.D. Abrahams (Eds) *Overland : Hydraulics and Erosion Mechanics*. UCL Press, London, 105–145.

Sharma, M.L., Gander, G.A. and Hunt, S.G. 1980. Spatial variability of infiltration in a watershed. *Journal of Hydrology*, 45, 101–122.

Simanton, J.R., Hawkins, R.H., Mohseni-Saarvai, M. and Renard, K.G. 1996. Runoff curve number variation with drainage area, Walnut Gulch, Arizona. *Transactions of the American Society of Civil Engineers*, 39(4), 1391–1394.

Singh, V.P. 1996. *Kinematic Wave Modelling in Water Resources*. Wiley, Chichester, 1399pp.

Sivapalan, M., Beven, K.J. and Wood, E.F. 1987. On hydrologic similarity. 2. A scale model of storm runoff production. *Water Resources Research*, 23, 2266–2278.

Sklash, M.G. 1990. Environmental isotope studies of storm and snowmelt runoff generation. In M.G. Anderson and T.P. Burt (Eds) *Process Studies in Hillslope Hydrology*. Wiley, Chichester, 401–435.

Slatyer, R.O. 1965. Measurements of precipitation interception by an arid zone plant community (*Acacia aneura*). *UNESCO Arid Zone Research*, 25, 181–192.

Smith, R.E. 1981. Rational models of infiltration hydrodynamics. In V.P. Singh (Ed.) *Modeling Components of Hydrologic Cycle*. Water Resource Publications, Littleton, CO., 107–126.

Smith, R.E. and Woolhiser, D.A. 1971. Overland flow on an infiltrating surface. *Water Resources Research*, 7(4), 899–913.

Smith, R.E. and Parlange, J.-Y. 1978. A parameter efficient infiltration model. *Water Resources Research*, 14, 533–538.

Smith, R.E., Goodrich, D.C., Woolhiser, D.A. and Unkrich, C.L. 1995. KINEROS – a kinematic runoff and erosion model. In V.P. Singh (Ed.) *Computer Models of Watershed Hydrology*. Water Resource Publications, Littleton, CO, 697–732.

Smith, R.E., Corradini, C. and Melone, F. 1999. A conceptual model for infiltration and redistribution in crusted soils. *Water Resources Research*, 35, 1385–1393.

Snelder, D.J. and Bryan, R.B. 1995. The use of rainfall simulation tests to assess the influence of vegetation density on soil loss on degraded rangelands in the Baringo District, Kenya. *Catena*, 25, 105–116.

Solé-Benet, A., Calvo, A., Cerda, A., Lazaro, R., Pini, R. and Barbero, J. 1997. Influences of micro-relief patterns and plant cover on runoff related processes in badlands from Tabernas (SE Spain). *Catena*, 31, 23–38.

Sorriso-Valvo, M., Bryan, R.B., Yair, A., Iovino, F. and Antronico, L. 1995. Impact of afforestation on hydrological response and sediment production in a small Calabrian catchment. *Catena*, 25, 89–104.

Soto, B. and Diaz-Fierros, F. 1998. Runoff and soil erosion from areas of burnt scrub: comparison of experimental results with those predicted by the WEPP model. *Catena*, 31, 257–270.

Steenhuis, T.S., Winchell, M., Rossing, J., Zollweg, J.A. and Walter, M.F. 1995. SCS runoff equation revisited for variable source runoff areas. *Journal of Irrigation and Drainage Engineering, ASCE*, 121, 234–238.

Taha, A., Gresillon, J.M. and Clothier, B.E. 1997. Modelling the link between hillslope water movement and stream flow: application to a small Mediterranean forest watershed. *Journal of Hydrology*, 203, 11–20.

Tayfur, G. and Kavvas, M.L. 1998. Areally-averaged overland flow equations at hillslope scale. *Hydrological Science Journal*, 43, 361–378.

Thiéry, J.M., d'Herbès, J.M. and Valentin, C. 1995. A model simulating the genesis of banded vegetation patterns in Niger. *Journal of Ecology*, 83, 497–507.

Thornes, J.B. 1994. Catchment and channel hydrology. In A.D. Abrahams and A.J. Parsons (Eds) *Geomorphology of Desert Environments*. Chapman & Hall, London, 257–287.

Thornes, J.B., Shao, J., Diamond, S., McMahon, M. and Hawkes, J.C. 1996. Testing the Medalus model. *Catena*, 26, 106–156.

Todini, E. 1996. The ARNO rainfall-runoff model. *Journal of Hydrology*, 175, 339–382.

Torri, D., Colica, A. and Rockwell, D. 1994. Preliminary study of the erosion mechanisms in a biancana badland (Tuscany, Italy). *Catena*, 23, 281–294.

Turner, J.V., Bradd, J.M. and Waite, T.D. 1991. The conjunctive use of isotopic techniques to elucidate solute concentration and flow processes in dryland salinized catchments. In *International Symposium on the Use of Isotopes in Water Resources Development*. IAHS.

Valentin, C. and Bresson, L.M. 1992. Morphology, genesis and classification of surface crusts in loamy and sandy soils. *Geoderma*, 55, 225–245.

Valentin, C. and d'Herbès, J.M. 1999. Niger tiger bush as a natural water harvesting system. *Catena*, 37, 231–256.

Vandervaere, J.-P., Peugeot, C., Vauclin, M., Jaramillo, R.A. and Lebel, T. 1997. Estimating hydraulic conductivity of crusted soils using disc infiltrometers and mintensiometers. *Journal of Hydrology*, 188–189, 203–223.

Vereecken, H., Diels, J., Van Orshoven, J., Feyen, J. and Bouma, J. 1992. Functional evaluation of pedotransfer functions for the estimation of soil hydraulic properties. *Soil Science Society of America Journal*, 56, 1371–1378.

Wainwright, J. 1996a. Hillslope response to extreme storm events: the example of the Vaison-la-Romaine event. In M.G. Anderson and S.M. Brooks, (Eds) *Advances in Hillslope Processes*. Wiley, Chichester, 997–1026.

Wainwright, J. 1996b. Infiltration, runoff and erosion characteristics of agricultural land in extreme storm events, SE France. *Catena*, 26, 27–47.

Wang, Z., Feyen, J., Van Genuchten, M. Th. and Nielsen, D.R. 1998. Air entrapment effects on infiltration rates and flow instability. *Water Resources Research*, 34, 213–222.

Western, A., Grayson, R.B., Blöschl, G., Willgoose, G.R. and McMahon, T.A. 1999. Observed spatial organisation of soil moisture and its relation to terrain indices. *Water Resources Research*, 35(3), 797–810.

White, S., Garćia-Ruíz, J.M., Marti, C., Valero, B., Errea, M.P. and Gomez-Villar, A. 1997. The 1996 Biescas Campsite disaster in the central Spanish Pyrenees and its temporal and spatial context. *Hydrological Processes*, 11, 1797–1812.

Wigmosta, M.S., Vail, L.W. and Lettenmaier, D.P. 1994. A distributed hydrology-vegetation model for complex terrain. *Water Resources Research*, 30, 1665–1680.

Wilcox, B.P., Wood, M.K. and Tromple, J.M. 1988. Factors influencing infiltrability of semi-arid mountain slopes. *Journal of Range Management*, 41, 197–206.

Wilcox, B.P., Newman, B.D., Brandes, D., Davenport, D.W. and Reid, K. 1997. Runoff from a semi-arid ponderosa pine hillslope in New Mexico. *Water Resources Research*, 33, 2301–2314.

Willgoose, G.R. and Kuczera, G. 1995. Estimation of subgrid scale kinematic wave parameters for hillslopes. *Hydrological Processes*, 9, 469–482.

Williams, J. and Bonell, M. 1988. The influence of scale of measurement on the spatial and temporal variability of the Philip infiltration parameters – an experimental study in an Australian savannah woodland. *Journal of Hydrology*, 104, 33–51.

Woolhiser, D.A. and Goodrich, D.C. 1988. Effects of storm rainfall intensity patterns on surface runoff. *Journal of Hydrology*, 102, 335–354.

Woolhiser, D.A., Smith, R.E. and Giraldez, J.V. 1996. Effects of spatial variability of saturated hydraulic conductivity on Hortonian overland flow. *Water Resources Research*, 32, 670–678.

Wösten, J.H.M., Finke, P.A. and Jansen, M.J.W. 1995. Comparison of class and continuous pedotransfer functions to generate soil hydraulic characteristics. *Geoderma*, 66, 227–237.

Yair, A. 1992. The control of headwater area on channel runoff in a small arid watershed. In A.J. Parsons and A.D. Abrahams (Eds) *Overland Flow: Hydraulics and Erosion Mechanics*. UCL Press, London.

Yair, A. and Danin, A. 1980. Spatial variations in vegetation as related to the soil moisture regime over an arid limestone hillside, northern Negev, Israel. *Oecologia*, 47, 83–88.

Yair, A. and Lavee, H. 1985. Runoff generation in arid and semi-arid zones. In M.G. Anderson and T.P. Burt (Eds) *Hydrological Forecasting*. Wiley, Chichester, 183–220.

Ye, W., Bates, B., Viney, N.R., Sivapalan, M. and Jakeman, A.J. 1997. Performance of conceptual rainfall–runoff models in low-yielding ephemeral catchments. *Water Resources Research*, 33, 153–166.

Yu, B. 1998. Theoretical justification of SCS method for runoff estimation. *Journal of Irrigation and Drainage Engineering, ASCE*, 124, 306–309.

Zhu, T.X., Cai, Q.G. and Zeng, B.Q. 1997. Runoff generation on a semi-arid agricultural catchment: field and experimental studies. *Journal of Hydrology*, 196, 99–118.

4   Sediment Dynamics of Ephemeral Channels

IAN REID

Department of Geography, Loughborough University, UK

4.1   INTRODUCTION

The compressive tectonics of the Tertiary period of Earth history, particularly those of the Miocene epoch, have produced a Mediterranean landform which is typified by high peaks and steep slopes. The character of many of the rivers that issue from such a landscape, especially that of headwaters, is governed by both the contortions of folds and the placement of nappes, and by the ways in which these have affected fluvial and glacial incision. Channels are often steep and there is considerable coupling with adjacent hillsides to form an integrated sediment transfer system. However, within this broad framework, smaller-scale extensional and trans-tensional motions have produced a different landscape. These have had their own effect upon the drainage system, providing local conduits (e.g. the rift structures of peninsular Greece; Collier et al., 1995) or beheading rivers (e.g. the trans-Jordanian rivers that now sink to the Dead Sea rather than to the Mediterranean; Gerson et al., 1984) and causing the complete reversal of drainage (Frostick and Reid, 1989).

Despite the prevalence of considerable structural complexity and the littoral proximity of high mountains along much of both the northern and southern shores, a number of large rivers such as the Rhône, the Po, the Ebro and the Nile pour into the Mediterranean Sea. They have been encouraged to do so by events not so distant in time. These have included the isolation of the Mediterranean through tectonic closure of the connection with the Atlantic and the dramatic lowering of base-level through evaporation which ensued – a lowering which produced the 'salinity crisis' of the Upper Miocene (Butler et al., 1995). More recently, there have been significant, if lesser, fluctuations in sea-level that were driven by global changes in climate during the Pleistocene (Butzer, 1962; Adamson, 1982). However, although these rivers fall to the Mediterranean, they are a product of climates that are far from Mediterranean in character, one branch of the Nile rising in equatorial central Africa and the Rhone being dominated by a snow-melt regime that reflects seasonality of climate in the Alps.

Perhaps more typical of the region are the small- to medium-sized rivers which have comparatively high longitudinal gradients and which are often coarse-grained (Macklin et al., 1995). They rise *within* the area classified as having a Mediterranean climate, with its strong seasonality. Their annual flow regime ranges widely in response to regional patterns of rainfall and snow-melt and includes those that are perennial or near-perennial, those that are winter-seasonal, and those that are ephemeral, where flow is storm-induced and where the channel remains dry for most of the year. This spectrum of flow regimes provides a commensurably

Dryland Rivers: Hydrology and Geomorphology of Semi-arid Channels. Edited by L.J. Bull and M.J. Kirkby.
© 2002 John Wiley & Sons, Ltd.

broad range in the magnitude and frequency of sediment transport in drainage basins that are otherwise similar in size, relief and so on.

As always, there is a need for much more information about the processes operating in such streams, but the number of sediment transport monitoring stations within the region is surprisingly low considering the obvious problems of reservoir 'siltation' (Zachar, 1982; Roquero, 1991) and the difficulties of maintaining road and rail communications (Schick, 1973). One factor discouraging a proliferation of such stations has been the highly intermittent nature of sediment transport events in many of the region's rivers and the need to justify the expense of long-term monitoring against a backcloth of slow and uncertain data acquisition. However, there is now emerging a body of information that gives some degree of confidence in a number of generalisations about patterns of sediment flux and their degree of predictability (Milliman and Syvitski, 1992; Woodward, 1995). Some of this comes from an assessment of rapid reservoir siltation (e.g. Lahlou, 1988) or the short-term monitoring of sediment flux (Palanques et al., 1990). Some comes from studies of lake cores and other stratigraphical evidence that can be used to provide a historical or prehistorical perspective, which usually implicates human activity as an important influence on landscape and the accelerated erosion that often follows (e.g. van Andel et al., 1986; Flower et al., 1989).

4.2   BEDLOAD TRANSPORT

4.2.1   Ephemeral Streams

Despite the discrete nature of flow events in ephemeral streams, there has been a growing interest in sediment flux in areas with dryland climates throughout the Old and New Worlds. This is driven, in part, by a need to understand the impact on sedimentation of extensive human interference with land cover (e.g. deforestation; Lisle and Madej, 1992), as well as other significant disturbances of the natural system, such as the hydraulic mining of alluvial placers (Simons and Ruh-Ming, 1982). Where radical changes in land-cover are historical, interest in understanding sedimentary processes and fluxes has been increased by the rapid infilling of reservoir lakes in situations where dams have been constructed recently to satisfy a present-day increase in agricultural demand for water (Roquero, 1991). There has also been an increasing awareness of the need to derive realistic estimates of bedload transport as a means of assessing the sustainability of potentially damaging operations such as the mining of alluvium for building aggregates (Kondolf, 1995).

The longest systematic record of flood flows in ephemeral rivers comes from Walnut Gulch in southern Arizona (Renard and Keppel, 1966), which is outside areas that experience a Mediterranean climate. In setting up the research station, the US Agricultural Research Service was interested primarily in the rainfall–runoff relations of semi-arid rangelands. Although some sediment transport data have been collected (Renard and Lane, 1975; Renard and Laursen, 1975), these have not been the focus of attention. Interest in sediments has been directed rather at understanding the hydraulic conductivity of the channel alluvium and its impact on transmission losses (Murphey et al., 1972). Indeed, it is this aspect of ephemeral channel sediments, rather than their transportability, that has often drawn the attention of those hydrologists and geomorphologists fascinated by flash floods (e.g. Burkham, 1970; Dunkerley, 1992; Anderson et al., 1998).

Within the Mediterranean region itself, Thornes (1976) provided some early data about the nature of some Spanish coarse-grained ephemeral streams, focusing on channel-bed sediments as they change downstream. Although the trends he found were subtle rather than clear-cut illustrations of patterns such as classical downstream fining, they drew his attention to the impact of tributaries on the trunk stream, the significance of which had also become apparent to Frostick and Reid (1977) after examining the microstratigraphy of Kenyan sand-bed ephemeral channel fills. Ahead of these studies, Schick (1970) had started monitoring sediment transport in the 1960s through a tracing programme on the Nahal Yael in the southern Negev. This was a deliberate attempt to overcome the disincentive to set up monitoring programmes that prevails in drylands because of the infrequency of events – Schick's (1988) persistence provides a record that has accumulated at less than one event per year. In the context of ephemeral rivers, his study provides a useful end-member of an environmental spectrum, although the hyper-arid climate of its setting and its location just beyond the boundary of the Mediterranean region that is usually drawn (Macklin et al., 1995) makes it more Saharo-Arabian in character. This long-term work by Schick on the transport of river-bed material has been complemented by Hassan (1990), who instituted a number of clast-tracing programmes in a couple of ephemeral streams of Judea, further north in Israel. Here, rainfall is more frequent and annual totals are higher at 300–500 mm. So, for example, Hassan reported three bedload generating events on the Nahal Hebron during a year's monitoring. He also showed that channel-bed disturbance during floods can involve scour and fill and that this induces individual clasts to move between surface and buried positions in the channel bed from flood to flood, so affecting their speed of passage downstream.

However, there was no live-bed information about bedload sediment transport in the Mediterranean region until Taconni and Billi (1987) installed a semi-automatic monitoring station on a near-perennial stream in Tuscany. This was followed soon after by Laronne et al. (1992) who, in 1989, placed a fully automatic station on an ephemeral gravel-bed stream in the northern Negev. This station employed three Birkbeck-type bed-slot samplers (Reid et al., 1980), giving the first information of its kind for the Mediterranean and for ephemeral streams – a continuous record of hydraulics and bedload flux during flash floods (Reid et al., 1995).

The Yatir database indicated that unit bedload flux is several orders of magnitude higher than had been established in gravel-bed streams elsewhere, with peak levels of 7 kg m^{-1} s^{-1} recorded at modest levels of boundary shear stress of 36 Pa (Laronne and Reid, 1993). Because the record for the Yatir hinted that such a programme was vital to answering some of the pressing problems of sedimentation in this kind of environmental setting, a similar station was installed nearby on the Nahal Eshtemoa. The Eshtemoa is larger than the Yatir, and five Birkbeck samplers have been deployed across the width of the channel since 1990 (Figure 4.1). Although the channel and the bed sediments differ in being wider and coarser, respectively, the record that has been obtained so far (Figure 4.2) exhibits a remarkable likeness to that of the Yatir. Indeed, when differences in bed material grain-size are taken into account in a dimensionless plot, the data of the two streams fit neatly into one envelope that describes bedload flux as a function of excess shear stress (Reid et al., 1998a).

This bedload–shear stress response is remarkably simple over three orders of magnitude of bedload. It more or less coincides with that predicted by the capacity transport equations developed by a number of engineers and physical scientists over a number of decades. So, although there is a small degree of scatter in the observed bedload, the unit flux can be approximated reasonably well using the 1948 version of the one-dimensional equation of Meyer-Peter and Müller (Figure 4.3).

Birkbeck-type slot samplers have been deployed on both the Yatir and the Eshtemoa in part because the unpredictability of flow – especially those that are capable of generating bedload – invites a fully automatic system. This, and the independent functioning of each sampler, has brought the added benefit of synchronised measurements of bedload flux at a number of cross-channel locations and has allowed, for the first time, an assessment of cross-stream bedload

Figure 4.1   The five fully-automatic Birkbeck-type bedload slot samplers on the Nahal Eshtemoa being overrun by the bore of a flash flood

hydraulics during the rapidly changing conditions of a flash flood – though the lessons need not be confined to ephemeral channels. Powell et al. (1999) show clearly the effects of sidewall drag in reducing boundary shear in the vicinity of the stream banks. This has the effect of halving bedload flux in this region relative to the channel centre. However, there is no demonstrable change in the influence of the sidewall as the aspect ratio of the flow changes from 30 to 6, suggesting less difference in the behaviour of wide and narrow channels than has previously been speculated.

4.2.2   Near-perennial Streams

The more recent installation of a monitoring station towards the western end of the Mediterranean in the Catalan Coastal Ranges provides hard-won live-bed data at the other end of the annual flow-regime spectrum – the near-perennial rivers. Two automatic Birkbeck

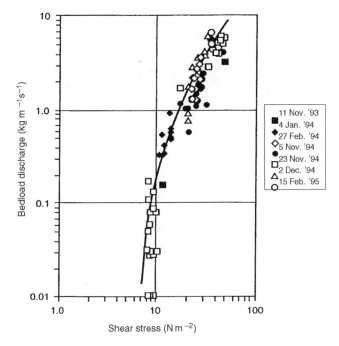

Figure 4.2 Bedload discharge as a function of shear stress on the ephemeral Nahal Eshtemoa, Israel. Data for individual flashfloods are differentiated. The curve is $i_b = 0.027(\tau - 7)^{1.52}$. Zero transport rates are placed arbitrarily at 0.01 kg m^{-1} s^{-1}. (After Reid et al., 1998a)

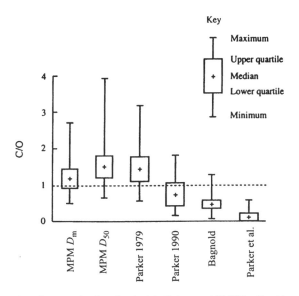

Figure 4.3 Dispersion diagram of ratios of calculated/observed (C/O) bedload transport rates for Nahal Yatir, Israel, using a number of established bedload functions. (After Reid et al., 1996)

Figure 4.4 The two fully-automatic Birkbeck-type bedload slot samplers on the near-perennial Río Tordera, NE Spain. Flow is from left to right. The I-beam carries a truck and chain-winch to facilitate removal of the inner collecting boxes when full of sediment. (Photograph courtesy of Celso García)

samplers and their attendant sensors have been set into the gravel-bed of the Río Tordera (Figure 4.4; García and Sala, 1997). The database that has been generated serves to highlight the significant behavioural differences of perennial and ephemeral streams of the region.

The measured flux rates of bedload are much smaller – at least an order of magnitude lower than the Yatir–Eshtemoa over the same range of shear stress. This, together with a critical value for shear in excess of 50 Pa (compared with 7 Pa for Eshtemoa and 3 Pa for Yatir) is related in part to the coarser nature of the surface-bed sediment (54 mm, compared with 16 and 6 mm). However, Figure 4.5 indicates a high degree of unpredictability, which has been attributed to the variable nature of the source areas on the bed of the stream (García et al., 1999, 2000). Decimetre to metre-scale patches of comparatively fine bed material are the principal source of bedload under moderate flows, but these fluctuate in their grain size, their areal extent, and their interconnectedness, depending on flood history. Occasionally, the armour layer consisting of much coarser sediment, between and on which the finer patches rest, is disrupted and this adds further to the scatter that is evident in the relation between bedload and shear stress.

Besides the extremely large scatter in the plot of bedload against contemporary shear stress in the Río Tordera, curves representing previously developed bedload equations commonly used in river engineering are seen to be largely divorced from the envelope of empirical data. This unpredictability of bedload flux in perennial gravel-bed streams has fascinated sedimentologists for several decades and is yet to be fully explained.

The complex bedload response pattern of the Río Tordera is matched by that of Virginio Creek, a near-perennial gravel-bed stream south of Firenze in Tuscany on which a vortex-tube

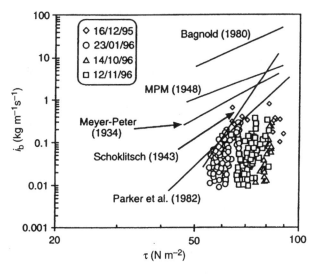

Figure 4.5 Bedload transport as a function of boundary shear stress, flood by flood, on the Río Tordera and predicted relations using a number of established bedload functions. (After García and Sala, 1997)

Figure 4.6 The vortex tube bedload sampler on the near-perennial Virginio Creek, Tuscany, Italy. Flow is from right to left. Sediment take-off to a set of rotating sieves and a weighing hopper is on the far bank. (Andrew Brayshaw for scale)

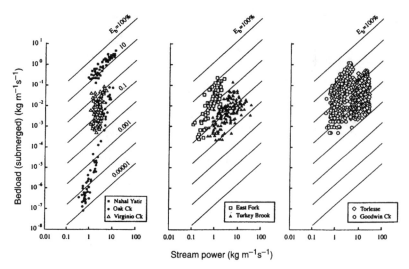

Figure 4.7 Bedload as a function of stream power in one ephemeral (Yatir), four perennial or near-perennial (Oak Ck, Turkey Bk, Torlesse Str. and Virginio Ck) and one seasonal (Goodwin Ck) gravel-bed streams and in the gravelly sand-bed East Fork River. Note that intervals where transport rates are zero are not shown. $E_b = 100 i_b / (\tan \alpha)$ in which $\tan \alpha = 0.63$ is Bagnold's (1973) percentage bedload transport efficiency index. (After Reid and Laronne, 1995)

bedload slot-sampler was installed (Figure 4.6; Taconni and Billi, 1987). Here, there is a clear demonstration of pulsation in the pattern of transport that is not matched by changes in hydraulic conditions. The resulting degree of scatter in the bedload–stream power relation of this stream is evident in Figure 4.7 where it can be seen that there is virtually no trend in bedload versus shear stress, at least over the range of hydraulic conditions for which there are data.

Lenzi et al. (1999) summarise the bedload record obtained with a complex fixed structure on the Rio Cordon, a small mountain torrent with a step-pool profile in the Dolomites. Flow here is perennial, though summer discharge is very low. The station separates the coarse (>20 mm) fraction of the load as sediment is carried over a grid in the stream bed and the accumulating volume is determined by scanning repeatedly with ultrasonic sensors. One of the interesting observations is that the hydraulic threshold of entrainment of coarse material appears to be consistently lower than that associated with the cessation of transport for material of the same calibre during the flow recession of each flood wave.

Further insight into the workings of dryland near-perennial streams comes from the Arbúcies River. This is a gravelly sand-bed stream that is tributary to the Río Tordera. Bedload has been monitored here using portable sampling equipment rather than a fixed installation (Figure 4.8; Batalla and Sala, 1995). Despite the small calibre of the bed material, there is a surprisingly indeterminate relation between sediment flux and water discharge, especially towards low flows. This indeterminacy is attributed by Batalla and Sala to the variable influence of migrating bedforms and to the development and breaching of a surface armour during the passage of flood waves. Some of the complexity may also be due to the complicated cross-sectional geometry of the channel. However, although the absence of a simple sediment flux–hydraulics relation may be surprising, given the fine-grained nature of much of the bed material, the behaviour of the Arbúcies River is not inconsistent with that of the celebrated gravelly sand-bed East Fork River (Figure 4.7; Leopold and Emmett, 1997).

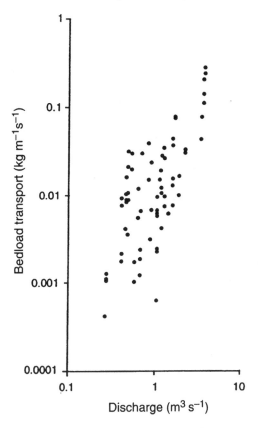

Figure 4.8 Bedload transport as a function of water discharge in the near-perennial gravelly sand-bed Arbúcies River, Catalonia, northeast Spain. (After Batalla and Sala, 1995)

4.2.3 Comparison of Ephemeral and Near-perennial Streams

The simplicity of bedload response in ephemeral gravel-bed channels has been attributed to several factors, the most significant of which is a lack of armour layer development (Reid and Laronne, 1995). Laronne et al. (1994) have postulated that the tendency towards size-selective entrainment that is a common feature of perennial gravel-bed rivers (e.g. Ashworth and Ferguson, 1989; Komar and Carling, 1991) is counteracted by the abundant supply of sediment that comes from the poorly vegetated hillslopes which characterise the semi-arid end of the climate spectrum. This concurs with the model of Dietrich et al. (1989) which illustrates that starving a channel of sediment leads to progressive armouring and vice versa. It is corroborated by the fieldwork of Lisle and Madej (1992) who show that deforestation in the catchment of Redwood Creek, northern California, has brought an abundant supply of sediment to the channel and a consequent reduction in the significance of the armour layer. The beds of the Yatir and Eshtemoa are neutrally or normally graded, i.e. each has an armour/subarmour median grain-size ratio of unity or less (Reid et al., 1999). By comparison, the near-perennial Tordera, with 80% of its water catchment covered by evergreen oak, has an armour/subarmour ratio of 3.4, while the value for Virginio Creek is 2.1.

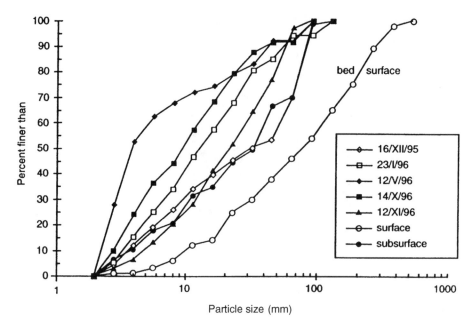

Figure 4.9 Bed material and bedload grain-size distributions, flood by flood, in the near-perennial Río Tordera, Catalonia, northeast Spain. (After García et al., 1999)

García et al. (1999) have added fresh insight into the processes that produce the scatter in the bedload flux–shear stress relation of an armoured gravel-bed channel. They have shown that patches of the bed having different grain-size distributions provide bedload differently as a function of flood magnitude and flood history. The size distribution of the mobilised sediment changes from flood to flood and they relate this to changes in the area and calibre of the source patches that have been exploited (Figure 4.9). In addition to this complex process, Taconni and Billi (1987) allude to a periodicity in bedload flux that has become recognised as characteristic of bedload discharge in perennial gravel-bed rivers by providing detailed time-series data of bedload flux in Virginio Creek (Figure 4.10).

The variable degree of independence of bedload and contemporary hydraulics has been attributed to a host of factors, many inferred, such as the translation of kinematic waves of sediment (Gomez, 1991). In two ephemeral streams of Judea, Church and Hassan (1992) throw some light on the effect of particle interlock on the release and transport of individual grains. The relation between distance of travel and size of clasts, flood by flood, is indistinct if all tracer pebbles (locked and free) are taken into account. However, if only the pebbles that are free before each flood are assessed, the anticipated inverse travel distance–size relation becomes much clearer (Figure 4.11). Church and Hassan suggest that the control which bed structure has on the release of clasts is more significant in smaller events, during which transport stage is closer to critical.

This might help to explain the uncharacteristically less-determinate pattern of bedload of the low magnitude, non-flashy, 27 February 1994 flood on the neighbouring Nahal Eshtemoa (Figure 4.12; Reid et al., 1998a). However, in general, the Eshtemoa's bedload response to changing hydraulic conditions is simple and clear-cut, as shown by the other sedigraphs of Figure 4.12. At this dry end of the Mediterranean spectrum, poorly vegetated water catchments

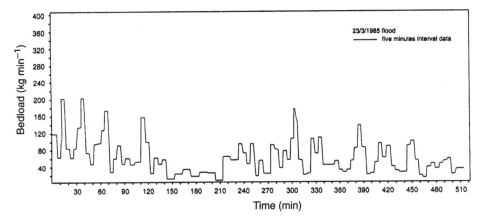

Figure 4.10 Five-minute interval bedload flux on the near-perennial Virginio Creek, Tuscany, Italy, showing pulses. (After Taconni and Billi, 1987)

provide an abundance of material of all calibres, either through sheet wash or gully extension, and this counteracts the selective entrainment that encourages armouring of the channel bed. Microforms and other structures are observed to be present but not dominant and Shields's entrainment function has a value at the lower end of the range reported for gravel-bed rivers (Buffington and Montgomery, 1997) of circa 0.03.

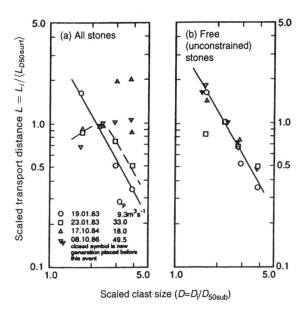

Figure 4.11 Scaled transport distance of clasts seeded in the bed of the Nahal Hebron, Israel, as a function of scaled clast size. The clasts are differentiated according to their degree of interlock before each of three flash floods. L_i is the mean transport distance of a clast size-group; $L_{D_{50surf}}$ is the transport distance of the size-group containing the median, D_{50}, of the bed surface layer; D_{50sub} is the median grain size of the subsurface layer. (After Church and Hassan, 1992)

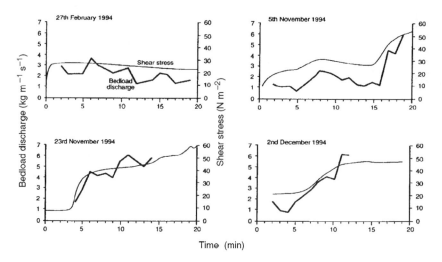

Figure 4.12   Time series of channel-average bedload flux and contemporary channel-average boundary shear stress during four flash floods in the ephemeral Nahal Eshtemoa, Israel. (After Reid et al., 1998a)

4.3   SUSPENDED SEDIMENT TRANSPORT

As might be expected in environmental settings where vegetation cover is sparse, drylands produce high concentrations of suspended sediment (Langbein and Schumm, 1958; Walling and Kleo, 1979). Indeed, it is from these environments that record levels of up to 68% solids have been measured in flash flood flows (Bondurant, 1951; Beverage and Culbertson, 1964; Lekach and Schick, 1982). However, for the Mediterranean region in particular, Woodward et al. (1992) draw attention to the role of lithotype underlying a water catchment in facilitating erosion. Generally, the exposure of silts and clays, sometimes sands and sandstones, by Alpine tectonics and the fact that the landform is comparatively young, with steep hillslopes often coupled directly to the channel system, lead to high concentrations of suspended sediment. In fact, in a specific example involving salt tectonics in the Dead Sea trough, Gerson (1977) was able to show that Pleistocene lake silts and clays that are being disturbed by contemporary uplift of the Sdom diapir are feeding local streams so that sediment concentrations are often in excess of 50%.

Another factor which provides an explanation for high suspended sediment loads is the change in vegetation cover that comes with forest and woodland removal and its replacement with grazed pasture or arable (Thornes, 1985). It has been estimated that half of the woodland clearance in the Mediterranean region had occurred by the end of the Roman Period (Tomaselli, 1977), giving early impetus to accelerated erosion. However, for a number of Moroccan lake basins, Flower et al. (1989) have examined sediment cores to reveal a significant impact on sediment delivery that arises through twentieth-century woodland clearance, indicating that, locally, the effect of landscape change continues to be important.

The pattern and the range in magnitude of suspended sediment transport can be illustrated from studies carried out across a climate gradient in Israel. In a remarkably detailed and impressive early study of six ephemeral streams that drain towards the Mediterranean from the Judean highlands in northern Israel, where annual rainfall is circa 600 mm, Negev (1969) reveals a reasonably common relation between suspended sediment load and water discharge.

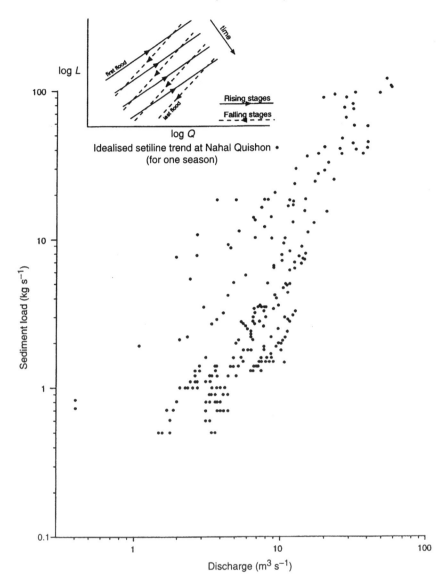

Figure 4.13 Suspended sediment load as a function of water discharge for five discrete flash floods in the ephemeral Nahal Quishon, northern Israel. The inset is a schematic interpretation of the non-stationary and hysteretic relation of the two variables through successive flash floods. (After Negev, 1969)

The sampling interval was hourly during rising flood stage, increasing during flow recession. The scatter in the plot of suspended sediment versus water discharge is modest for four of the streams, giving a rationale for fitting least-squares rating curves. However, for the Nahal Quishon, which outfalls to the Mediterranean at Haifa, the scatter is much greater (Figure 4.13). Negev disaggregates the plot, flood by flood, and shows, convincingly, the phenomenon of sediment supply exhaustion as the season progresses (Figure 4.13, inset), a pattern that was to be corroborated by later studies of perennial systems (Walling, 1977). This non-stationary process is used to explain the lesser degree of predictability of suspended sediment concentration in the case of the amalgamated dataset of this particular ephemeral stream.

Concentrations of suspended sediment are seen to be high in the coastal streams of Israel when they are compared with those typical of perennial systems. The range shown by Negev's (1969) dataset is 400–2000 mg l^{-1}. However, by shifting significantly down the steep climate gradient that characterises the Levant, and by moving into the Hebron Hills of southern Judea on the northern fringes of the Negev Desert where annual rainfall is 280 mm and the vegetation is sparse, the ephemeral Nahal Eshtemoa provides an even more dramatic contrast. For this upland water catchment, the record provides no values of suspended sediment concentration less than 1600 mg l^{-1} and maxima of 229 000 mg l^{-1}. The mean value lies at around 34 000 mg l^{-1} (Figure 4.14; Powell et al., 1996; Alexandrov et al., in review). The suspended sediment–water discharge plot for Nahal Eshtemoa contains considerable scatter. This may reflect, among a host of factors, the changing size and location of the source areas of runoff in a setting where rainfall is often delivered by spatially discrete convective storms, each of which may wet only part of the water catchment (Bull et al., 2000).

4.4 SEDIMENT YIELD

The suspended sediment concentrations of flash floods in Levantine streams do not set record levels by world standards, but they reflect the very high potential for soil erosion in a type-environment that has precious little soil. They also signal the significant problems associated with managing water resources by impoundment in a situation where rapidly increasing population and economic development are causes of more than commensurate increases in the demand for water.

Inbar (1992) compiled data on sediment yield for a number of drainage basins in Mediterranean-type environmental settings of both the Old and New Worlds. He shows a

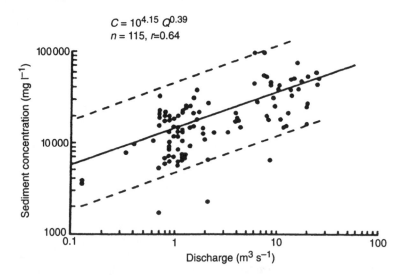

Figure 4.14 Suspended sediment concentration as a function of water discharge during flash floods in the ephemeral Nahal Eshtemoa, Israel. (After Powell et al., 1996)

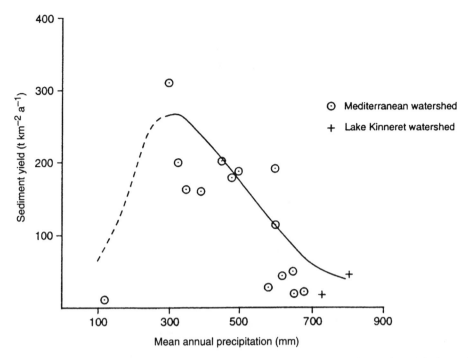

Figure 4.15 Langbein–Schumm relation between specific sediment yield and annual precipitation and cross-plot values established for drainage basins in northern Israel. Yield is derived from the measurement of suspended sediment only. (After Inbar, 1992)

reasonable degree of conformity between specific yields of basins in northern Israel and the classic yield–rainfall relation developed by Langbein and Schumm (1958; Figure 4.15). There is a similar conformity with yields established for selected basins that lie across a climate gradient in Spain (e.g. Sala, 1983; López-Bermúdez, 1979). However, Inbar draws out a distinction between the specific sediment yields of these Old World water catchments and those of the New World where more recent human disturbance of land cover may be a major factor behind significantly higher values at any given level of effective annual rainfall.

The record of the Nahal Eshtemoa provides an indication of the extremely spasmodic nature of the events that contribute to annual yield in dryland environments. Here, somewhat unusually, the sediment yield data consist not only of suspended load (usually the sole component measured) but also of bedload. Of passing interest is that the average *bedload* yield of $39\,t^{-1}\,km^{-2}\,a^{-1}$ is twice that of the entire *suspended* load of the Río Tordera in Catalonia at the wet end of the Mediterranean spectrum where annual rainfall is double (Sala, 1983). In the northern Negev, inter-annual variation is dictated by the capricious nature of discrete rainfall events (Figure 4.16; Powell et al., 1996). Transport stage is not infrequently subcritical and bedload does not always make a contribution. In addition, the number of events varies widely from year to year. This high degree of inter-annual variation is illustrated spectacularly by a cumulation of the runoff and sediment yield components (Figure 4.17; Reid et al., 1998b). In the (admittedly short) four-year record, the first year provides 33% of the runoff and 33% of the sediment yield; the equivalent figures for the fourth year are 61% and 65%; this leaves only 6% and 2% for the two intervening years taken together.

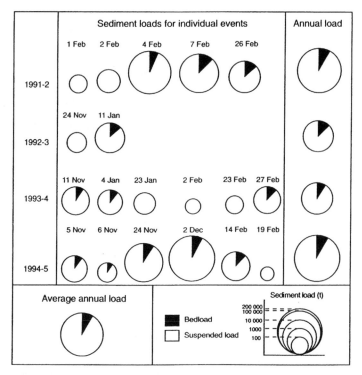

Figure 4.16 Annual, average annual, and storm-event suspended load and bedload sediment yields of the 119 km² water catchment of the ephemeral Nahal Eshtemoa, Israel. (After Powell et al., 1996)

There can be no doubt that the specific yields of almost all drainage basins in regions with Mediterranean-type climate reflect the liberation of material, both coarse and fine, by accelerated erosion and that this reflects, in turn, the impact of human interference with land cover and landscape. Inbar's (1992) study, by comparing New and Old Worlds, was designed to show that the immediate effect of human occupation is ameliorated with time. However, it is almost certainly the case that even the drainage basins of the Mediterranean region itself are currently exporting amounts of material in excess of early to middle Holocene equivalents that lay in a landscape which was yet to be ravaged by both agriculture and domesticated grazing animals (van Andel et al., 1986). This is particularly the case where the exposed lithotype encourages gully extension and the development of badlands, good examples of which have been documented all over the region and especially in Tunisia (De Ploey, 1974), Italy (Rendell, 1986), Greece (Woodward et al., 1992) and Spain (Vandekerckhove et al., 1998).

However, an interesting study of channel change illustrates, in reverse, the effect of reducing sediment production through the imposition of conservation measures. Rozin and Schick (1996) trace the dramatic changes in channel form of the Nahal Hoga on the coastal plain of central Israel (Figure 4.18). Through an analysis of aerial photographs for a number of epochs between 1945 and 1992, they show a channel transformed from braided to single-thread as changes in land use favour soil conservation and as runoff ratios fall from around 3 to less than 1%. The implications for changes in sediment yield are enormous. In highlighting human impact, this study shows that an already wide regional spectrum of stream behaviour has been widened considerably as the landscape has become variously and variably artificial.

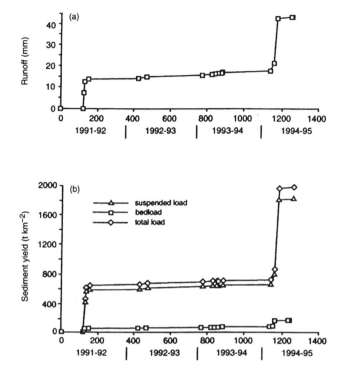

Figure 4.17   Cumulative (a) runoff and (b) sediment yield (suspended and bedload) from the first four years of record on the ephemeral Nahal Eshtemoa, Israel. (After Reid et al., 1998b)

4.5 CONCLUSIONS

Ephemeral streams in drylands, and especially in the Mediterranean, have been shown to produce record levels of bedload flux during in-bank flows that exert only moderate levels of boundary shear stress. The flux is at least an order of magnitude higher than that in perennial streams of similar size, with rates reaching up to six orders of magnitude higher than those of counterparts in humid environments. The explanation lies in the abundant supply of sediment of all calibres that is delivered both from the sparsely vegetated side-slopes of the water catchment and from the gullies that propagate in the vicinity of the channel. The size-selective transport that would encourage the development of an armour layer on the stream bed in the supply-limited conditions typical of a vegetated terrain is counteracted, and armouring is weak or non-existent. Consequently, bedload flux is at or near capacity, is shown to be a simple function of excess shear stress (in contrast with armoured near-perennials), and is predicted reasonably well by previously developed one-dimensional transport equations. This holds significant promise for successful channel engineering and suggests that margins of error in calculating the likelihood of sustaining the channel bed over a reasonable range of expected flows can be specified at moderate levels without danger of serious under- or overestimation.

Suspended sediment loads are high in dryland ephemeral streams, though not the highest recorded world wide. Besides having a significant impact on the density of the flow and there-

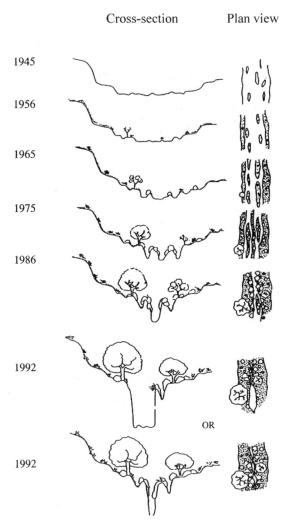

Figure 4.18 Stages in the development of the channel of Nahal Hoga, Israel, following the imposition of soil and water conservation measures in the water catchment. The cross-sections have a schematic vertical exaggeration. (After Rozin and Schick, 1996)

fore on the hydraulic behaviour of the streams, such high concentrations produce high sediment yields and bring about measurable reductions in reservoir capacity, flood by flood. For example, Reid et al. (1998b) show that for a notional small reservoir with a capacity of 1M m^3 at the outfall of the 119 km^2 Eshtemoa water catchment, the half-life of the impoundment would be only 12 years as a result of sediment accumulation. Zachar (1982) reports on a number of similarly sized Algerian reservoirs which were silted completely within 20 years of dam closure. Similar rapid losses of storage are also reported for Tunisian and Moroccan reservoirs (Ghorbel and Claude, 1977; Lahlou, 1988), all of which mean eventual loss of function or costly maintenance through periodic flushing (Tolouie et al., 1993).

The lessons are all too clear for river engineers and managers of water resources: sediment transport in ephemeral streams is an environmental nuisance of major proportions. Leopold (1992) has drawn attention to the fact that several ancient cultures, such as the Levantine Nabateans, recognised this by constructing headwater check dams. The processes that set dryland sediment transport several orders of magnitude above its counterparts in more humid environments need to be fully understood so that the problem can be ameliorated, if possible, by land management within the water catchment.

ACKNOWLEDGEMENTS

I pay particular tribute to colleagues who have worked with me in studying ephemeral streams since I first stepped into one in 1974: Lynne Frostick, Jonathan Laronne, Mark Powell, Lev Meirovitch, John Layman, Yitshak Yitshak, Julia Alexandrov and Celso García. My gratitude also goes out to those who have assisted either in the field or through invaluable discussions over a number of years, particularly Asher Schick and Marwan Hassan.

REFERENCES

Adamson, D.A. 1982. The integrated Nile. In M.A.J. Williams and D.A. Adamson (Eds) *A Land between Two Niles*. Balkema, Rotterdam, 221–234.

Alexandrov, Y., Laronne, J.B. and Reid, I. (in review). Suspended sediment concentration and its variation with water discharge in a dryland ephemeral channel, northern Negev, Israel. *Journal of Arid Environments*.

Anderson, N.J., Wheater, H.S., Timmis, A.J.H. and Gaongalelwe, D. 1998. Sustainable development of alluvial groundwater in sand rivers of Botswana. In H.S. Wheater and C. Kirby (Eds) *Hydrology in a Changing Environment*, Vol. II. J. Wiley & Sons, Chichester, 367–376.

Ashworth, P.J. and Ferguson R.I. 1989. Size-selective entrainment of bed load in gravel bed streams. *Water Resources Research*, 25, 627–634.

Bagnold, R.A. 1973. The nature of saltation and of 'bedload' transport in water. *Proceedings of the Royal Society of London*, A332, 473–504.

Batalla, R.J. and Sala, M. 1995. Effective discharge for bedload transport in subhumid Mediterranean sandy gravel-bed river (Arbúcies, North-East Spain). In E.J. Hickin (Ed.) *River Geomorphology*. J.Wiley & Sons Ltd, Chichester, 93–103.

Beverage, J.P. and Culbertson, J.K. 1964. Hyper-concentrations of suspended sediment. *Proceedings of the American Society of Civil Engineers, Journal of Hydraulics Division*, 90, HY6, 117–128.

Bondurant, D.C. 1951. Sedimentation studies at Conchas Reservoir in New Mexico. *Transactions of the American Society of Civil Engineers*, 116, 1292–1295.

Buffington, J.M. and Montgomery, D.R. 1997. A systematic analysis of eight decades of incipient motion studies, with special reference to gravel-bedded rivers. *Water Resources Research*, 33, 1993–2029.

Bull, L.J., Kirkby, M.J., Shannon, J. and Hooke, J.M. 2000. The impact of rainstorms on floods in ephemeral channels in southeast Spain. *Catena*, 38, 191–209.

Burkham, D.E. 1970. *Depletion of streamflow by infiltration in the main channels of the Tucson Basin, southeastern Arizona*. United States Geological Survey Water Supply Paper 1939.

Butler, R.W.H., Lickorish, W.H., Grasso, M., Pedley, H.M. and Ramberti, L. 1995. Tectonics and sequence stratigraphy in Messinian basins, Sicily: constraints on the initiation and termination of the Mediterranean salinity crisis. *Geological Society of America Bulletin*, 107, 425–439.

Butzer, K.W. 1962. Coastal geomorphology of Majorca. *Annals of the American Association of Geographers*, 52, 191–212.

Church, M. and Hassan, M.A. 1992. Size and distance of travel of unconstrained clasts on a streambed. *Water Resources Research*, 28, 299–303.

Collier, R.E.L., Leeder, M.R. and Jackson, J.A. 1995. Quaternary drainage development, sediment fluxes and extensional tectonics in Greece. In J. Lewin, M.G. Macklin and J.C. Woodward (Eds) *Mediterranean Quaternary River Environments*. Balkema, Rotterdam, 31–44.

De Ploey, J. 1974. Mechanical properties of hillslopes and their relation to gullying in central semi-arid Tunisia. *Zeitschrift für Geomorphologie*, 21, 177–190.

Dietrich, W.E., Kirchner, J.W., Ikeda, H. and Iseya, F. 1989. Sediment supply and the development of the coarse surface layer in gravel-bedded rivers. *Nature*, 340, 215–217.

Dunkerley, D.L. 1992. Channel geometry, bed material, and inferred flow conditions in ephemeral stream systems, Barrier Range, Western N.S.W., Australia. *Hydrological Processes*, 6, 417–433.

Flower, R.J., Stevenson, A.C., Dearing, J.A., Foster, I.D.L., Airey, A., Rippey, B., Wilson, J.P.F. and Appleby, P.G. 1989. Catchment disturbance inferred from paleolimnological studies of three contrasted sub-humid environments in Morocco. *Journal of Paleolimnology*, 1, 293–322.

Frostick, L.E. and Reid, I. 1977. The origin of horizontal laminae in ephemeral stream channel-fill. *Sedimentology*, 24, 1–9.

Frostick, L.E. and Reid, I. 1989. Climatic versus tectonic controls of fan sequences: lessons from the Dead Sea, Israel. *Journal of the Geological Society London*, 146, 527–538.

García, C., Laronne, J.B. and Sala, M. 1999. Variable source areas of bedload in a gravel-bed stream. *Journal of Sedimentary Research*, 69, 27–31.

García, C., Laronne, J.B. and Sala, M. 2000. Continuous monitoring of bedload flux in a mountain gravel-bed river. *Geomorphology*, 34, 23–31.

García, C. and Sala, M. 1997. Aplicación de fórmulas de transporte de fondo a un río de gravas: comparación con las tasa reales de transporte obtenidas en el Río Tordera. *Ingegnería del Agua*, 5, 59–72.

Gerson, R. 1977. Sediment transport for desert watersheds in erodible materials. *Earth Surface Processes*, 2, 343–361.

Gerson, R., Grossman, S. and Bowman, D. 1984. Stages in the creation of a large rift valley – geomorphic evolution along the southern Dead Sea Rift. In J.T. Hack and M. Morisawa (Eds.) *Tectonic Geomorphology*. George Allen & Unwin, 53–73.

Ghorbel, A. and Claude, J. 1977. Mesure de l'envasement dans les retinues de sept barrages en Tunisie: estimation des transports solides. *International Association Hydrological Sciences Publication*, 122, 219–232.

Gomez, B. 1991. Bedload transport. *Earth Science Reviews*, 31, 89–132.

Hassan, M.A. 1990. Scour, fill and burial depth of coarse material in gravel bed streams. *Earth Surface Processes and Landforms*, 15, 341–356.

Inbar, M. 1992. Rates of fluvial erosion in basins with a Mediterranean type climate. *Catena*, 19, 393–409.

Komar, P.D. and Carling, P.A. 1991. Grain sorting in gravel-bed streams and the choice of particle sizes for flow-competence evaluations. *Sedimentology*, 38, 489–502.

Kondolf, G.M. 1995. Managing bedload sediment in regulated rivers: examples from California, USA. In J.E. Costa, A.J. Miller, K.W. Potter and P.R. Wilcock (Eds) *Natural and Anthropogenic Influences in Fluvial Geomorphology*. American Geophysical Union, Geophysical Monograph 89, 165–176.

Lahlou, A. 1988. The silting of Moroccan dams. *International Association of Hydrological Sciences Publication*, 174, 71–77.

Langbein, W.B. and Schumm, S.A. 1958. Yield of sediment in relation to mean annual precipitation. *Transactions American Geophysical Union*, 20, 637–641.

Laronne, J.B., Reid, I. Yitshak, Y. and Frostick, L.E. 1992. Recording bedload discharge in a semiarid channel. In D.E. Bogen, D.E. Walling and T.J. Day (Eds) *Erosion and Sediment Transport Monitoring Programmes in River Basins*. *International Association of Hydrological Sciences Publication* No. 210. Institute of Hydrology, Wallingford, 79–86.

Laronne, J.B. and Reid, I. 1993. Very high rates of bedload sediment transport by ephemeral desert rivers. *Nature*, 366, 148–150.

Laronne, J.B., Reid, I., Yitshak, Y. and Frostick, L.E. 1994. The non-layering of gravel stream beds under ephemeral flood regimes. *Journal of Hydrology*, 159, 353–363.

Lekach, J. and Schick, A.P. 1982. Suspended sediment in desert floods in small catchments. *Israel Journal of Earth Sciences*, 31, 144–156.

Lenzi, M.A., D'Agostino, V. and Billi, P. 1999. Bedload transport in the instrumented catchment of the Rio Cordon. Part I: Analysis of bedload records, conditions and thresholds of bedload entrainment. *Catena*, 36, 171–190.

Leopold, L.B. 1992. Base level rise: gradient of deposition. *Israel Journal of Earth Sciences*, 41, 57–64.

Leopold, L.B. and Emmett, W.W. 1997. Bedload and river hydraulics – inferences from the East Fork River, Wyoming. *United States Geological Survey Professional Paper* 1583.

Lisle, T.E. and Madej, M.A. 1992. Spatial variation in armouring in a channel with high sediment supply. In P. Billi, R.D. Hey, C.R. Thorne and P. Tacconi (Eds) *Dynamics of Gravel-Bed Rivers*. John Wiley & Sons Ltd., Chichester, 277–291.

López-Bermúdez, F. 1979. Inundaciones catastrôficas, precipitaciones torrenciales y erosión en la provincia de Murcia. *Papeles del Dipartimento de Geografia, Universida de Murcia*, 49–60.

Macklin, M.G., Lewin, J. and Woodward, J.C. 1995. Quaternary fluvial systems in the Mediterranean basin. In J. Lewin, M. G. Macklin and J.C. Woodward (Eds) *Mediterranean Quaternary River Environments*. Balkema, Rotterdam, 1–25.

Meyer-Peter, E. and Müller, R. 1948. Formulas for bedload transport. *International Association of Hydraulic Structures Research Report, Second Meeting, Stockholm*, 39–64.

Milliman, A.C. and Syvitski, J.P.M. 1992. Geomorphic/tectonic control of sediment discharge to the ocean: the importance of small mountainous rivers. *Journal of Geology*, 100, 525–544.

Murphey, J.B., Lane, L.J. and Diskin, M.H. 1972. Bed material characteristics and transmission losses in an ephemeral stream. Hydrology and water resources in Arizona and the Southwest. *Proceedings 1972 Meeting Arizona Section, American Water Association and the Hydrology Section, Arizona Academy Science, Prescott, Arizona*, 2, 455–472.

Negev, M. 1969. *Analysis of data on suspended sediment discharge in several streams in Israel*. Israel Hydrological Service, Hydrological Paper no. 12.

Palanques, A., Plana, F. and Maldonado, A. 1990. Recent influence of man on the Ebro margin sedimentation system, northwestern Mediterranean Sea. *Marine Geology*, 95, 247–263.

Powell, D.M., Reid, I., Laronne, J.B. and Frostick, L.E. 1996. Bed load as a component of sediment yield from a semiarid watershed of the northern Negev. In D.E. Walling and B. Webb (Eds) *Erosion and Sediment Yield: Global and Regional Perspectives. Proceedings of the Exeter Symposium*, July 1996. International Association of Hydrological Sciences Publication, 236, 389–397.

Powell, D.M., Reid, I. and Laronne, J.B. 1999. Hydraulic interpretation of cross-stream variations in bedload transport. *Journal of Hydraulic Engineering*, 125, 1243–1252.

Reid I. and Laronne, J. B. 1995. Bedload sediment transport in an ephemeral stream and a comparison with seasonal and perennial counterparts. *Water Resources Research*, 31, 773–781.

Reid, I., Laronne, J.B. and Powell, D.M. 1995. The Nahal Yatir bedload database: sediment dynamics in a gravel-bed ephemeral stream. *Earth Surface Processes and Landforms*, 20, 845–857.

Reid, I., Laronne, J.B. and Powell, D.M. 1998a. Flashflood and bedload dynamics of desert gravel-bed streams. *Hydrological Processes*, 12, 543–557.

Reid, I., Laronne, J.B. and Powell, D.M. 1999. Impact of major climate change on coarse-grained river sedimentation – a speculative assessment based on measured flux. In A.G. Brown and T.A. Quine (Eds) *Fluvial Processes and Environmental Change*. J. Wiley & Sons, Chichester, 105–115.

Reid, I., Layman, J.T. and Frostick, L.E. 1980. The continuous measurement of bedload discharge. *Journal of Hydraulic Research*, 18, 243–249.

Reid, I., Powell, D.M. and Laronne, J.B. 1996. Prediction of bedload transport by desert flash-floods. *Journal of Hydraulic Engineering, American Society of Civil Engineers*, 122, 170–173.

Reid, I., Powell, D.M. and Laronne, J.B. 1998b. Flood flows, sediment fluxes and reservoir sedimentation in upland desert rivers. In H. Wheater and C. Kirby (Eds) *Hydrology in a Changing Environment Volume II*. J. Wiley & Sons Ltd, Chichester, 377–386.

Renard, K.G. and Keppel, R.V. 1966. Hydrographs of ephemeral streams in the Southwest. *Proceedings of the American Society of Civil Engineers, Journal of Hydraulics Division*, 92, 33–52.

Renard, K.G. and Lane, L.J. 1975. Sediment yield as related to a stochastic model of ephemeral runoff. Present and prospective technology for predicting sediment yields and sources. *Proceedings Sediment-yield Workshop, USDA Sedimentation Laboratory, Oxford, Mississippi*, 1972, 253–263.

Renard, K.G. and Laursen, E.M. 1975. Dynamic behaviour model of ephemeral streams. *Proceedings of the American Society Civil Engineers, Journal of Hydraulics Division*, 101, 511–528.

Rendell, H.M. 1986. Soil erosion and land degradation in southern Italy. In R. Fantechi and N.S. Margaris (Eds) *Desertification in Europe*. Reidel, Dordrecht, 184–193.

Roquero, E. 1991. Le Barrage de Nijar ou d'Isabel II (Almeria, Espagne), une étude de l'influence de la géomorphologie sur l'accéleration de l'ensablement. *Zeitschrift für Geomorphologie Supplement Band*, 83, 9–16.

Rozin, U. and Schick, A.P. 1996. Land use change, conservation measures and stream channel response in the Mediterranean/semiarid transition zone: Nahal Hoga, southern Coastal Plain, Israel. In D.E. Walling and B. Webb (Eds) *Erosion and Sediment Yield: Global and Regional Perspectives. Proceedings of the Exeter Symposium*, July 1996. International Association of Hydrological Sciences Publication, 236, 427–444.

Sala, M. 1983. Fluvial and slope processes in the Fuirosos basin, Catalan Ranges, north east Iberian coast. *Zeitschrift für Geomorphologie*, 27, 393–411.

Schick, A.P. 1970. Desert floods. *Symposium on the Results of Research on Representative Experimental Basins*. International Association Scientific Hydrologists/UNESCO, 478–493.

Schick, A.P. 1973. Alluvial fans and desert roads – a problem in applied geomorphology. *International Symposium on Geomorphological Processes and Process Combinations, Göttingen, Proceedings*, 418–425.

Schick, A.P. 1988. Hydrologic aspects of floods in extreme arid environments. In V.R. Baker, R.C. Kochel, and P.C. Patton (Eds) *Flood Geomorphology*. John Wiley & Sons, New York, 189–203.

Simons, D.B. and Ruh-Ming, Li 1982. Application of flow routing techniques to river regulation. In R.D. Hey, J.C. Bathurst and C.R. Thorne (Eds) *Gravel-bed Rivers*. J.Wiley & Sons Ltd, Chichester, 583–599.

Taconni, P. and Billi, P. 1987. Bed load transport measurements by vortex-tube trap on Virginio Creek, Italy. In C.R. Thorne, J.C. Bathurst and R.D. Hey (Eds) *Sediment Transport in Gravel-bed Rivers*. John Wiley & Sons Ltd., Chichester, 583–606.

Thornes, J.B. 1976. *Semi-arid erosional systems: case studies from Spain*. London School of Economics Geographical Paper 7.

Thornes, J.B. 1985. The ecology of erosion. *Geography*, 70, 222–235.

Tolouie, E., West, J.R. and Billam, J. 1993. Sedimentation and desiltation in the Sefid-Rud reservoir, Iran. In J. McManus and R.W. Duck (Eds) *Geomorphology and Sedimentology of Lakes and Reservoirs*. J. Wiley & Sons, Chichester, 125–138.

Tomaselli, R. 1977. Degradation of the Mediterranean maquis. In *Mediterranean Forests and Maquis: Conservation and Management*. Man and the Biosphere Technical Notes 2, UNESCO, Paris.

van Andel, T.H., Runnels, C.N. and Pope, K.O. 1986. Five thousand years of land use and abuse in the southern Argolid, Greece. *Hesperia*, 55, 103–128.

Vandekerckhove, L., Poesen, J., Oostwoud Wijdenes, D. and de Figueiredo, T. 1998. Topographical thresholds for ephemeral gully initiation in intensively cultivated areas of the Mediterranean. *Catena*, 33, 271–292.

Walling, D.E. 1977. Limitations on the rating curve technique for estimating suspended sediment loads, with particular reference to British rivers. *International Association of Hydrological Sciences Publication*, 122, 34–48.

Walling, D.E. and Kleo, A.H.A. 1979. Sediment yields of rivers in areas of low precipitation: a global view. In *The Hydrology of Areas of Low Precipitation. Proceedings of the Canberra Symposium*, December 1979. International Association Hydrological Sciences Publication, 128, 479–493.

Woodward, J.C. 1995. Patterns of erosion and suspended sediment yield in Mediterranean river basins. In I.D.L. Foster, A.M. Gurnell and B.W. Webb (Eds) *Sediment and Water Quality in River Catchments*. John Wiley & Sons, Chichester, 365–389.

Woodward, J.C., Lewin, J. and Macklin, M.G. 1992. Alluvial sediment sources in a glaciated catchment: the Voidomatis Basin, northwest Greece. *Earth Surface Processes and Landforms*, 17, 205–216.

Zachar, D. 1982. Soil erosion. *Developments in Soil Science*. Elsevier, Amsterdam.

5 Modelling Event-based Fluxes in Ephemeral Streams

JULIE SHANNON[1], ROY RICHARDSON[2] AND JOHN THORNES[1]

[1] Department of Geography, Kings College London, UK
[2] Philip Williams & Associates, Ltd. (PWA), Corte Madera, USA

5.1 INTRODUCTION

As part of the MEDALUS III European Union Project on ephemeral channel behaviour (Brandt and Thornes, 1996), a joint investigation of ephemeral channels as highways for the products of land degradation has been undertaken by a team from the Universities of Leeds, Portsmouth, Murcia, Barcelona and Kings College London. The main objective, developed on the basis of earlier work, was to address the issue generally neglected in land degradation studies and models, of how the sediment produced by catchment erosion in the upland phase was evacuated from the basin by flows in ephemeral channel systems. Work was concentrated on the Nogalte basin in the province of Murcia, southeast Spain, building on an earlier application of the MEDRUSH model, by the Leeds and Kings College London teams, described in Kirkby et al. (1996).

In this chapter we report the efforts to develop models suitable for modelling flow and sediment transport under the peculiar conditions of ephemeral channel behaviour, as outlined in Section 5.2, Ephemeral channel characteristics and flow events. These special modelling requirements are outlined in Section 5.3, using the Rambla de Nogalte as a test case. Finally the conceptual framework and modelling approaches are discussed, especially the balance of effort between distributed deterministic models and stochastic models based on a range of approaches from random walk to diffusion-based strategies.

5.2 EPHEMERAL CHANNEL CHARACTERISTICS AND FLOW EVENTS

Ephemeral channels in arid and semi-arid environments differ greatly from their perennial counterparts in humid temperate environments. The climatic and flow characteristics are so different that existing models of flow in perennial channels are unsuitable. Here, a brief overview of the main characteristics of ephemeral channels and the climatic conditions in semi-arid areas will be given, as a more in-depth discussion precedes this chapter in Chapters 3 and 4.

Dryland Rivers: Hydrology and Geomorphology of Semi-arid Channels. Edited by L.J. Bull and M.J. Kirkby.
© 2002 John Wiley & Sons, Ltd.

5.2.1   Runoff Generation

Runoff generation was considered in Chapter 3. Fluvial processes in drylands are driven by precipitation, therefore it is critical to understand the temporal and spatial variability of rainfall events (Graf, 1988). The Mediterranean experiences low annual rainfall with totals of less than 500 mm and most of the rain is delivered in short storms with high intensities, mainly in the winter months. Storm cells tend to be limited in their areal extent, with storm cells being less than 10 to 14 km in diameter (Rossi and Siccardi, 1990; Míro-Granada and Gelabert, 1974), and so it is rare for all of larger catchments to be affected by the same rainfall event. The areal discontinuity of rainfall events or spottiness was studied by Sharon (1972) in Israel, where, for catchments of a few hundred square kilometres the percentage of the catchment receiving intensive rainfall on a certain day could be as low as 20%. Storm cells also tend to migrate across a catchment as they deliver rainfall. As a result, catchments experience variability both spatially and temporally in areas affected by rainfall. Rainfall intensities have also been observed to change very sharply during storms (Bull et al., 2000; Gutiérrez et al., 1998), further emphasising the unpredictable nature of storm events in semi-arid environments. Generally, because the spatial variability in rainfall intensities and totals is combined with variability in land use and soil properties, the prediction of runoff generating areas is very difficult, having implications on the timing and incidence of flows into the main channel and hydrograph propagation downstream. As a result of rainfall spatial variability, different parts of the catchment will flow at different times, thus prediction and monitoring is extremely difficult; also, the asynchronous flow pattern means that catchments are rarely responding to the same event (Graf, 1988; Walters, 1990). This variability in both geomorphological and climatological parameters becomes increasingly important as the catchment size increases (El-Hames and Richards, 1994). Antecedent conditions also play a significant role in the response of a catchment to a storm event (White et al., 1997), influencing both hillslope runoff and main channel hydrograph propagation. The variability in catchment response, and the unpredictable nature of storm cells, mean that main channel hydrographs are less predictable than those in perennial channels (Reid and Frostick, 1997).

Figure 5.1 shows the rainfall intensities and estimated peak discharges recorded for a storm on 29 September 1997 in the Rambla de Nogalte catchment in southeast Spain. The timing of the peak intensities varied throughout the catchment, indicating a northeasterly migration of the storm cell (Bull et al., 2000). Intensities of over 200 mm h^{-1} were recorded at some gauges while other gauges recorded intensities of less than 100 mm h^{-1}. Bull et al. (2000) report that high intensities were measured throughout the catchment, however only in localised areas of the catchment were high peak discharges estimated and geomorphic changes observed.

Variability of peak discharges and geomorphic change can provide a useful insight into the pattern of rainfall intensities (Gutiérrez et al., 1998) and runoff processes (Bull et al., 2000) operating in semi-arid catchments. Variability in runoff production throughout a catchment has important implications on stream flow in the main channel and hydrograph characteristics.

5.2.2   Stream Flow and Hydrograph Characteristics

Singh (1997) divides factors affecting streamflow hydrographs into the following categories: watershed conditions; storm precipitation dynamics; infiltration; and antecedent conditions. As already mentioned, in semi-arid catchments all of these categories exhibit high temporal and spatial variability, leading to difficulties when trying to estimate stream hydrographs for a given rainfall event.

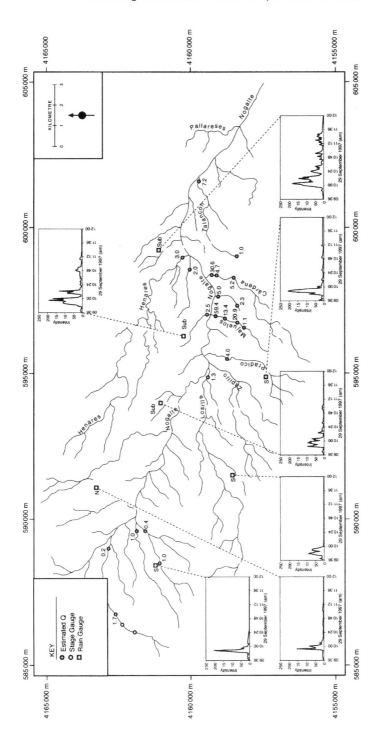

Figure 5.1 Variability in rainfall intensity and timing of peak intensity for a storm recorded in the Rambla de Nogalte, southeast Spain, 29 September, 1997. Adapted from Bull et al. (2000, p. 198)

Peebles et al. (1981), described the flow characteristics of ephemeral channels as being in response to surface runoff as the groundwater level is usually far below the surface. Flows are generated by storm rainfall events of brief duration in which large volumes of surface runoff water enter the channel in a short period of time causing flash floods. Peak flow rate is usually reached almost instantaneously as a result of the quick runoff and because ephemeral flood waves form a steep wave front early in their journey downstream. The steep flood wave occurs, at least in part, as a result of the high infiltration rates into the permeable stream bed at the wave front and decreases in the upstream direction, so that the leading edge of the wave becomes steeper as it moves downstream (Pilgrim et al., 1988; Smith, 1972). As the peak travels faster than the leading edge of the flood wave, the result is that they almost coincide and produce a shock front (Pilgrim et al., 1988).

Generally, hydrographs of ephemeral channels are characterised by steep rising limbs, with an almost instantaneous rise to peak flow and a steep recession limb. The recession limb is of longer duration than the rising limb, therefore almost the entire hydrograph consists of the recession curve. Floods in semi-arid environments are also characterised by their short duration, often less than a few hours. Figure 5.2 shows the rainfall-runoff record for the Rambla de Torrealvilla in southeast Spain. The flashy nature of the hydrographs is evident from the short time to peak discharge and steep rising and falling limbs. Figure 5.3 shows two hydrographs for another catchment in southeast Spain, the Rambla de Nogalte. These two stage hydrographs for the upper and lower sections of a channel reach also exhibit steep rising and falling limbs. The distance between the gauges and different times to peak discharge indicate that the flood peak was travelling at approximately $1\,\mathrm{m\,s}^{-1}$. The characteristic short-lived nature of flow events in ephemeral channels is also evident in Figure 5.3, where the flow in the channel lasted approximately 1 hour.

Overall the most important aspect of semi-arid hydrographs that sets them apart from those in perennial channels, is how they are transformed as they propagate downstream. Hydrograph propagation is affected by three important factors: variability of inputs into the main channel from subcatchment flows, losses due to infiltration into the bed and banks of the channel and

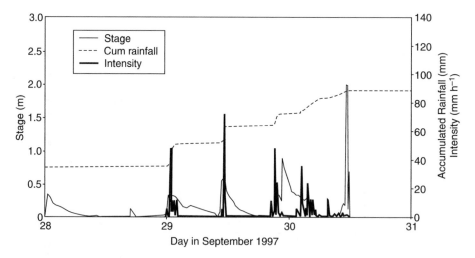

Figure 5.2 Rainfall and runoff recorded in the Rambla de Torrealvilla, 29 September 1997. Adapted from Bull et al. (2000, p. 199)

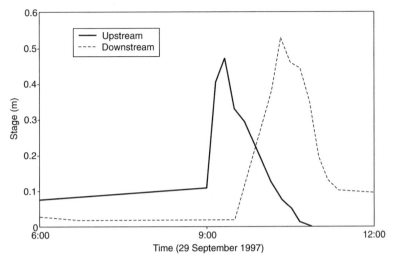

Figure 5.3 Upstream and downstream stage hydrographs for flow recorded in the Rambla de Nogalte, 29 September 1999

the geomorphic structure of the network (Rodriguez-Iturbe, 1994). As already outlined in the previous section, runoff can be generated over small areas of the catchment, and so tributary inflow may occur when the rest of the channel system is dry (Thornes, 1977). The combination of asynchronous flow and transmission losses in semi-arid catchments, results in flow volumes and hydrograph peaks decreasing downstream due to transmission losses and then suddenly increasing again where there is active tributary inflow, as described by Thornes (1977). The exact timing and pattern of flows and hydrograph propagation is dependent upon the complex nature of the catchment, channel and rainfall conditions for each individual flow event. Transmission losses can have a dramatic effect on the hydrograph shape and volume and will be considered in the next section.

5.2.3 Transmission Losses

Transmission losses are one of the main features that separate ephemeral channels from perennial channels. Water that infiltrates into the permeable bed and banks of the channel is known as transmission loss, it is part of the stream flow at one location but is lost from the flow before reaching some location downstream (Butcher and Thornes, 1978), resulting in a reduction in the hydrograph volume and peak discharge downstream.

Transmission losses are affected by a complex interaction of variables such as:

'(1) soil-moisture characteristics of the channel sediments and bank material; (2) initial moisture distribution in the channel sediments; (3) physical structure of the channel materials, including the presence of a surface crust, cracks, and/or stratification; (4) depth of stream flow and its rate of change with time; (5) duration of stream flow; (6) surface area wetted by the flow; (7) sediment content of the flow; (8) channel erosion and deposition; (9) chemical composition of surface and subsurface water and channel sediments; and (10) temperature of surface and subsurface water.' (Freyberg, 1983, p. 599)

Transmission losses can have a marked effect on the water yield from a catchment, affecting both the shape and size of the runoff hydrograph (Babcock and Cushing, 1941; Renard and Keppel, 1966; Burkham, 1970; Renard, 1970; Jordan, 1977; Thornes, 1977) as well as the peak discharge (Renard and Keppel, 1966; Lane, 1982a). They also have the effect of steepening the flood wave as it propagates downstream (Butcher and Thornes, 1978). Thus, transmission losses are a vital component of any channel-routeing model and flood-warning system.

In addition to affecting downstream hydrographs, transmission losses help to support riparian vegetation and have important implications on groundwater recharge which is constantly being depleted in the Mediterranean by pressures from agriculture, industry and increasingly by tourism. Burkham (1970) points out that transmission losses and the subsequent recharge from ephemeral channel streams form a component of the groundwater system that has not been adequately defined. As all flood routeing methods rely on the solution of the continuity equation, and because transmission losses are a fundamental component, their estimation is vital (Lane et al., 1971). Generally, the estimation of output hydrographs from reaches is an essential development of water-resource systems in semi-arid environments (Sharma and Murthy, 1995).

There are several problems associated with the estimation of transmission losses from ephemeral channels. The infrequent and often extreme nature of flow events means that observation and measurements are fraught with difficulties. Also, the broad alluvial channel beds can abstract large quantities of water before it can be gauged further downstream (Lane, 1972).

Babcock and Cushing (1941) pioneered some of the early work into transmission losses and several approaches have been adopted since then. Most work has been carried out in the American southwest, Saudi Arabia and India, with considerably less data and literature on this phenomenon in the Mediterranean. Table 5.1 summarises the methods and findings of some studies into transmission losses in ephemeral channels.

Transmission losses reduce discharge whether or not there is a major contribution from subcatchments further downstream. They also affect sediment transport, alterations to the channel bed morphology and are a vital component of any flow routeing model.

5.2.4 Catchment and Channel Characteristics

The rainfall and hydrograph characteristics outlined above provide unique characteristics for semi-arid catchments and channels. As ephemeral channels always exhibit unsteady, non-uniform and often supercritical flow, they are not amenable to some of the generalisations that are made about perennial channels (Thornes, 1980). Channel flow is intermittent, and channels remain dry for most of the year. They exhibit an inverse relationship between the magnitude and frequency of events. As a consequence of the rainfall characteristics, subcatchment flows into the main channel are spatially and temporally variable, with downstream channel reaches experiencing flow while upstream reaches receive little or no flow from subcatchments. Flow into the main channel need not be restricted by the lack of storm cells delivering to a subcatchment. Anthropogenic factors such as check dams and terraces inhibit the flow of water and sediment into the main channel. Entire catchments rarely respond to the same event and asynchronous behaviour is characteristic (Butcher and Thornes, 1978).

Flows in the main channel and from subcatchments result in the channel bed morphology reflecting the flow history in the main channel. The main channel slope does not decrease downstream as in perennial channels, but it is spatially variable and depends on subcatchment fan formations and sediment deposition within the main channel. Generally, channel slopes in the Nogalte main channel range between 1 and 2 degrees, but can be higher where large fans increase the local channel slope at large subcatchment inputs (Figure 5.4). However, for most flows channel slope remains constant and relatively small, and it is variations in the geometry of flow and the volume of flow itself which are largely responsible for the energy relations that dominate channel behaviour (Thornes, 1980).

Channel width does not increase steadily downstream as in perennial channels. It is highly sinuous, ranging from very broad sections to very narrow confined sections of channel (Figure 5.5), again reflecting the spatial variability of the flow history and inputs from subcatchments (Thornes, 1977, 1979). Thornes (1980) noted that, in general, ephemeral channels have two sections, separated by a point of alluviation, whose location appears to be determined by the network characteristics of the system. The upstream section has coupled channels and low width to depth ratios. In the downstream section, below the point of alluviation, the cross-sections are different. The channel bed in non-cohesive materials (sands and gravels) has high rates of transmission loss, large subsurface storage potential and channels with relatively large width to depth ratios. One important feature is that the alluvial plane may be much wider than the most recently occupied channel and it has a slope of nearly constant gradient (Thornes, 1980).

Generally, the flow from subcatchments into the main channel can have a marked effect on both the local channel slope and width at large, active tributary junctions. The lower and higher magnitude events do not simply differ according to their discharge and frequency of occurrence. Lower magnitude events do not result in complete channel submersion on reaches where the channel is very broad, and the flows occupy the smaller braided channels that are incised into the main channel. Eventually the water flows 'over-bank' and ultimately the entire alluvial fill may be submerged. The larger discharge events that result in complete channel submersion occur less frequently, scouring the channel bed and causing mass sediment transport. After a large event that occupies the entire alluvial plain width, the flow breaks down into multiple channels by the emergence of bed features produced by the high flows, and during this phase saltation appears to be the dominant mode of transport. Alternatively, recession may be rapid, leaving behind a plain bed which is then reworked by subsequent smaller flows (Thornes, 1980). The main difference between rare large flood events and the more frequent low flow events is that morphology controls the flow patterns during low flow events, whereas flow controls morphology during higher discharge events. This morphological control of the low flow events is further developed in the discussion of stochastic modelling of flows.

The spatial variability in flow events means that flow histories in one part of the catchment may be very different from those in another. Of course subcatchments with different lithologies will behave very differently. Bull et al. (2000) found that land use, slope and soil type had a marked effect on runoff generation in subcatchments in the Nogalte basin, and thus flow into the main channel was highly localised even though the whole catchment experienced the storm event (Figure 5.1).

5.2.5 Sediment Transport

Sediment transport was discussed extensively in Chapter 4, and will be considered here only briefly with some modelling approaches expanded later in the chapter. Sediment transport depends upon the prevailing flow event and the channel bed material type and availability. The transport of sediment as waves, with the sudden deposition of material with little sorting as a result of reduced stream power due to transmission losses or at tributary junctions, is common. As a result of predominantly low flow conditions and a plentiful supply of material from hillslopes, sediment transport tends to be transport limited. Sediment in transport is prone to being deposited rapidly during hydrograph recession or as a result of a loss of stream power due

Table 5.1 Transmission losses and methods for estimation (adapted from Knighton and Nanson, 1997, p. 187)

Reference	Study area	Method	Predicted or measured transmission losses	Notes
Babcock and Cushing (1941)	Queen Creek, Arizona	Stage gauge measurements	32 100 acre-feet out of 64 300 acre-feet	Total amounts from events over a period from 12 Feb. 1940 to 18 Mar. 1941
Buono and Lang (1980)	Mojave River Basin, California	Stage gauge measurements	Loss as a percentage of the total flow, 77.2% and 88.2% measured in 1969 and 1978, respectively	Measured at The Forks, Afton
Burkham (1970)	Tucson Basin, Arizona	Inflow loss rate equations	47 000 acre-feet (70% of the total)	Total measured losses over period 1936–63
Dunkerley and Brown (1999)	Homestead Creek – Fowlers Creek Western NSW, Australia	Field measurements of the change in peak flow and wetted perimeter downstream, from an initial measured peak flow of $9.1 \, \mathrm{m^3 \, s^{-1}}$	Average 13.2% loss per km for sub-bankfull flow and 5–6.9% for overbank flow or bankfull flow	
Hughes and Sami (1992)	Goba River, South Africa	Moisture observations using neutron probe access tubes	75% and 22% of the flow, recorded from two different flow events	75% loss for an event on 3 Oct. 1989, and 22% on 14 Nov. 1989

Reference	Location	Method	Results	Comments
Jordan (1977)	Western Kansas	Differential equations	Average loss per valley mile is 1.3% of flow at upper gauging station	Results from a 4.1 mile experimental reach
Lane (1972)	Walnut Gulch, Arizona	Leaky reservoir models	Maximum discharge reduced by 50% and flow volume reduced by 35%	
Lane (1985)	Walnut Gulch, Arizona	Differential equations	2-year flood peak reduced by 2% (0.0142 mi^2 watershed), 30% (5.98 mi^2 watershed)	
Lane et al. (1971)	Walnut Gulch, Arizona	Regression equations	Between 0.08 and 81.91 acre-feet, corresponding outflow peak reduction 100% and 2.3% respectively	For events taking place on 11 Sep. 1966 and 11 Sep. 1964 respectively
Sharma et al. (1994)	Luni Basin, NW India	Differential equations	38×10^3 m^3 km^{-1} (rocky terrain; $L = 73$ km) 6500×10^3 m^3 km^{-1} (deep alluvium; $L = 34$ km)	Range in 15 reaches
Walters (1990)	Queen Creek, Arizona	Inflow loss rate equations	62.4×10^3 m^3 km^{-1}	Average of 10 events; $L = 32$ km

L = Length

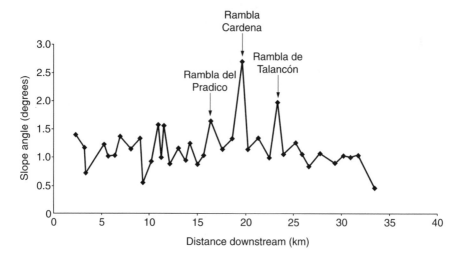

Figure 5.4 Channel slope variations downstream for the Rambla de Nogalte, southeast Spain, showing major tributary input

to transmission losses, indicating that channel behaviour and sediment transport are very sensitive to slight changes in the controlling parameters. Entrainment and sediment transport are explained by available stream power and the channel-bed morphology is essentially a response to the sediment behaviour of the channel. In the absence of entrainment, the channel boundary is fixed and conforms essentially to the configuration provided by the last available event (Thornes, 1980).

By the nature of the events in ephemeral channels, data on bedload and suspended sediment transport are extremely rare. Sediment transport in perennial gravel-bed rivers has received much more attention than the perennial counterparts due mainly to the infrequent events and need for expensive measuring, monitoring and logging equipment required to make accurate measurements of events occurring throughout the year. Any data that are collected should be

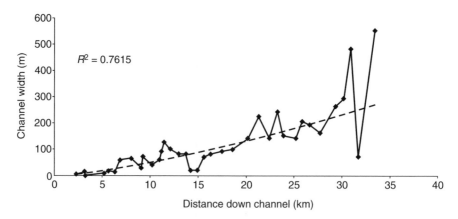

Figure 5.5 Channel width variations downstream for the Rambla de Nogalte, southeast Spain

considered in the context of all data available, that is, each event is unique to the environment and specific conditions under which it occurred.

Data from Israel have, however, provided an invaluable insight into the processes behind sediment transport in gravel-bed ephemeral channels. Information from the Nahel–Yatir ephemeral channel experimental sites (Laronne et al., 1994; Reid et al., 1995) have provided important information on how operating processes differ greatly from perennial channel counterparts (Reid et al., 1994), and therefore should be considered separately if they are to be understood.

Ephemeral channels also exhibit a much more efficient flow regime than perennial channels, making them very effective environments for the movement of all particle sizes, compared with perennial channels (Reid and Laronne, 1995).

5.3 MODELLING: IMPORTANT FEATURES AND PROBLEMS

Modelling the transfer of water and sediment is particularly difficult due to the lack of data available. The shortage of data can be attributed to rarity of measured events, the danger that accompanies any kind of measurement carried out manually and the loss of monitoring equipment. Some catchments, such as Walnut Gulch, Arizona, have been instrumented and monitored for well over 40 years (Osborn and Lane, 1969), providing data and knowledge about the processes operating during a flow event. Each flow event should be considered as an individual dataset and data from more than one event, for different catchments, are required if the rainfall–runoff processes are to be better understood.

The limited available rainfall–runoff records for the Mediterranean, coupled with a limited understanding of this catastrophic phenomenon, do not allow for precise risk management evaluations (Gutiérrez et al., 1998) or development of rainfall–runoff models in semi-arid environments. Model simulation of runoff hydrographs from arid and semi-arid catchments is a key to answering the problems of flood prevention, groundwater recharge through channels, reservoir sedimentation and channel stability both for present-day scenarios and for future climatic change scenarios in an area where the paucity of data is well known. The lack of available flow data results in the use of procedures for data generation. However, Srikathan and McMahon (1980) note that one of the major problems in applying data generation models is the modelling of zero flows. This is particularly important because during periods with no stream flow, no data on the catchment's antecedent soil moisture status are available (Ye et al., 1997), which is a crucial component of any rainfall–runoff model in semi-arid catchments.

As well as a lack of data for model parameters, thresholds for runoff are dependent upon catchment characteristics as well as rainfall intensity, and can only begin to be investigated through observations of different storm magnitudes. The non-linear relationship between rainfall and runoff adds complexity to the modelling process (Ye et al., 1997). Also, if flow and sediment transport are to be accurately predicted, then estimation of transmission losses is critical.

Generally, models are considered as

'. . . an efficient consolidation of available data, a positive framework for incorporating future data, a means for reducing the inherent physical indeterminacy of hydrological systems, and a basis for imbedding a priori hypotheses which allow for the possibility of occurrence of events that have not been observed historically.' (Kisiel et al., 1971, p. 1699)

Table 5.2 Summary of some of the research that has been carried out into flow and sediment transport in ephemeral channels

Area	References
Southwest USA	Burkham, 1976; Diskin and Lane, 1972; Drissel and Osborn, 1968; Graf, 1983a, 1983b; Keppel and Renard, 1962; Lane, 1982b; Lane and Renard, 1972; Lane et al., 1994b; Leopold et al., 1966; Murphey et al., 1972; Renard, 1970; Renard and Lane, 1975; Renard and Laursen, 1975; Schumm, 1961
Africa	Reid and Frostick, 1987
Australia	Dunkerley and Brown, 1999; Knighton and Nanson, 1994, 1997
Israel	Ben-Zvi et al., 1991; Greenbaum et al., 1998; Hassan, 1990; Hassan and Reid, 1990; Laronne et al., 1992, 1994; Laronne and Reid, 1993; Meirovich et al., 1998; Powell et al., 1996; Reid and Laronne, 1995; Reid et al., 1995, 1996, 1998; Schick, 1977
India	Sharma and Murthy, 1994a, 1994b, 1996a, 1996b, 1998
Saudi Arabia	Parissopoulos and Wheater, 1995; Sorman et al., 1997
Mediterranean	Bull et al, 2000; Butcher and Thornes, 1978; Conesa-García, 1995; Conesa-García and Alonso-Sarría, 1997; Faulkner, 1992; Gutiérrez et al., 1998; Harvey, 1984; Piñol et al., 1997; Thornes, 1976, 1980; White et al., 1997

However, the lack of observed data that tends towards a modelling approach also has the effect of making the task more difficult (Pilgrim et al., 1988).

Table 5.2 summarises some of the research that has been carried out into ephemeral channel flow and sediment transport, particularly in the American southwest, and to a lesser extent in Africa, Australia, India, Israel, Saudi Arabia and the Mediterranean.

It is generally accepted that more detailed information on river-bed topography is required in order to understand more about the spatially distributed process-form feedbacks in fluvial geomorphology (Naden, 1987; Richards, 1988; Lane et al., 1994b). The overall channel topography is important, particularly in channel systems that exhibit braiding, as it is this braided system that determines the flow patterns, amount of water transported and sediment transport and deposition in gravel-bed ephemeral channels. The aerial approach is a good way to obtain a representation of the overall pattern and the mesoscale channel characteristics both across and down channel. Cross-section profiles can give indicators about cross-stream profiles and processes, but little about the linkages and processes downstream, thus downstream profiles are equally important (Lane et al., 1994b).

The remainder of this chapter will consider different modelling approaches, both deterministic and stochastic, for the transfer of water and sediment in a gravel-bed ephemeral channel in southeast Spain, the Rambla de Nogalte.

5.4 CASE STUDY: RAMBLA DE NOGALTE

5.4.1 Catchment and Channel Characteristics

The Rambla de Nogalte is situated in the southeast on the border of Murcia and Almería. It is 171 km^2 with a channel length of approximately 33 km (Figure 5.6). The geology of the area is

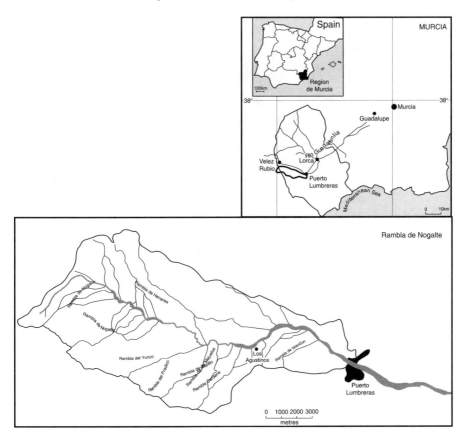

Figure 5.6 Catchment diagram of Rambla de Nogalte showing its location in southeast Spain

mainly phyllites and mica schists. To the south of the boundary fault on the Guadalentín graben the channel becomes unconfined and distributes onto a very large fan where the geology is predominantly Tertiary marls and Quaternary deposits. Land use in the catchment is a mixture of almond and olive cultivation on suitable slopes; and shrubs, matorral and abandoned land on hillslopes unsuitable for cultivation (Figure 5.7). There is some wheat cultivation on the flood terraces within the main channel.

The main channel is a broad, gravel-bed river. Downstream the channel width varies greatly from just a few metres to over 300 m in places (Figure 5.5). The channel bed has a complex morphology with riparian vegetation and the main channel is dissected by a series of braided channels that carry the lower flow events that rework the channel bed in between larger flood events.

The channel has a history of catastrophic flooding, current anthropogenic alterations and large-scale engineering works, which make it an important study area for past as well as future flood events and the impact of anthropogenic activity. In 1973 the channel suffered a devastating flood, in which hundreds of people in the town of Puerto Lumbreras were killed. This is the largest event on record for this channel at the town of Puerto Lumbreras. Table 5.3 shows the discharge and recurrence intervals of other events recorded in the Nogalte.

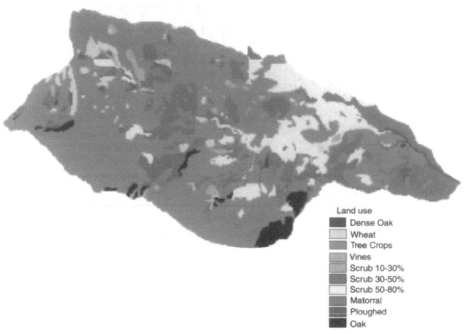

Figure 5.7   Land use map of the Rambla de Nogalte

Table 5.3   Discharge events and recurrence intervals for the Rambla de Nogalte measured at Puerto Lumbreras (adapted from Conesa-García, 1995, p. 175)

Discharge (m s^{-1})	No. of events	Times yr^{-1}	Recurrence interval
106.0	1	0.17	6.00
36.3	3	0.50	2.00
7.0	9	1.50	0.67
0.9	20	3.30	0.30

5.4.2   Flow Events and Channel Change

The most recent considerable event that caused some reworking of certain parts of the catchment and main channel occurred in September 1997. A maximum discharge of 59.4 m^3 s^{-1} was estimated in the Rambla de los Majuelos (Figure 5.1), and the overall event was estimated to have a recurrence interval of seven years (Bull et al., 2000).

Monthly totals ranging from 0 to over 200 mm have been recorded in the area. Yearly rainfall totals are in the region of 300–350 mm. Figure 5.8 shows the monthly rainfall totals for Tonosa, Velez Rubio from 1967 to 1994. The wettest months tend to be September, October and November, with very little or no rainfall in July and August.

Anthropogenic factors that influence flow events include engineering structures and farming. Engineering structures, such as check dams which exist in many of the subcatchments as well as

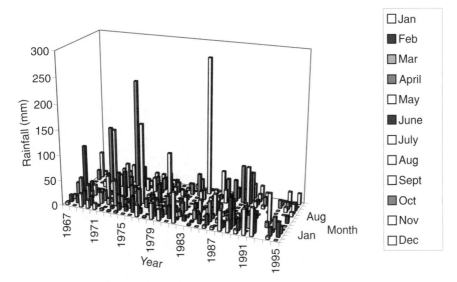

Figure 5.8   Rainfall for Velez Rubio, 1967 to 1994

a few in the main channel, alter the routeing of hydrographs and interrupt the transfer of sediment to the main channel from subcatchments, as well as in the main channel. When the subcatchment check dams are not full, sediment is trapped behind them and virtually clear water is released downstream. This has the effect of providing an amount of water with greater energy just below the check dam, and field evidence indicates that this produces massive scouring of the dam itself and the main channel just below it (Figure 5.9). If check dams are

Figure 5.9   Scouring of the channel bed in front of a large check dam in the Rambla Cárdena

Figure 5.10   (a) Large borrow pit in the main channel as a result of gravel extracted for motorway construction. Figure (b) shows a different borrow pit, and the effect that the extraction of bed material has on channel storage, resulting in surface ponding. The maximum depth of the water was estimated to be approximately 2 m

full to capacity and not properly maintained, which is generally the case in the Nogalte, then the check dam now takes on the effect of producing a larger waterfall structure over which the flood wave propagates. Gutiérrez et al. (1998) attributed the rapid rise to peak flood discharge in the Arás campsite disaster to the large amount of sediment load released when the check dams failed. This caused massive sediment deposition on the Arás alluvial fan, where the Las Nieves campsite was located. Check dams, therefore, have major implications for channel management including hydrograph propagation, sediment transfer and delivery to the main channel.

Other human-induced channel changes have resulted from motorway construction in the area. Large areas of the channel have been dug out to take advantage of the gravel of which the main channel is composed. This has resulted in large borrow pits within the main channel and areas with virtually no vegetation cover. This, in turn, has major implications for an advancing flood wave down the main channel (Figure 5.10).

Therefore, once the climate, land use, soils and geology have been accounted for, anthropogenic factors play an important role in the transfer of water and sediment through an ephemeral catchment system and cannot be ignored.

5.4.3 Catchment Monitoring

The catchment has been instrumented with seven tipping bucket rain gauges, one automatic weather station and two pressure transducer stage-level recorders located at the top and bottom of a short channel reach that has been chosen to study flow. This system of observations has already provided data and an insight into processes operating within an ephemeral channel catchment, both in terms of climate and stream flow. However, it has also proved to be inadequate in terms of describing storm cell sizes and locations because the rain gauge network is not dense enough, but the density of rain gauges that would be required is not feasible. One event has been recorded in the channel. In September 1997, the catchment experienced high rainfall intensities and flow in parts of the main channel. In this storm event, rainfall intensities were very high and varied across the catchment. Rainfall intensity alone, however, did not account for areas of the catchment that produced high runoff discharges. Land use, soils and slopes were important in determining runoff-producing areas and subcatchments that flowed into the main channel (Bull et al., 2000). Flow was recorded in the channel reach. These data not only provide input and output hydrographs from a channel reach, but a measure of the decrease in discharge down the channel as a result of transmission losses.

The remainder of this chapter will concentrate on some modelling approaches that can be adopted in order to study the transfer of water and sediment in ephemeral channels. We turn now to a discussion of the main modelling approaches used to study the transfer of water and sediment in ephemeral channels.

5.5 FLOOD ROUTEING IN EPHEMERAL CHANNELS USING A TWO-DIMENSIONAL FINITE ELEMENT MODEL

5.5.1 Introduction to Computational Fluid Dynamics

Modelling of flood flows, sediment transport and channel morphology in ephemeral systems can be approached through an accurate description of flow hydraulics. Recent advances in the application of computational fluid dynamics (CFD) to the study of river flows has shown the potential for high resolution hydrodynamic modelling in attaining such results (Bates et al.,

1992, 1997; Anderson and Bates, 1994; Bates and Anderson, 1993; Nicholas and Walling, 1997; Hodskinson, 1996).

Computational fluid dynamic codes allow the simulation of detailed spatial patterns of flow properties through a predefined geometry. Hydraulic-modelling approaches in geomorphology have traditionally been based upon a finite difference solution of the one-dimensional St Venant equations (Cunge et al., 1980; Sammuels, 1983, 1990; Fread, 1985) which consider the mass continuity (equation (5.1a)) and conservation of momentum (equation (5.1b)) between two consecutive cross-sections Δx apart. These are first-order partial differential equations in Q and h which can be solved using a finite difference procedure such as the Preissmann (1961) or Abbott and Ionesco (1967) schemes (Bates et al., 1997).

$$\frac{\partial Q}{\partial t} + \frac{\partial (Q^2/A)}{\partial x} + g.A\left[\frac{\partial h}{\partial x} + S_f\right] = 0 \tag{5.1a}$$

$$\frac{\partial Q}{\partial t} + \frac{\partial A}{\partial t} = 0 \tag{5.1b}$$

where Q = discharge $(L^3 T^{-1})$

t = time (T)

A = flow area (L^2)

x = distance in downstream direction (L)

g = acceleration due to gravity $(L\,T^{-2})$

h = depth of flow (L)

S_f = friction slope

Advances in computing power and computational fluid dynamics have meant that CFD applications to river geomorphology can now involve the application of two-dimensional finite element techniques, which treat the computational domain as a continuous field rather than a series of cross-sections and draws on the flexibility of the finite element space discretisation to represent complex topographies with a minimum number of elements (Bates et al., 1997). The finite element mesh effectively constitutes an independent Digital Terrain Model and therefore all flow effects over topography are inherently included.

Such codes solve the second-order partial differential equations for depth-averaged flow derived from the three-dimensional Navier–Stokes equations. This results in an equation for two-dimensional mass continuity (equation (5.2a)) and the two-dimensional finite element St Venant equations for conservation of momentum (equations (5.2b) and (5.2c)).

$$\frac{\partial h}{\partial t} + \frac{\partial (uh)}{\partial x} + \frac{\partial (vh)}{\partial y} = 0 \tag{5.2a}$$

$$\frac{\partial u}{\partial t} + u\frac{\partial u}{\partial x} + v\frac{\partial u}{\partial y} + g\left(\frac{\partial h}{\partial x} + \frac{\partial a_0}{\partial x}\right) - \frac{\partial^2 u}{\partial x^2} - \frac{\partial^2 u}{\partial y^2} + \frac{gu}{C^2 h}\sqrt{u^2 + v^2} = 0 \tag{5.2b}$$

$$\frac{\partial v}{\partial t} + u\frac{\partial v}{\partial x} + v\frac{\partial v}{\partial y} + g\left(\frac{\partial h}{\partial y} + \frac{\partial a_0}{\partial y}\right) - \frac{\partial^2 v}{\partial x^2} - \frac{\partial^2 v}{\partial y^2} + \frac{gv}{C^2 h}\sqrt{u^2 + v^2} = 0 \tag{5.2c}$$

where x = distance in the x-direction (longitudinal to flow direction) [L]

u = **horizontal flow velocity in the** x**-direction** $[L\,T^{-1}]$

y = distance in the y-direction (lateral to flow direction) [L]

v = **horizontal flow velocity in the** y**-direction** $[L\,T^{-1}]$

t = time [T]

g = acceleration due to gravity $[L\,T^{-2}]$

h = **water depth** [L]

a_0 = elevation of the profile bottom [L]

C = Chezy roughness coefficient $[L^{2/3}\,T^{-1}]$

The models solve for the three unknowns (highlighted in bold) u, v and h using varying applications of the finite element scheme. To simulate the effect of turbulent momentum transfer on flow properties, CFD models generally employ a mean flow concept, which averages instantaneous velocities through time to produce mean values, and include Reynolds stress terms which are added to the governing equations to represent velocity fluctuations.

5.5.2 Simulation of Flood Inundation

Until recently, the application of finite element models has been limited to fairly small-scale problems such as detailed analysis of flow around structures (King and Norton, 1978), at river confluences (Niemeyer, 1979) and at channel bends (Hodskinson, 1996). These problems have tended to involve scenarios where the whole solution domain is inundated during a simulation. Partially wet elements within the model are either included or excluded as a whole, thereby leading to an irregular flow boundary based on the element geometry rather than the physical topography of the problem area (Figure 5.11). Erroneous flow velocities may occur with relatively small depth changes in such schemes (Bates et al., 1997). The finite element method as originally developed is therefore of limited use for studies with initially dry domains.

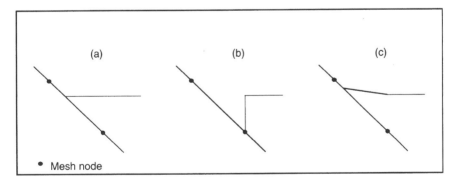

Figure 5.11   (a) Free surface position on a partially wet element. (b) Standard RMA-2 interpolation of a free surface and (c) Interpolation achieved using the enhanced wetting and drying algorithm of King and Roig (1988). (Adapted from Bates et al., 1997)

However, recent developments in computing power and algorithms related to flow over initially dry conditions has allowed the application of two-dimensional finite element methods to flow problems over larger scales with inundation scenarios. A significant amount of work has been undertaken in this field related to flood inundation (Bates et al., 1992, 1997; Gee et al., 1990; Feldhaus et al., 1992; Bates and Anderson, 1993). Research in this area (Lynch and Gray, 1980; King and Roig, 1988; King and Norton, 1978; Gee et al., 1990; Bates et al., 1992; Baird et al., 1992; Bates and Anderson, 1993; Anderson and Bates, 1994) has led to the development of generalised two-dimensional finite element codes that have the ability to model the inundation of initially dry areas. Two-dimensional finite element modelling of inundation flows has been shown to be viable with realistic results faithfully reproducing field data (Bates et al., 1997).

5.5.3   Application of CFD to the Rambla de Nogalte, Spain

There are currently a number of CFD codes available for simulating two-dimensional flows. The work reported here uses the RMA-2 model, originally developed for the US Army Corps of Engineers. RMA-2 solves the depth-averaged shallow water equations for a mesh comprising six-node triangular and eight-node quadrilateral elements using a fully implicit Galerkin weighted residual technique. The numerical integration for the Galerkin procedure is performed iteratively using a Newton–Raphson type solver (Norton et al., 1973). At each stage in the iteration procedure RMA-2 fully assembles the solution matrices. RMA-2 includes a specific algorithm for inundation of initially dry areas (King and Roig, 1988) which simulates a smooth transition between wet and dry states. This is achieved using a domain coefficient which represents the portion of a mesh element used in the simulation and represents more accurately the simulated water boundary.

The results reported here describe the preliminary stages of a study to simulate flood flows through an initially dry ephemeral channel. A 2-km reach of the Rambla de Nogalte, between established stage-gauge recording stations, was selected as the study site for the simulation. A computational mesh was established for the study site (Figure 5.12) using a geo-referenced aerial photograph of the reach. Height information was added to the mesh using an extensive ground survey of the site (approximately 400 points), and sediment data was collected to give roughness information ($D_{50} = 2.0$ mm).

On 29 September 1997 a flood event was gauged at the study site (Figure 5.3). From the upstream gauged stage record for this flow event, an effective discharge was calculated (Figure 5.13). The effective discharge is the discharge which transports the greatest load of sediment during the period of record (Andrews, 1980). The calculation of effective discharge involved integrating a flow–frequency histogram (above the threshold condition for sediment transport) with a sediment rating curve to produce a histogram of sediment load as a function of discharge. The effective discharge is represented by the peak of the histogram and is, in this case, interpreted as being the most significant flow in terms of sediment transport and therefore channel morphology. The sediment rating curve was constructed using the Bagnold stream power equation (Bagnold, 1980).

For the event observed on 29 September 1997, an effective discharge of $5.0 \, \mathrm{m^3 \, s^{-1}}$ was calculated. As a first stage in modelling this flood event, a steady-state flow simulation of the effective discharge was performed. The Limerinos equation (Limerinos, 1970) was used to calculate Manning n friction factor for the main channel ($= 0.0023$ for a discharge of $5.0 \, \mathrm{m^3 \, s^{-1}}$) and 0.05 was chosen for the floodplain. Depth-averaged velocity vectors for the steady-state simulation are shown in Figure 5.14.

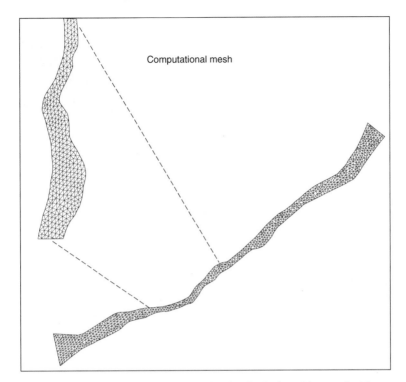

Figure 5.12 Finite element mesh of the Rambla de Nogalte, Spain, for a 2-km reach at the gauged site (slope = 0.023)

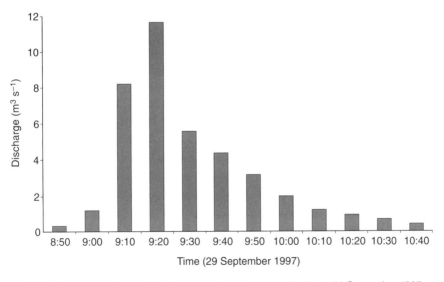

Figure 5.13 Discharge hydrograph for flood event gauged at study site on 29 September 1997

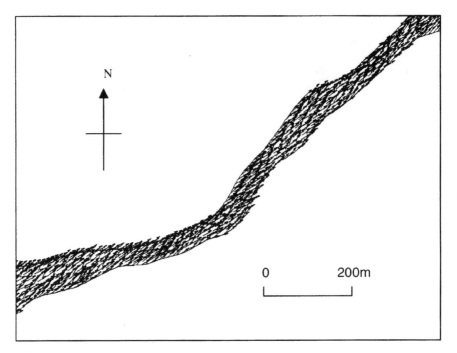

Figure 5.14   Depth-averaged velocity vectors for a 'steady-state' two-dimensional simulation of a $5.0\,m^3\,s^{-1}$ event at the Nogalte gauging site

The application of the two-dimensional finite element schemes to simulating flood inundation has been demonstrated by many researchers in this field. Here, the qualitatively successful simulation of a steady-state flow condition for a relatively small reach of the Rambla de Nogalte has highlighted the potential of such applications in this environment and further work is now required to simulate a fully dynamic flood wave propagation through such a reach with the aim of applying the resulting detailed hydraulic data to successful sediment transport and channel morphology modelling.

5.6   STOCHASTIC MODELLING OF EPHEMERAL CHANNEL FLOWS

5.6.1   Conceptual Framework

A major barrier to modelling hydrological processes in semi-arid areas is the lack of understanding and model representation of the distinctive features and processes associated with runoff generation in these regions, along with the paucity of field data (Zhu et al., 1999).

Ephemeral channels relate more to processes in braided channels during low flow conditions. As a result, morphology is a critical control on both process type and its rate of operation and, as such, should be a detailed component of any geomorphological study (Lane, 1998), whether this be a boundary condition in numerical modelling, laboratory- or field-based studies. Advances in spatially distributed numerical modelling require a spatially distributed representation of form as a boundary condition in the study of process; this is more feasible as the ability of computers to handle large amounts of data increases. Also, numerically modelling the interaction between form and process will allow for the investigation and sensitivity analysis of different boundary conditions and the effect on river channel form (Lane, 1998).

Channel morphology has an important effect upon the ability of a reach to transport sediment. It can be argued that this also applies to hydrograph propagation as, during lower flows in ephemeral channels, the sediment transport capacity is reduced to a minimum, as a result of flows with low capacities for transport, and the channel-bed morphology can be considered as a boundary condition for the event in question, providing the necessary spatially distributed morphology dataset for a reach and the initial conditions that will determine the flow paths and overall hydrograph propagation. Changes in downstream morphology in ephemeral channels are important for the propagation of hydrographs, especially during lower discharge events where the linkages of braided channels and the bed morphology are instrumental in the passage of a flood wave on an event basis.

Ephemeral channels have different process mechanisms during different discharge events; different magnitude discharges have implications for modelling; and different flow levels or domains can be conceptualised. The more frequent lower flow events do not submerge the whole channel width, but occupy the braided channels that have been formed by medium discharge, within bank flow events, and can be thought of as the lower discharge channel system (Figure 5.15). The geometry of the channel bed is reworked to different degrees according to the frequency and magnitude of events. Dunkerley and Brown (1999) distinguish between sub-bankfull flow events and bankfull and shallow overbank events in terms of estimating transmission losses (Table 5.1). They state that instead of porosity and the depth of channel fill being the controlling factors for transmission losses, local aspects of channel geometry, pool abstraction and interfilament abstraction play a more important role when considering

Figure 5.15 Section of the main channel of the Rambla de Nogalte during a low discharge flow event

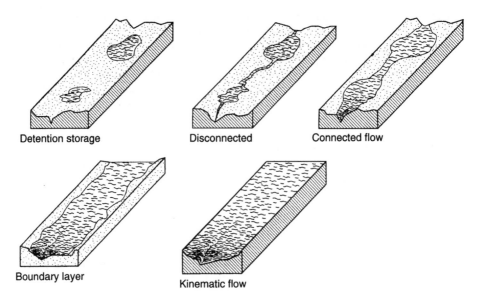

Figure 5.16   Conceptual ephemeral channel flow domains

transmission losses from lower sub-bankfull flow events. That is, the bed and bank characteristics become relatively less important and local channel-bed morphology becomes increasingly
important (Dunkerley and Brown, 1999).

Figure 5.16 shows the schematic flow domains for an ephemeral channel according to
different discharge levels in the channel; and Table 5.4 outlines the main features and processes
operating in each flow domain.

Detention storage allows greater opportunity for infiltration into the bed of the channel and
eventually reduces the transfer of water down the channel to zero, it is water in closed depressions that eventually evaporates or is infiltrated into the ground. The lower flow events (disconnected and connected flow, Figure 5.16) can be considered to be controlled by the channel-
bed morphology and do not usually exhibit the shear stresses required to rework the existing
channel-bed material (Dunkerley and Brown, 1999), except perhaps for some reworking of the
channel-bed fines. However, these lower magnitude flows occur most frequently, with transmission losses consuming the flow such that hydrographs may not propagate very far downstream.
Nevertheless, an understanding of these low flow events is important if the long-term recharge
of aquifers is to be estimated, and in order to gain a better understanding of flow processes in
ephemeral channels.

As the discharge in the channel increases, the flow in the channel becomes continuous at the
boundary layer and, where major reworking of the channel-bed material takes place, we call the
flow domain 'kinematic'.

The spatial variation in rainfall and runoff leads to variation of lateral input hydrographs to
the main channel. This variability of lateral inputs distinguishes ephemeral channel forms,
particularly in terms of the dumping of sediment as waves or alluvial fans and sets them
apart from their perennial counterparts. This should be included in models of ephemeral
channel flow and sediment transport. Figure 5.17 shows how a catchment can be considered
in terms of modelling the variability and propagation of channel flows.

Table 5.4 Characteristics of ephemeral channel flow domains

Flow domain	Frequency	Type of flow	Comments
Detention storage	Detention storage will occur for most rainfall events. Storage duration will depend on volume of rainfall and antecedent channel conditions	No flow – only ponded water in channel depressions	No reworking of the channel bed and no channel flow. Ponding results in channel abstracting all surface water. Channel bed morphology determines the location of detention storage
Disconnected flow	Depending on antecedent conditions and rainfall amounts, usually 1–3 times/year	Low flow as a result of detention storage pools overflowing	Onset of hydrograph propagation downstream, but at slow speeds. No reworking of the channel bed and transmission losses according to local bed morphology and conditions
Connected flow	Same as for disconnected flow, but higher rainfall totals and longer duration rainfall. 1–3 times/year	Low flow that occupies braided channels, deeper flow depths than disconnected flow	Hydrograph propagation speeds increase. Flow and transmission losses are still controlled by channel-bed morphology, but there may be some reworking of finer channel-bed sediments
Boundary layer flow	Much less frequent, but will depend on channel width, every 10–50 years	Bankfull flow that just submerges a channel reach. Friction from channel bed can influence hydrograph propagation speeds	Major reworking of the channel-bed material, where evidence of all smaller flows is destroyed. Sediment transport and deposition as waves. Transmission losses according to the channel dimensions and infiltration characteristics
Kinematic flow	Same as boundary layer flow, however more rainfall totals required, every 10–50 years	Same as boundary layer flow, but flow depth is deeper therefore friction from channel bed does not influence hydrograph propagation	Same as for boundary layer flow, except the possibility of higher transmission losses if over bank flow occurs

- - - - - - Indicates boundary between flows that are controlled by the channel-bed morphology, and those that perform major channel-bed reworking.

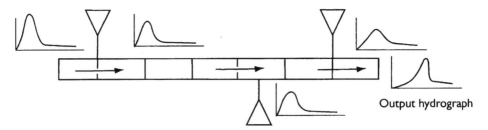

Figure 5.17 Schematic diagram showing how the catchment can be considered. Input hydrographs are routed down the main channel with the output hydrograph produced at the catchment outlet

For routeing of water at the lower flow domains, information on the channel-bed morphology is required. A high-resolution DEM shown in Figure 5.18 for a section of the main channel at the tributary with the Rambla Cárdena has been developed; however, this only covers a section of approximately 1 km in length. Deterministic routeing according to the channel-bed morphology is possible for this section; however, to make a DEM at this resolution for the whole 33 km of the channel is not feasible due to time, financial and computing restraints. As the data available for the rest of the main channel is lower resolution surveyed cross-section data, a

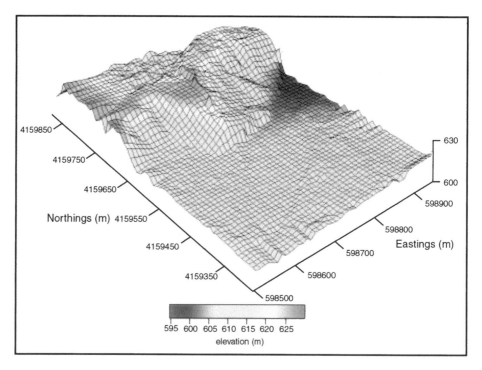

Figure 5.18 10-m DEM of a 1-km length reach of the Rambla de Nogalte at the junction of the Rambla Cárdena

deterministic approach is unsuitable. A stochastic approach is, on the other hand, much more appropriate because of the uncertain elements of the bed topography. The concept of the random walk can be used to route the low discharge events through the main channel with variations of the subcatchment inputs.

5.6.2 Random Walk Modelling of Channel Flows

Determination of flow directions on a cellular grid is not new to hillslope or channel routeing. Tarboton (1997) reviewed work that has been carried out on cellular flow directions, and presented a new procedure for calculating flow directions from grid-based DEMs. Findings showed that the best procedure should be a single flow direction that follows the steepest slope of the eight neighbouring cells. However, as already mentioned, obtaining a DEM for the whole length of a channel is unrealistic and other procedures need to be adopted.

Murray and Paola (1994, 1997) developed a simple cellular computer model for braided streams in which the complex physical laws that govern processes in braided streams are embedded in the model via a series of rules that are based on abstractions of the physics governing the dynamic flow behaviour. This approach means that only the most basic processes need to be investigated, thus simplifying the modelling procedure and concentrating on the important aspects of the flow dynamics. Another simplified procedure is that of the random walk.

In random walk models the central concept is that each step is determined by the previous state (position) and that the specific path taken is determined randomly.

Interest focuses on average path travel distances and times, and these can be used when determining transmission losses and sediment transport. For example, the number of people travelling into a city each day to go to work from all directions needs to be estimated in order to set up a transport network that can cope with them. The actual route of each individual is of little relevance; it is the behaviour of the whole system of people that is of interest. Knowing what each person does exactly on one day would generate too much data, and travellers often change their mode of transport due to extenuating circumstances. If all of this information was used to develop a transport network it would be incredibly time consuming with little improvement of the system knowledge (Morley and Thornes, 1972).

Observing the mass behaviour of a system and assigning probabilities to the system provides enough information with which to set up bus, rail and road networks for example. In the case of water in ephemeral channels, we can say that we are not interested in the specific routes taken by parcels of water according to the channel-bed morphology, but we are interested in an ensemble of parcels and how they behave in a channel reach, where they start, where they end and the average or most probable path taken by them. In other words, where a hydrograph begins, where it ends or diminishes and the average time taken for that process. The micro-behaviour of an individual particle defies meaning in the practical sense and it is the behaviour of the ensemble of parcels that is of interest (Harr, 1925).

The movement of water through a channel reach can be compared to the movement of water through a porous medium, which can be considered by two approaches; the construction of an ordered continuum with the motion of particles responding exactly to the governing laws (deterministic), or a medium wherein the exact position is not predictable but a good approximation can be derived as to where it is most likely to be (probabilistic). This can be applied to water in a channel reach. If the bed morphology elevations are available, then the deterministic laws of flow can determine the exact position of water (as is the case where we have a high-

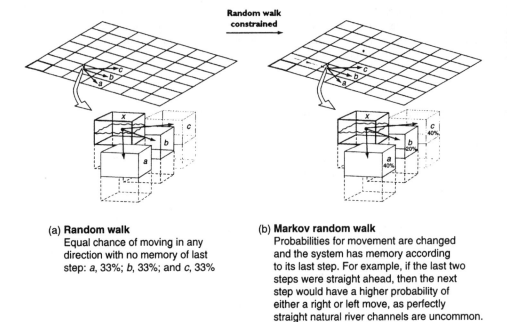

(a) **Random walk**
 Equal chance of moving in any
 direction with no memory of last
 step: *a*, 33%; *b*, 33%; and *c*, 33%

(b) **Markov random walk**
 Probabilities for movement are changed
 and the system has memory according
 to its last step. For example, if the last two
 steps were straight ahead, then the next
 step would have a higher probability of
 either a right or left move, as perfectly
 straight natural river channels are uncommon.

Figure 5.19 (a) and (b) Examples of random and Markov random walks as used in the channel flow routeing

resolution DEM). However, if bed elevation data is unavailable, then approximations of where water is most likely to be can be used as a surrogate for deterministic routeing procedures.

By adopting a random walk strategy for a simplified routeing procedure for the whole channel and catchment on an event base, the problem is being viewed in its simplest form, and constraints can be progressively imposed on the random behaviour in order to mimic the processes more realistically. In this case the amount of data required and computational power is somewhat reduced compared to a fully deterministic approach.

Gradually the behaviour of a walk is constrained from completely random to Markovian (Figure 5.19(a) and (b)). As the behaviour is constrained, more knowledge of the system is required and this can be thought of as gradually imposing process laws on the behaviour of the system instead of it being completely random. Assuming that moves are considered to be straight forward, forward left and forward right, the probability of these steps can be changed in order to simulate different types of behaviour – that is, more sinuous flow could be simulated by assigning higher probabilities to the forward left and forward right directions (Figure 5.19(a) and (b)). This is how knowledge of the fluvial processes can be introduced into the random walk. For example, in a completely random walk, water in a channel depression may move forward, but realistically this should not happen. If, for example, information from the previous state could be incorporated into the random walk (it then becomes a Markov process), then we could check to see if water moved during the last step. If it did not we could then determine that it is in a channel depression and therefore is unlikely to move until the discharge in the channel increases.

Figure 5.20 shows simple model output for random walks through a channel reach with different probability rules. In the first case, water has an equal probability of moving to any of

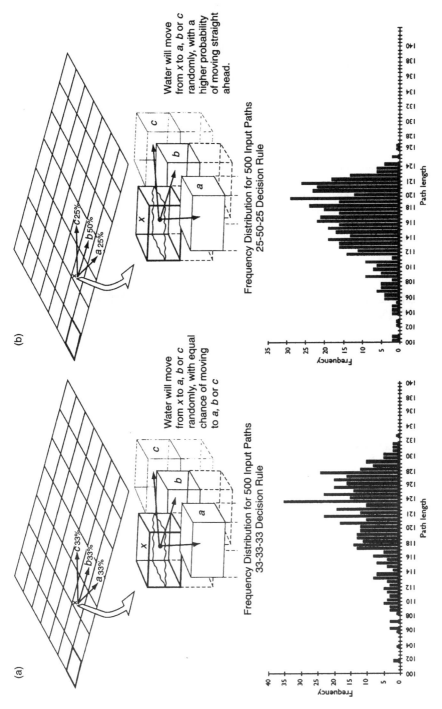

Figure 5.20 (a) Frequency distributions of path lengths for a 33–33–33 decision rule; (b) shows that for a 25–50–25 decision rule. Note the larger path lengths are obtained in (a) as opposed to (b), where there is a higher probability of the walk moving left or right and therefore taking a longer, more sinuous

the three cells in front (left–middle–right). The output is a distribution of the path lengths for 500 random walks, through a reach 100 units (or cells) in length. The path length is simply the incremented distance from each move (1 unit for a straight-ahead move and 1.41 for a diagonal move). Thus the minimum and maximum path lengths can be 100 and 141 respectively.

It can be seen that the distribution for the 33–33–33 rule base ranges from just over 101 units up to about 133 units (Figure 5.20(a)), and that for the 25–50–25 rule base is from 100 to just over 126 units (Figure 5.20(b)). By changing the rule base to one that promotes more straight-ahead movement (or flow), the maximum path length is reduced, and would therefore increase the speed with which water would propagate down the channel.

These simple model runs show that by changing the rule base of the output path lengths, which could be considered as a surrogate for hydrograph propagation, different channel processes can be simulated. For example, rougher channel reaches may have higher probabilities of 'no movement' from cells, which simulates the process of Detention storage, and less sinuous channels could have higher probabilities of forward movement.

This work is ongoing at present, but the use of random and Markov random walks could be powerful in an environment where the lack of data is the most problematic component of any modelling that is undertaken.

5.7 MODELLING SEDIMENT TRANSPORT IN EPHEMERAL CHANNELS

For all the reasons outlined earlier in this chapter and in other chapters, the measurement of sediment transport and production present substantial difficulties. The problem can be addressed either as an issue of single-particle transport or the movement of ensembles of particles called sediment waves or slugs.

Adopting the single-particle approach, single particles are moving intermittently through a complex maze of bedforms and secondary channels, rather like soil particles moving by overland flow across rough terrain. These successive steps occur under radically different flow conditions, encouraging the adoption of the random walk approach described in the previous section, or by analogy with a queue in which particles arrive at the back of a queue and depart from its front edge. If arrival and departure rates can be probabilistically described, then queue theory (Thornes, 1971) might provide a suitable analogy and methodology for dealing with the phenomenon.

Another approach is to accept that particles move as wave phenomena in (dynamically) unstable bedforms, where the behaviour of individual particles is constrained by the behaviour of the group as a whole. Dealing with huge numbers of particles requires us to adopt a more appropriate strategy. We discuss here two possibilities for approximating the mass transfer of sediments, a diffusion model and a kinematic wave model.

The diffusion approach can be regarded as the end member of the probabilistic methods described in the previous section. Imagine that individual particles move from reach to reach along a channel, with the reaches being length dx. Then assume further that the net flux from reach r to reach $r + 1$ in a time dt is proportional to the difference in the number of particles $N(r, t) - N(r + 1, t)$, where $N(r, t)$ is the number of particles in the rth reach at time t. If the constant of proportionality D is the same for all pairs of sites, then it follows that N changes according to the rule $N(r, t + dt) = N(r, t) + D\,dt$

$$N(x, t + dt) = N(r, t - 1) + D\,dt[N(x + dx, t) + N(x - dx, t) - 2N(x, t)] \qquad (5.3)$$

If one allows the time step and lattice size to shrink to zero in a compatible way, then in the limit this becomes the diffusion equation

$$\frac{DN}{dt} = \frac{D\,d^2N}{dt^2} \tag{5.4}$$

This is, in fact, the Forward, Kolmogorov diffusion equation for the density function of a stochastic process (Cox and Miller, 1965, p. 215 et seq). Reece (1986) expands the equation as follows:

$$\frac{d^2y}{dx^2} = \frac{y_{i+1} - y_i)(x_{i+1} - x_i) - (y_i - y_{i-1})(x_i - x_{i-1})}{(x_{i+1} - x_{i-1})/2} \tag{5.5}$$

and this provides a suitable algorithm for coding into a programme. This approach has been used in the simulations that follow. The general procedure is to start with an initial condition that includes the number of particles in each reach, section and especially, of course, the first section. Given that the Rambla de Nogalte is 33 km long, each reach in the hypothetical model is presumed to be 1 km long and divided into 100-m sections. The choice of 100 m is on the basis of seeking to represent a channel junction fan as the first section in a reach. Each time iteration is assumed to occur at the annual maximum peak flow, when we assume there are no constraints to sediment transport. A better approach would be to allow *each* flow event to transport particles between reaches selectively, so that a few events would transfer larger particles between reaches, and a few would transfer smaller particles. Assuming that D is a function of discharge and available particle size, transmission losses would affect propagation through the reaches, and selective sizes would appear in the outcome in different sections. We have not yet experimented along these lines, but two examples illustrate that the approach is promising. The first demonstrates the simulation of a wave of sediments furnished in the first section of the reach through the reach. With an initial value of 1000 particles in the first section, the wave is seen to propagate and alternate down the reach with a D value of 0.5 (Figure 5.21). The output

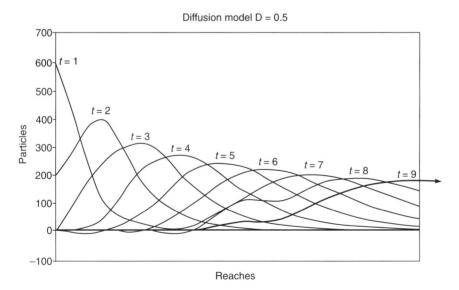

Figure 5.21 Wave propagation down a reach at successive discrete time intervals

Table 5.5 Distribution of sediment units by space and time. Reading down a column gives the sedigraph for that reach. Reading across the rows gives the reach-wise distribution through time of the sediment units at the selected time step

	reach = 1	reach = 2	reach = 3	reach = 4	reach = 5	reach = 6	reach = 7	reach = 8	reach = 9
$t = 1$	600	120	24	4.8	0.96	0.192	0.0384	0.01	0.0015
$t = 2$	200	400	152	44.8	11.84	2.944	0.704	0.16	0.037
$t = 3$	0	240	312	158.4	59.52	19.2	5.6448	2	0.4116
$t = 4$	0	40	232	264	156.8	69.44	26	9	3
$t = 5$	0	0	72	216	233.28	152.64	76	32	12
$t = 6$	0	0	8	91.2	200.64	211.456	148	80	37
$t = 7$	0	0	0	19.2	101.76	187.392	194	142	83
$t = 8$	0	0	0	1.6	101.76	107.2	176	181	138
$t = 9$	0	0	0	0	30.08	39.36	109.67	167	171

can be envisaged in matrix form, where the columns are the reaches and the rows are successive times (Table 5.5). By observing a single column through all time, we obtain the flux of sediments through a reach or a section of a reach. Figure 5.22 is a plot of a column as a section in a reach through time. By plotting the contents of a row in the matrix, we observe the particle number distribution at an instant in time throughout the system. Clearly, if particular columns (locations) represent the tributary junctions, then they can be provided with a time-varying number of particles, reflecting the respective inputs from the tributary. If these results seem fanciful, then they should be compared (but not in magnitude terms) with the sediment waves described (and modelled) in the Fly River in Papua New Guinea by Pickup et al. (1983).

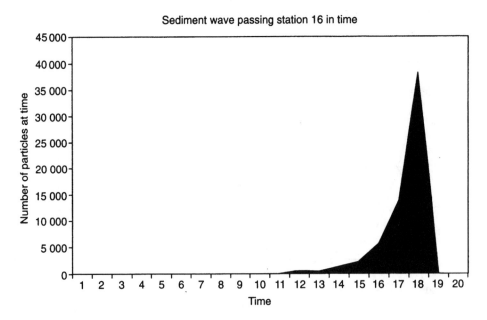

Figure 5.22 A plot of sediment in reach 16 through time (i.e. sedigraph)

Recall that the key assumption is that the transport is driven by the second difference of the number of particles in spatially successive sections. Call this the concentration of particles and then the parallel with other diffusion models is immediately transparent. A section with a large concentration of particles adjacent to a section with very few particles will produce a large transfer either upstream or downstream or both. The combination of upstream and down-stream is illustrated in Figure 5.23, where an initial wave is at $x = 0$. The function at the end of the reach combines with another wave originating at the tributary at $x = 500$.

By the definition of the problem, the waves are travelling at $100/\Delta t\,\mathrm{m\,yr^{-1}}$ if we assume that they are moved by the annual peak flow, and they are moving at a rate proportional to concentration (the number of particles/reach) and to the diffusion coefficient.

If the velocities of movement are represented by straight lines in the $x - t$ plane, then we can see from Figure 5.23 that the forward- and backward-moving waves intersect to produce the megawave, given their starting point at the respective junctions. The solid circle at the inter-section marks the point at which the backward-travelling wave from 5 meets the forward-travelling wave from 0. The forward wave covers greater distance (x) in a shorter time than the backward-travelling wave. In reality, of course, the bed comprises a whole spectrum of bed waves of different sizes, travelling at different velocities, especially when the waves are made up of particles of different sizes that are moving at different velocities.

These basic ideas are embedded in the theory of traffic flow and hence the traffic analogy, first used by Lighthill and Whitham (1955). They modelled long shallow waves in water, for exam-ple in canals, which was subsequently introduced by Langbein and Leopold as a potential model for the spacing of particle sizes on bar surfaces (1968). A useful introduction to these ideas is sketched in Leopold et al. (1964, p. 1066) and in Thornes and Brunsden (1977). Detailed mathematical treatment is given in Abbott (1966). Our treatment follows Hberman (1977).

Imagine traffic flowing along a highway (Figure 5.24) and that the traffic density ρ is given by the number of cars per unit length of road. Then adopt three process laws:

Law 1. The velocity of cars is a function of their density (ρ) according to

$$\mu = \lambda(1/\rho - 1/\rho_{\max}) \qquad \text{for } \rho \geq \rho_c \tag{5.6}$$

where μ is the velocity and $\rho_{\max}$ is the maximum possible velocity (as in Figure 5.25(a)) at a high density (just less than bumper to bumper); as the spacing becomes less and less, the velocity increases up to some speed, $\mu_{\max}$, beyond which the law does not hold.

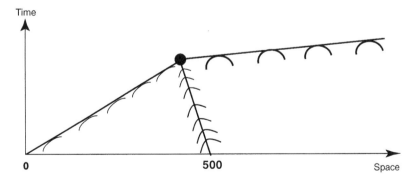

Figure 5.23 Convergence of waves propagating in two channels and converging on a junction. Larger waves are produced and move off downstream (to the right)

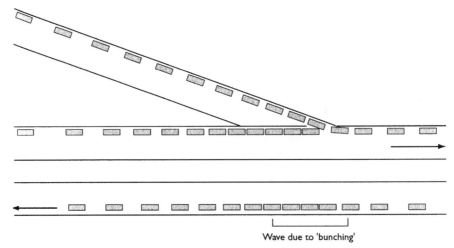

Wave due to 'bunching'

Figure 5.24　Schematic diagram of traffic flowing on a highway

Law 2. Let the flow rate be the product of concentration and velocity

$$Q = \mu\rho \tag{5.7}$$

On the motorway with high speed and high density, the throughput of vehicles is at a maximum.

Law 3. This is the conservation law:

$$\partial\rho/\partial t - \partial q(\rho)/\partial x = 0 \tag{5.8}$$

In other words, the change in density in time equals the difference in input and output to a reach of the motorway. The conservation equation can be expanded by the chain rule of calculus

$$\partial q(\rho)/\partial x = \partial(a) \cdot \partial\rho/\partial x \tag{5.9}$$

so that,

$$\partial\rho/\partial t = \partial q/\partial\rho \cdot \partial\rho \, \partial x \tag{5.10}$$

with $\partial\rho/\partial x$ the incoming traffic at the junctions.

This is a partial differential equation because ρ depends on x and t. In order to solve this equation, we require initial conditions (how the traffic is distributed at time, $t = 0$) and parameter values. We seek to solve the above equation to find how ρ varies with t.

Hberman (1977, pp. 301–305) suggested that the equation could be solved in one of the following ways: (i) writing it as a simple differential equation that can be integrated using the usual operations of calculus, or (ii) linearising the partial differential equation by assuming that the initial density remains constant (i.e. $\rho(x, 0) = \rho$) and independent of x. The density remains constant because all the cars are moving at the same speed. By introducing a perturbation that is a function of x, $(\rho(x, 0) = f(x)$, i.e. by letting the density vary at specific values of x (junctions where vehicles are entering) and then expanding, using a Taylor Series Expansion to give the linearised equation

$$\partial\rho_i/\partial t + \partial q(\rho_0)/\partial\rho \cdot \partial\rho_i/\partial x \tag{5.11}$$

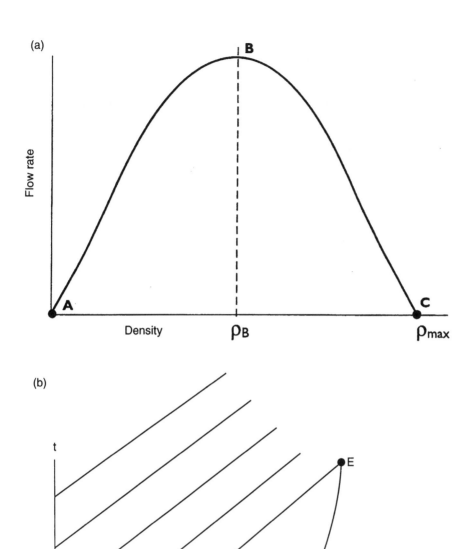

Figure 5.25 (a) The relationship between traffic density and flow rate. (b) A set of trajectories of waves moving at the same speed. At E, a faster moving wave meets the wave started at D

This is solved by Hberman (1977, p. 303) by looking at the partial differential equation moving in a co-ordinate system with velocity c.

$$\rho_i(x, t) = f(x - ct) \tag{5.12}$$

So the density is propagating as a wave with speed c. This is expressed by the relationship between flow rate and density, shown in Figure 5.25, in which the behaviour is shown as a curve with flow rate rising from zero at A and increasing to a maximum at B, with a density of ρ_B. The continuing increase in density causes the traffic flow rate to fall steeply until, at C, they are bumper to bumper, with a velocity of zero and a maximum density ρ_{max}. It is a common experience that, in driving in fast-moving motorway traffic, waves develop, so we are continually joining the back of a wave and leaving the front. Such waves of denser traffic may be stationary (as near a traffic inflow or an accident on the other carriageway) or they move negatively (back up the road in a direction opposite to the traffic flow) or positively (with the traffic down the motorway). If we are unlucky we move along the motorway in a dense patch.

The speeds of the wave are the gradients of the tangents to the curve in Figure 5.25(a). Between A and B on the rising limb, they are positive. At C the wave of traffic is stationary and on the falling limb negative. The equation describing the speed of the wave (called the celerity), not to be confused with the speed of the traffic (μ), is related to the flux–density curve by

$$c = \partial q(\rho_0)/\partial \rho \tag{5.13}$$

indicating that the wave speed depends on the density. If a wave starts along the road (Figure 5.25(b)) at point D when it is assumed to be zero, it has a location x. Then a line can be drawn as $x = ct$, representing the successive positions in time and space of the higher density wave of traffic. The lines shown in the figure are lines of constant traffic density and wave celerity. They are called characteristics and the Method of Characteristics is a third method of obtaining the position of waves in time and space. We shall not develop this idea further (see Abbott, 1966).

In the case of a near uniform flow, $dx/dt = c$, so all the lines are now parallel, but not necessarily straight. For example, if the wave is accelerating, the lines will steepen (as at E in Figure 5.25(b)). We have now come full circle in our diffusion modelling. We saw that one sediment wave can overtake or back up into another, where the characteristics intersect. In traffic flow, this is usually associated with disastrous consequences.

Wooding adopted the traffic analogy (1965), so the peak water wave moves proportional to its height, $v = f(h)a$, as did Nye (1960) in his modelling of ice waves moving through a glacier. In our diffusion example, given the movement of one step in unit time, the velocity is dx/dt, where dx is the length of the basic step that the particles are assumed to move (maximum 100-m section if not size dependent), but if the quantity moved (defined by the diffusion coefficient) is a function of the value Q in the section (as we have modelled it), then larger waves will move more sediment *per event*, leading to a faster movement of the larger wave, i.e. $dq/dx = f(\rho x_1, -\rho x_2)$, as required by the kinematic wave theory.

The traffic analogy is attractive for sediment movement because, in transport studies, engineers are dealing with different traffic loads arising from different sources in the network and moving through the network to one (or more often several) destinations. The stream–order system is like the road–order system, with the higher order links receiving sediments from lower and lower-order links in a network system, in which each link is an input–output system that can be modelled by a cascade of kernel functions, as in hydrological systems.

Another approach is to regard the links as states in the systems, through which sediment units (particles, tonnes) are moving. Then the transition from one link to another can be viewed as a Markov transition matrix, whose entries are the probabilities of movement of the sediment

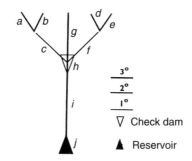

	a	b	c	d	e	f	g	h	i	j
a	0.7	0	0.3	0	0	0	0	0	0	0
b	0	0.7	0.3	0	0	0	0	0	0	0
c	0	0	0.8	0	0	0	0	0.2	0	0
d	0	0	0	0.7	0.3	0	0	0	0	0
e	0	0	0	0	0.7	0.3	0	0	0	0
f	0	0	0	0	0	0.8	0	0.2	0	0
g	0	0	0	0	0	0	0.7	0.3	0	0
h	0	0	0	0	0	0	0	0.9	0.1	0
i	0	0	0	0	0	0	0	0	0.9	0.1
j	0	0	0	0	0	0	0	0	0	1

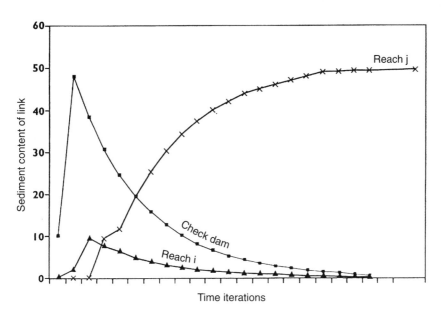

Figure 5.26 Schematic diagram of stream network (top) together with a possible transition matrix (middle). The graph shows a set of time realisations for a check dam, reach below a check dam (reach i) and reservoir (called reach j), which is an absorbing state

unit from reach x to reach $x + 1$, in a downstream direction. Thornes (in press) has used this concept to conceptualise the movement of sediments through a stream network subject to check dams to argue for the optimum location of check dams in the system, as shown in Figure 5.26. Each change of state (reach) is a step transition, so the model is a discrete-time Markov process, in which we have to assume that the transitions occur at fixed times. In practice, the transition times should themselves be spaced according to some predetermined distribution giving rise to the continuous-time Markov chain process. The problem has been solved analytically by Rodriguez-Iturbe and Valdes (1979) and applied by Conesa-García and Alonso-Sarría (1997). They relate the Instantaneous Unit Hydrograph (IUH) to the Hortonian characteristics of the network to produce the Geomorphological Unit Hydrograph (GUH). In a set of papers Rodriguez-Iturbe has developed the theory to demonstrate how the Hortonian laws influence the GUH and how the IUH can thereby be predicted. The GUH can then be coupled to the rainfall amounts and intensities to produce the distributional characteristics of families of GUHs in different rainfall regimes. Finally the theory is used to derive the frequency distribution of peak flows (Rodriguez-Iturbe, 1994).

In our use of the Markov transition matrix for sediment transfer, we follow the usual procedure of pre-multiplying the transition matrix (defined empirically on the basis of channel characteristics – flow and slope) with the vector of initial state occupancies (as in Figure 5.26) to yield the state occupancies (reach-sediment concentrations) at any step or after the commencement of the simulation. In this way, sediment waves introduced into first-order channels can be 'tracked' through the system, rather as the flood of vehicles leaving a set of seaside resorts at the end of a sunny afternoon can be used to predict the evening rush hour into a nearby major city.

5.8 CONCLUSIONS

Generally, there are great advantages to modelling the short-term dynamics of ephemeral channels, and certain modelling approaches outlined in this chapter can begin to overcome some of the problems associated with the lack of data in this area. However, there is still a lack of understanding of the rainfall–runoff processes in semi-arid environments and it is only through observation of these events, and the geomorphological change associated with them, that a better understanding of the processes will be obtained. The evaluation of paleofloods combined with modelling techniques is certainly a way forward in semi-arid rainfall–runoff modelling (Bull et al., 2000; Martínez-Goytre et al., 1994; Pilgrim *et al.*, 1988).

REFERENCES

Abbott, M.B. 1966. *An Introduction to the Methods of Characteristics*. Thames & Hudson, London, 242pp.

Abbott, M.B. and Ionesco, F. 1967. On the numerical computation of nearly horizontal flows. *Journal of Hydraulics Research*, 5, 97–117.

Anderson, M.G. and Bates, P.D. 1994. Initial testing of a two-dimensional finite element model for floodplain inundation. *Proceedings of the Royal Society of London Series A*, 444, 149–159.

Andrews, E.D. 1980. Effective and bankfull discharge of streams in the Yampa Basin, Western Wyoming. *Journal of Hydrology*, 46, 310–330.

Babcock, H.M. and Cushing, E.M. 1941. Recharge to ground-water from floods in a typical desert wash, Pinal County, Arizona. *EoS Transactions of the American Geophysical Union*, 23, 49–56.

Bagnold, R.A. 1980. An empirical correlation of bed load transport rates in flumes and natural rivers. *Proceedings of the Royal Society of London Series A*, 372, 453–473.

Baird, L., Gee, D.M. and Anderson, M.G. 1992. Ungauged catchment modelling II. Utilisation of hydraulic models for validation. *Catena*, 19, 33–42.

Bates, P.D. and Anderson, M.G. 1993. A two-dimensional finite element model for river flow inundation. *Proceedings of the Royal Society of London Series A*, 440, 481–491.

Bates, P.D., Anderson, M.G., Baird, L., Walling, D.E. and Simm, D. 1992. Modelling floodplain flows using a two-dimensional finite element model. *Earth Surface Processes and Landforms*, 17, 575–588.

Bates, P.D., Anderson, M.G., Hervouet, J.-M. and Hawkes, J.C. 1997. Investigating the behaviour of two-dimensional finite element models of compound channel flow. *Earth Surface Processes and Landforms*, 22, 3–17.

Ben-Zvi, A. Massoth, S. and Schick, A. P. 1991. Travel time of runoff crests in Israel. *Journal of Hydrology*, 122, 309–320.

Brandt, J. and Thornes, J.B. (Eds) 1996. *Mediterranean Desertification and Land Use*. John Wiley, Chichester; 554pp.

Bull, L.J., Kirkby, M.J., Shannon, J. and Hooke J.M. 2000. The impact of rainstorms on floods in ephemeral channels in southeast Spain. *Catena*, 38(3), 191–209.

Buono, A. and Lang, D.J. 1980. Aquifer recharge from the 1969 and 1978 floods in the Mojave River Basin, California. *United States Geological Survey Water Resources Investigations, Open File Report*, 80–207: 1–25.

Burkham, D.E. 1976. Flow from small watersheds adjacent to the study reach of the Gila River Phreatophyte Project, Arizona. *United States Geological Survey Professional Paper*, 655–I, 1–19.

Burkham, D.E. 1970. Depletion of streamflow by infiltration in the main channels of the Tucson Basin, Southeastern Arizona. *United States Geological Survey Water Supply Paper*, 1939-B, 1-36.

Butcher, G.C. and Thornes, J.B. 1978. Spatial variability in runoff processes in an ephemeral channel. *Zeitschrift für Geomorphologie Supplement Band*, 29, 83–92.

Conesa-García, C. 1995. Torrential flow frequency and morphological adjustments of ephemeral channels in South-East Spain. In E.J. Hickin, (Ed.) *River Geomorphology*. John Wiley & Sons Ltd.

Conesa-García, C. and Alonso-Sarría, F. 1997. Stochastic matrices applied to the probabilistic analysis of runoff events in a semi-arid stream. *Hydrological Processes*, 11, 297–310.

Cox, D.R. and Miller, H.D. 1965. *The Theory of Stochastic Processes*. Methuen & Co. Ltd, London.

Cunge, J.A., Holly, F.M. and Verwey, A. 1980. *Practical Aspects of Computational River Hydraulics*. Pitman, London, 420pp.

Diskin, M.H. and Lane, L.J. 1972. A basinwide stochastic model for ephemeral stream runoff, in Southeastern Arizona. *Hydrological Sciences Bulletin*, 17(1), 61–76.

Drissel, J.C. and Osborn, H.B. 1968. Variability in rainfall producing runoff from a semiarid rangeland watershed, Alamogordo Creek, New Mexico. *Journal of Hydrology*, 6, 194–201.

Dunkerley, D. and Brown, K. 1999. Flow behaviour, suspended sediment transport and transmission losses in a small (sub-bank-full) flow event in an Australian desert stream. *Hydrological Processes*, 13, 1577–1588.

El-Hames, A.S. and Richards, K.S. 1994. Progress in arid-lands rainfall-runoff modelling. *Progress in Physical Geography*, 18(3), 343–365.

Faulkner, H. 1992. Simulation of summer storms of differing recurrence intervals in a semiarid environment using a kinematic routing scheme. *Hydrological Processes*, 6, 397–416.

Feldhaus, R., Huttges, J., Brockhau, T. and Rouve, G. 1992. Finite element simulation of flow and pollution transport applied to a part of the River Rhine. In R.A. Falconer, K. Shiono, and R.G.S. Matthews, (Eds) *Hydraulic and Environmental Modelling; Estuarine and River Waters*. Ashgate Publishing, Aldershot, 323–334.

Fread, D.L. 1985. Channel routing. In M.G. Anderson and T.P. Burt (Eds) *Hydrological Forecasting*. John Wiley & Sons, Chichester, 437–503.

Freyberg, D.L. 1983. Modelling the effects of a time-dependent wetted perimeter on infiltration from ephemeral channels. *Water Resources Research*, 19, 559–566.

Gee, D.M., Anderson, M.G. and Baird, L. 1990. Large scale floodplain modelling, *Earth Surface Processes and Landforms*, 15, 512–523.

Graf, W.L. 1983a. Flood-related channel change in an arid-region river. *Earth Surface Processes and Landforms*, 8, 125–139.

Graf, W.L. 1983b. Variability of sediment removal in a semiarid watershed. *Water Resources Research,* 19(3), 643–652.

Graf, W.L. 1988. *Fluvial Processes in Dryland Rivers.* Springer-Verlag.

Greenbaum, N., Margalit, A., Schick, A. P., Sharon, D. and Baker, V.R. 1998. A high magnitude storm and flood in a hyperarid catchment, Nahel Zin, Negev Desert, Israel, *Hydrological Processes,* 12, 1–23.

Gutiérrez, F., Gutiérrez, M. and Sancho, C. 1998. Geomorphological and sedimentological analysis of a catastrophic flash flood in the Aras drainage basin (Central Pyrenees, Spain). *Geomorphology,* 22, 265–283.

Harr, M.E. 1925. *Mechanics of Particulate Media.* McGraw-Hill.

Harvey, A.M. 1984. Geomorphological response to an extreme flood: a case from Southeast Spain. *Earth Surface Processes and Landforms,* 9, 267–279.

Hassan, M. A. 1990. Observations of desert flood bores. *Earth Surface Processes and Landforms,* 155, 481–485.

Hassan, M.A. and Reid, I. 1990. The influence of microform bed roughness elements on flow and sediment transport in gravel bed rivers. *Earth Surface Processes and Landforms,* 15, 739–750.

Hberman, R. 1977. *Mathematical Models, Mechanical Vibrations, Population Dynamics and Traffic Flow (An Introduction to Applied Mathematics).* Prentice Hall, Englewood Cliffs, New Jersey.

Hodskinson, A. 1996. Computational fluid dynamics as a tool for investigating separated flow in river bends, *Earth Surface Processes and Landforms,* 21, 993–1000.

Hughes, D.A. and Sami, K. 1992. Transmission losses to alluvium and associated moisture dynamics in a semi-arid ephemeral channel system in Southern Africa. *Hydrological Processes,* 6, 45–53.

Jordan, P.R. 1977. Streamflow transmission losses in western Kansas. *Journal of Hydraulic Engineering American Society of Civil Engineers,* 103(HY8), 905–919.

Keppel, R.V. and Renard, K.G. 1962. Transmission losses in ephemeral stream beds. Proceedings of the American Society of Civil. Engineers, *Journal of the Hydraulics Division,* 88(HY-3), 59–68.

King, I.P. and Norton, W.R. 1978. Recent applications of RMAs finite element models for two-dimensional hydrodynamics and water quality, *Proceedings of the Second International Conference on Finite Elements in Water Resources.* Pentech Press, London, 81–99.

King, I.P. and Roig, L.C. 1988. Two-dimensional finite element models for floodplains and tidal flats. In K. Niki and M. Kawahara (Eds) *Proceedings of an International Conference on Computational Methods in Flow Analysis, Okayama, Japan,* 711–718.

Kirkby, M.J., Baird, A.J., McMahon, M.D., Mitchell, P.L., Shao, J., Sheehy, J.E., Thornes, J.B. and Woodward, F.I. 1996. The MEDALUS slope catena model: a physically based process model for hydrology, ecology and land degradation interaction. In C.J. Brandt and J.B. Thornes (Eds) *Mediterranean Desertification and Land Use.* John Wiley & Sons, Chichester, 87–108.

Kisiel, C.C., Duckstein, L. and Fogel, M.M. 1971. Analysis of ephemeral flow in aridlands. *Journal of Hydraulic Engineering American Society of Civil Engineers,* 97(HY10), 1699–1717.

Knighton, A.D. and Nanson, G.C. 1994. Flow transmission along an arid zone anastomosing river, Cooper Creek, Australia. *Hydrological Processes,* 8, 137–154.

Knighton, A.D. and Nanson, G.C. 1997. Distinctiveness, diversity and uniqueness in arid zone river systems. In D.S.G. Thomas (Ed.) *Arid Zone Geomorphology. Process, Form and Change in Drylands.* John Wiley & Sons, Chichester.

Lane, L.J. 1972. A proposed model for flood routing in abstracting ephemeral channels. *Proceedings of the 1972 Meetings of the Arizona Section – American Water Resources Association and the Hydrology Section, Arizona Academy of Sciences. May 5-6, 1972. Prescott, Arizona,* 439–453.

Lane, L.J. 1982a. Distributed model for small semiarid watersheds. *American Society of Civil Engineers Journal of Hydraulic Engineering,* 108(HY10), 1114–1131.

Lane, L.J. 1982b. Development of a procedure to estimate runoff and sediment transport in ephemeral streams. *Recent Developments in the Explanation and Predictions of Erosion and Sediment Yield, Proceedings of the Exeter Symposium,* July 1982. *IAHS Publication* 137, 275–282.

Lane, L.J. 1985. Estimating transmission losses. *Proceedings, Development and Management Aspects of Irrigation and Drainage Systems, Irrigation Division, American Society of Civil Engineers, San Antonio, Texas,* 106–113.

Lane, S. N. 1998. Hydraulic modelling in hydrology and geomorphology: a review of high resolution approaches. *Hydrological Processes*, 12, 1131–1150.

Lane, S.N., Chandler, J.H. and Richards, K.S. 1994a. Developments in monitoring and modelling small-scale river bed topography. *Earth Surface Processes and Landforms*, 19, 349–368.

Lane, L.J., Diskin, M.H. and Renard, K.G. 1971. Input–output relationship for an abstracting ephemeral stream channel system. *Journal of Hydrology*, 13, 22–40.

Lane, L.J., Nichols, M.H., Hernandez, M. and Osterkamp, W.R. 1994b. Variability in discharge, stream power, and particle-size distributions in ephemeral-stream channel systems. In *Variability in Stream Erosion and Sediment Transport, Proceedings of the Canberra Symposium, December 1994. IAHS Publication No. 224*, 335–342.

Lane, L.J. and Renard, K.G. 1972. Evaluation of a basin-wide stochastic model for ephemeral runoff from semiarid watersheds. American Society of Agricultural Engineers. In K.S. Richards (Ed.) *River Channels, Environment and Process*. Basil Blackwell, 15(2), 280–283.

Langbein, W. and Leopold L.B. 1968. Riverchannels bars and dunes – the theory of kinematic waves. *United States Geological Survey Professional Paper*, 122L, 11–20.

Laronne, J.B. and Reid, I. 1993. Very high rates of bedload sediment transport by ephemeral desert rivers. *Nature*, 366, 148–150.

Laronne, J.B., Reid, I., Yitshak, Y. and Frostick, L.E. 1992. Recording bedload discharge in a semiarid channel, Nahel Yatir, Israel. In J. Bogen, D.E. Walling and T.J. Day (Eds) *Erosion and Sediment Monitoring Programs in River Basins. International Association Hydrological Sciences Publication*, 210, 79–86.

Laronne, J.B., Reid, I., Yitshak, Y. and Frostick, L.E. 1994. The non-layering of gravel streambeds under ephemeral flow regimes. *Journal of Hydrology*, 159, 353–363.

Leopold, L.B., Emmett, W.W. and Myrick, R.M. 1966. Channel and hillslope processes in a semiarid area of New Mexico. *US Geological Survey Professional Paper*, 352G, 193–253.

Leopold L.B., Wolman, M.G. and Miller, J.P. 1964. *Fluvial Processes in Geomorphology.*, Freeman, San Francisco.

Lighthill, M.H. and Whitham, G.B. 1955. On kinematic waves: 1 Flood movement in long rivers. *Proceedings of the Royal Society of London Series A*, 299, 281–136.

Limerinos, J.T. 1970. Determination of the Manning coefficient from measured bed roughness in natural channels. *Geological Survey Water-Supply Paper*, 1898-B. US Government Printing Office, Washington, DC, 20402.

Lynch, D.R. and Gray, W.G. 1980. Finite element simulations of flow deforming regions. *Journal of Computational Physics*, 36, 135–153.

Martinez-Goytre, J. House, P.K. and Baker, V.R. 1994. Spatial variability of small-basin paleoflood magnitudes for a southeastern Arizona mountain range. *Water Resources Research*, 30, 1491–1501.

Meirovich, L., Ben-Zvi, A., Shentsis, I. and Yanovich, E. 1998. Frequency and magnitude of runoff events in the arid Negev of Israel. *Journal of Hydrology*, 207, 204–219.

Miró-Granada, R. and Gelabert, I. 1974. *Les cuves catastrophiques sur le Mediterranee occidentale*. IAHS/AISH Publication 112.

Morley, C.D. and Thornes, J.B. 1972. A Markov Decision Model for network flows *Geographical Analysis*, 4, 180–193.

Murphey, J.B., Lane, L.J. and Diskin, M.H. 1972. Bed Material Characteristics and Transmission Losses in an Ephemeral Stream. Hydrology and Water Resources in Arizona and the southwest. *Proceedings 1972 Meeting Arizona Section, American Water Association and the Hydrology Section, Arizona Academy of Science*. Prescott, Arizona, 2, 455–472.

Murray, A.B. and Paola, B. 1994. A cellular model of braided rivers. *Nature*, 37, 54–57.

Murray, A.B. and Paola, B. 1997. Properties of a cellular braided-stream model. *Earth Surface Processes and Landforms*, 22, 1001–1025.

Naden, P.S. 1987. Modelling gravel-bed topography from sediment transport. *Earth Surface Processes and Landforms*, 12, 353–367.

Nicholas, A.P. and Walling, D.E. 1997 Modelling flood hydraulics and overbank deposition on river floodplains. *Earth Surface Processes and Landforms*, 22, 59–77.

Niemeyer, G. 1979. Efficient simulation of non-linear steady flow. *Journal of Hydraulics Division American Society of Civil Engineers*, 105, 185–196.

Norton, W.R., King, I.P. and Orlob, G.T. 1973. *A finite element model for lower granite reservoir*. A report prepared for the U.S. Army Corps of Engineers, Walla Walla District, Washington, Water Resources Engineers, Walnut Creek, California, 105pp.

Nye, J.F. 1960. The response of a glacier to changes in the rate of nourishment and wastage. *Proceedings of the Royal Society of London Series A*, 275, 87–112.

Osborn, H.B. and Lane, L.J. 1969. Precipitation runoff relations for very small semiarid rangeland watersheds. *Water Resources Research*, 5, 419–425.

Parissopoulos, G.A. and Wheater, H.S. 1995. Simulation of groundwater recharge from flash floods – the case of Wadi Tabalah. In N.X. Tsiourtis (Ed.) *Water Resources Management Under Drought or Water Shortage Conditions*. Balkema, Rotterdam, 169–177.

Peebles, W.R., Smith, R.E. and Yakowitz, S.J. 1981. A leaky reservoir model for ephemeral channel flow recession. *Water Resources Research*, 173, 628–636.

Pickup, G., Higgins, R.J. and Grant, I. 1983. Modelling sediment transport as a moving wave. The transfer and deposition of mining waste. *Journal of Hydrology*, 60, 282–301.

Pilgrim, D.H., Chapman, T.G. and Doran, D.G. 1988. Problems of rainfall–runoff modelling in arid and semiarid regions. *Hydrological Sciences Journal*, 33(4), 379–400.

Piñol, J., Beven, K. and Freer, J. 1997. Modelling the hydrological response of Mediterranean catchments, Prades, Catalonia. The use of distributed models as aids to hypothesis formulation. *Hydrological Processes*, 11, 1287–1306.

Powell, D.M., Reid, I., Laronne, J.B. and Frostick, L.E. 1996. Bed load as a component of sediment yield from a semiarid water shed of the northern Negev. Erosion and sediment yield. *Global and Regional Perspectives (Proceedings of the Exeter Symposium, July 1996)*. IAHS Publication, 236, 389–397.

Preissmann, A. 1961. Propagation des intumescences dans les canaux et rivieres, *1er Congres de l'Association Francaise de Calcul, Grenoble*, 433–442.

Reece, G. 1986. *Microcomputer Modelling by Finite Differences*. London: Macmillan Education Ltd, 126pp.

Reid, I. and Frostick, L.E. 1987. Flow dynamics and suspended sediment properties in arid zone flash floods. *Hydrological Processes*, 1, 239–253.

Reid, I. and Frostick, L.E. 1997. Channel form, flows and sediments in deserts. In D.S.G. Thomas (Ed.) *Arid Zone Geomorphology*, 2nd Edn. John Wiley & Sons, 205–231.

Reid, I. and Laronne, J.B. 1995. Bed load sediment transport in an ephemeral stream and a comparison with seasonal and perennial counterparts. *Water Resources Research*, 31(3), 773–781.

Reid, I., Laronne, J.B. and Powell, D.M. 1995. The Nahel Yatir bedload database: sediment dynamics in a gravel-bed ephemeral stream. *Earth Surface Processes and Landforms*, 20, 845–857.

Reid, I., Laronne, J.B. and Powell, D.M. 1998. Flash-flood and bedload dynamics of desert gravel-bed streams. *Hydrological Processes*, 12, 543–557.

Reid, I., Powell, D.M. and Laronne, J.B. 1996. Prediction of bed-load transport by desert flash floods. *Journal of Hydraulic Engineering*, 122(3), 170–173.

Reid, I., Powell, D.M., Laronne, J.B. and Garcia, C. 1994. Flash floods in desert rivers: studying the unexpected. *Eos, Transactions American Geophysical Union*, 75(39), 452–453.

Renard, K.G. 1970. The Hydrology of Semiarid Rangeland Watersheds. *United States Department of Agriculture, Agricultural Research Service*, ARS 41–162, 1–26.

Renard, K.G. and Keppel, R.V. 1966. Hydrographs of ephemeral streams in the southwest. *ASCE Journal of Hydraulic Engineering*, 92(HY2), 33–52.

Renard, K.G. and Lane, L.J. 1975. Sediment yield as related to a stochastic model of ephemeral runoff, present and prospective technology for predicting sediment yields and source. *Proceedings of the Sediment Yield Workshop USDA Sediment Laboratory (Sedimentation Lab), Oxford*, MS, 1972, 253–263.

Renard, K.G. and Laursen, E.M. 1975. Dynamic behaviour of ephemeral streams. *American Society of Civil Engineers Journal of Hydraulic Engineering*, 101(HY5), 511–529.

Richards, K.S. 1988. Fluvial geomorphology. *Progress in Physical Geography*, 12, 435–456.

Rodriguez-Iturbe, I. 1994. The geomorphological unit hydrograph. In K. Bevan and M.J. Kirkby (Eds) *Channel Network Hydrology*. John Wiley & Sons, Chichester, 43–69.

Rodriguez-Iturbe, I. and Valdes, J. 1979. The geomorphic structure of hydrologic response. *Water Resources Research*, 15(6), 1409–1420.

Rossi, F. and Siccardi, F. 1990. Coping with floods: the research policy of the Italian group for prevention from hydro-geological disasters. In F. Siccardi and R.L. Bras (Eds) *Natural Disasters in European Mediterranean Countries*, US Natural Science Foundation, Italian Natural Research Council, 395–414.

Sammuels, P.G. 1983. Computational modelling of flood flows in embanked rivers, *Proceedings of an International conference on the Hydraulic Aspects of Floods and Flood Control, BHRA*, 229–240.

Sammuels, P.G. 1990. Cross section location in one-dimensional models. In W.R. White (Ed.), *International Conference on River Flood Hydraulics*. John Wiley & Sons, Chichester, 339–350.

Schick, A.P. 1977. A tentative sediment budget for an extremely arid watershed in the Southern Negev. In D.O. Doehring (Ed.) *Geomorphology in Arid Regions*. A Proceedings Volume of the Eighth Annual Geomorphology Symposium held at the State University of New York at Binghampton, 23–24 September 1977, 139–163.

Schumm, S.A. 1961. The effect of sediment characteristics on erosion and deposition in ephemeral stream channels. *USGS Professional Paper*, 352C, 31–69.

Sharma, K.D. and Murthy, J.S.R. 1994a. Estimating transmission losses in arid regions. *Journal of Arid Environments*, 26(3), 209–219.

Sharma, K.D. and Murthy, J.S.R. 1994b. Estimating transmission losses in an arid region – a realistic approach. *Journal of Arid Environments*, 26(2), 107–112.

Sharma, K.D. and Murthy, J.S. 1995. Hydrological routing of flow in arid ephemeral channels. *Journal of Hydraulic Engineering*, 121(6), 466–471.

Sharma, K.D. and Murthy, J.S.R. 1996a. A conceptual sediment transport model for arid regions, *Journal of Arid Environments*, 33, 281–290.

Sharma, K.D. and Murthy, J.S.R. 1996b. Ephemeral flow modelling in arid regions. *Journal of Arid Environments*, 33, 161–178.

Sharma, K.D. and Murthy, J.S.R. 1998. A practical approach to rainfall-runoff modelling in arid zone drainage basins. *Hydrological Sciences Journal*, 43(3), 331–348.

Sharma, K.D., Vangani, N.S., Menenti, M., Huygen, J. and Vich, A. 1994. Spatiotemporal variability of sediment transport in arid regions. In *Variability in Stream Erosion and Sediment Transport. Proceedings of the Canberra Symposium, December 1994*. IAHS Publication 224, 251–258.

Sharon, D. 1972. The spottiness of rainfall in a desert area. *Journal of Hydrology*, 17, 161–175.

Singh, V.P. 1997. Effect of spatial and temporal variability in rainfall and watershed characteristics on stream flow hydrograph. *Hydrological Processes*, 11(12), 1649–1669.

Smith, R.E. 1972. Border irrigation advance and ephemeral flood waves. *Journal Irrigation and Drainage Division American Society of Civil Engineers*, Vol. 98, No. IR2, pp. 289–307.

Sorman, A.U., Abdulrazzak, M.J. and Morel-Seytoux, H.J. 1997. Groundwater recharge estimation from ephemeral streams. Case study: Wadi Tabalah, Saudi Arabia. *Hydrological Processes*, 11(12), 1607–1619.

Srikathan, R. and McMahon, T.A. 1980. Stochastic generation of monthly flows for ephemeral streams. *Journal of Hydrology*, 47, 19–40.

Tarboton, D.G. 1997. A new method for the determination of flow directions and upslope areas in grid digital elevation models. *Water Resources Research*, 33(2), 309–319.

Thornes, J.B. 1971. State, environment and attribute in scree slope studies. In D. Brunsden (comp.) *Slopes, Form and Process. Institute of British Geographers Special Publication*, 3, 49–63.

Thornes, J.B. 1977. Channel changes in ephemeral streams: observations, problems and models. In K.G. Gregory, (Ed.) *River Channel Changes*. John Wiley, Chichester.

Thornes, J.B. 1979. Fluvial processes. In C.E. Embleton and J.B. Thornes (Eds) *Processes in Geomorphology*. Allen & Unwin, London, 213–271.

Thornes, J.B. 1980. Structural instability and ephemeral channel behaviour. *Zeitschrift für Geomorphologie Supplement Band*, 36, 233–244.

Thornes, J.B. (in press) Applied geomorphology. *Proceedings of the VI National Reunion of the Spanish Geomorphological Society, Madrid, 19–20 Sept 2000*.

Thornes J.B. and Brunsden, D. 1967. *Geomorphology and Time*. Methuen, London.

Thornes, J.B. 1976. *Semi-arid Erosional System: Case Studies from Spain*. London School of Economics Geographical Paper, 7, 207pp.

Walters, M.O. 1990. Transmission losses in arid regions. *Journal of Hydraulic Engineering*, 116(1), 129–139.

White, S., García-Ruíz, J.M. and Gomez-Villar, A. 1997. The 1996 Biescas campsite disaster in the Central Spanish Pyrenees, and its temporal and spatial context. *Hydrological Processes*, 11(14), 1797–1812.

Wooding, R.A. 1965. A hydraulic model for the catchment-stream problem. I Kinematic wave theory. *Journal of Hydrology*, 3, 254–267.

Ye, W., Bates, B.C., Viney, N.R., Sivapalan, M. and Jakeman, A.J. 1997. Performance of conceptual rainfall-runoff models in low-yielding ephemeral catchments. *Water Resources Research*, 33(1), 153–166.

Zhu, T.X., Band, L.E. and Vertessy, R.A. 1999. Continuous modelling of intermittent stormflows on a semi-arid agricultural catchment. *Journal of Hydrology*, 226, 11–29.

6 Morpho-dynamics of Ephemeral Streams

JANET HOOKE AND JENNY MANT

Department of Geography, University of Portsmouth, UK

6.1 INTRODUCTION

This chapter is concerned with the morphology of ephemeral streams and the dynamics of channel form. It will examine the range of characteristics, the controls on form, the causes of change in form and the processes and time-scales of these changes. It focuses in particular on the role of single floods and sequences of events over time-scales of a few years to a few decades. This viewpoint provides a link between the short-term processes of Chapters 4 and 5 and the longer-term perspective of Chapter 3. It concentrates upon the main river channels flowing in valley floors, thus occupying a spatially intermediate position between gullies (Chapters 8 and 9) and alluvial fans (Chapter 7).

Flows occur in these channels when and where sufficient runoff is generated on contributing slopes and in tributary subcatchments. In turn, the channels become major zones of sediment transport. This sediment is supplied directly from hillslopes, by mass movements or by fluvial flows, or from channel erosion, often remobilizing previously deposited alluvium. The valley floors are also zones of major sediment storage, the location and extent of which alters on both short and longer time-scales. The valley floors generally contain channels, floodplain and terraces. The high relief of much of the Mediterranean means that floodplains and terraces are often of limited area but they are economically important as areas of agricultural activity. However, these zones are highly hazardous due to the unpredictability of flow events, although, conversely, the water supplied via the sudden and short-lived flows is essential for the direct supply to agricultural usage and for the replenishment of groundwater. Traditionally, much of the flow was utilised via various water-harvesting, storage and diversion schemes, although, since the introduction of alternative irrigation methods, these have fallen into disrepair.

Geomorphologists have traditionally focused upon process–form relations, and the driving forces of runoff and sediment production, but it is important – particularly within ephemeral channels where the legacy of a past flow may persist for many years – that the influence of morphology upon process is examined and that feedbacks are considered. A key issue in relation to the understanding of the impacts of events and the nature of changes in the channels of ephemeral streams is to determine how systematic are the changes and how far relationships can be detected. It has long been accepted that ephemeral channels tend to be in a non-equilibrium state and therefore unstable, with some propensity to sudden switches of characteristics, commonly termed 'metamorphosis' (Schumm and Lichty, 1963; Burkham, 1972). The

Dryland Rivers: Hydrology and Geomorphology of Semi-arid Channels. Edited by L.J. Bull and M.J. Kirkby.
© 2002 John Wiley & Sons, Ltd.

extent or lack of predictability of behaviour, change and characteristics has important implications for channel management.

A major problem in examining the issues outlined above is the dearth of adequate data and information relating to changes in ephemeral channels, and the geomorphological impacts of flood events. Given the sporadic nature of flood events in semi-arid environments, the chances are low that a significant event will occur in a monitored section within the reasonable lifetime of a research project. Various observations of individual floods have been made, but the number of publications giving details of morphological impacts or of changes are still very few. Even less information is available on the cumulative effects of events and on changes over decadal time-scales. World wide, the most abundant information available is from North America. Some interesting studies have been undertaken in Australia but mostly on systems with very different characteristics from those of the Mediterranean, and a few studies are available from southern Africa and South America. Even within the Mediterranean the distribution of studies is patchy, with most having been undertaken in Spain, southern France, Italy and Israel. A few case studies do exist for parts of the southern margin of the Mediterranean basin in north Africa, but heavy reliance is placed in this discussion on the experience from the western Mediterranean.

6.2 TYPES AND VARIABILITY OF CHANNEL MORPHOLOGY, SUBSTRATE AND VEGETATIVE COVER

Perennially, seasonally and ephemerally flowing channels have all been shown to be present within relatively small geographical areas of the Mediterranean region, and the Segura catchment in the Murcia province, southeast Spain, is one such example (Figure 6.1), despite having a relatively low average rainfall of c. 350 mm a^{-1} (Vidal-Abarca et al., 1992). In general, perennially flowing channels in this region originate beyond the Mediterranean margin, in higher and wetter areas. Conversely, ephemeral streams are mostly either small, short systems, including tributaries, or streams in the drier parts of the Mediterranean region where average rainfall is <400 mm a^{-1} or on particular lithologies which produce little continuous runoff.

Macklin et al. (1995, p. 18) stated that by 'taking the 500 m contour as the mountain-lowland boundary, it is clear that most of the Mediterranean basin is drained by steepland river systems'. Although there are a few large river systems with extensive lowland such as the Ebro, Po, Nile and Rhone, most streams conform to Macklin et al.'s assertion. The systems are thus characterised by small, steep upper catchments and a relatively long transfer zone (in Schumm's (1977) 3-zone system). These are often confined, with short distal zones, and end in fans which may or may not connect with another system or the coastal margin. A few of the larger subsystems emerge from the confines of the uplands into basins or the coastal margin and develop into broad, unconfined channels. Thus a major control on the fluvial morphology is the topography and tectonics influencing the degree of confinement and incision or entrenchment. A major division may therefore be made into confined and unconfined channels (see Figure 6.2(a) and (b).

A second major influence on channel characteristics is that of lithology and sediment supply. Lithology obviously influences the runoff regime, but it also influences the calibre and amount of sediment load supplied. In terms of the type of channel morphology which develops, through the direct influence of the sediment and indirectly through its feedback effects on vegetation, a major division may be made between the gravel-bed, cohesionless sedimentary channels and those with a supply of fine, cohesive material (Figure 6.2(c) and (d)). The former occur where

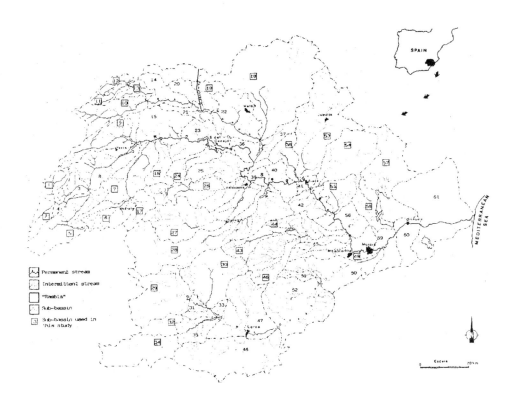

Figure 6.1   Channel network of the Segura catchment, Murcia province (Vidal-Abarca et al., 1992)

coarser material is supplied and have a tendency to form wide, braided and highly unstable channels, whereas those with cohesive material tend to be narrower, compound in form and rather more stable because of the higher resistance of the material. Thus, geomorphologically, the former occur in the hard rock areas of schist, limestone and where gravels predominate. The latter occur in marl, and other soft rock areas. Channels may also vary in type along their course, particularly as confinement and source areas vary. Thus in parts of the region, canyon-like reaches alternate with much wider sections, particularly where the fluvial systems cross the structure or grain of relief as is so common in the Mediterranean.

The characteristic components of the channel morphological system can each be examined. Given the above division of major contrasting types of system or reach, it is obviously difficult to generalise, but certain characteristics are frequently cited in the literature. Relative to perennial channels in more humid areas, ephemeral channels are generally rather wide and shallow. Channel morphology as such is often poorly defined, with low and indistinct channel banks. In many systems the channel widens very rapidly at the upstream end where hillslope tributaries reach the valley floor, or begins abruptly as a gully. Direct observations indicate that width may then increase more slowly downstream than for perennial systems and often jerkily, notably in relation to tributaries (Figure 6.3). Wolman and Gerson (1978) and Reid and

Figure 6.2 Photos of (a) unconfined, (b) confined

Figure 6.2   (c) Gravel-bed and (d) marl channels

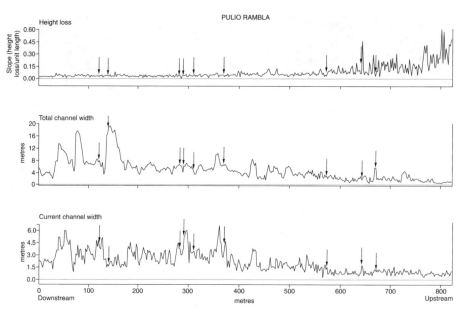

Figure 6.3   Width and slope variations downstream in Pulio Rambla (Thornes, 1976)

Frostick (1992) identify an asymptotic relationship of width to drainage area with a limit of c. 100 m reached at about 50 km^2 but relatively rapid increases at the upstream end. The lack of increase in width lower down the catchment is mainly due to the discontinuous nature of flow and high transmission losses in events. Analyses of hydraulic geometry have shown that at a cross-section ephemeral channels tend to have much greater increase in width with stage than perennial channels (Leopold et al., 1966). This low overall trend in the basin is often super-imposed on a very high variability of width, at one scale related to the structural controls on confinement but at a finer scale, the 'beaded' or autocorrelated type of form (Figure 6.3) described by Thornes (1976, 1980) and Alexander (1980). However, the extent to which this phenomenon is generally applicable is difficult to generalise due to the lack of systematic surveys and analyses of ephemeral channel morphology.

In terms of slope profiles and characteristic gradients, the predominance of steepland implies relatively high gradients. At the basin scale, a notable feature of the longitudinal profiles is the rapid shallowing of gradient near the origin then the lack of variability downstream, though with major discontinuities being related to tributaries, structural controls, headcuts or artificial controls such as check dams. One of the most comprehensive and systematic analyses of basin longitudinal profiles is that by Navarro Hervas (1991) for the Guadalentín basin. She confirms the uniform gradients for the main part of the course and the lack of concavity of the profiles, as commonly reported for semi-arid areas (Leopold et al., 1964; Thornes, 1980) except where there are major structural controls. The profiles of the Ramblas Torrealvilla and Nogalte, which have both been studied intensively in the MEDALUS programme and much quoted in this book, are shown in Figure 6.4. Navarro Hervas (1991) calculated the mean slope of the main channels of these two streams as 0.0246% and 0.028% respectively, in spite of their different topographic and lithological settings and total catchment size. However, within reaches there are very great differences in types of channel in variability of relief. Some gravel-bed channels are remarkably smooth in their profiles with little differentiation in form or sediment downstream. Other

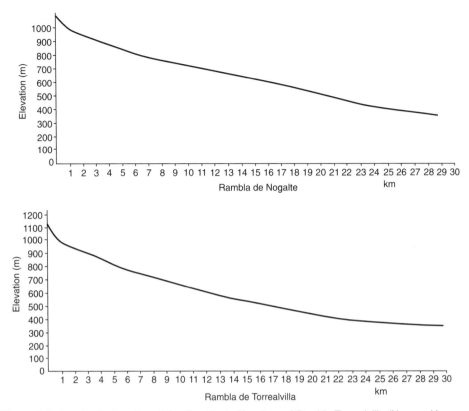

Figure 6.4 Longitudinal profiles of the Rambla de Nogalte and Rambla Torrealvilla (Navarro Hervas, 1991)

channels exhibit large scour holes succeeded by depositional lobes. Yet others exhibit step-pool topography in their upper reaches and a pool-riffle type relief in their middle courses.

Various classifications of channel patterns have been suggested in the literature. Examples of straight, meandering and braided patterns all occur on ephemeral streams in the Mediterranean. However, arguably few of the 'meandering' channels are true alluvial meanders, as many of them are confined and irregular in pattern and have low sinuosity.

As indicated above, a major influence upon channel characteristics and upon channel behaviour is that of sediment supply. Various types of sources have been identified in semi-arid systems but much debate surrounds the relative contribution of these sources and it is generally agreed that much more research is needed (Poesen and Hooke, 1997). The major types of sources are: mass movements, mainly landslides and debris flows; gullies; floodplain sources; and channel erosion of alluvium and colluvium. There is increasing evidence that it is the major linearities, notably gullies and tributary channels which contribute a high proportion of the sediment from the slopes (Poesen et al., 1996; Conesa-García, 1990; Llorens and Gallart, 1992). There is also much evidence that even where slope erosion is high, delivery of the sediment to the stream system may be low, with a lack of connectivity between the slope and channel (Parsons and Wainwright, 1995). The degree of connectivity is a crucial factor and itself may vary with event size or land use practices. A very significant factor influencing the dynamics and form of the channel is the timing of delivery of sediment load. This influences the sedigraph of the event and thus the effects of the event. For example, if a major sediment wave or input arrives after the peak discharge, then scouring is likely to take place initially, whereas if the peak sediment load is early in the event then there is unlikely to be excess capacity in the flow for erosion. Furthermore, if the flow in tributaries or gullies peaks after the main channel, or at times when no flow occurs in the main channel, then tributary fans and deposits may be formed into or across the main channel. In more arid areas, Schick and Lekach (1987) showed how such fans can even cause a dam in the main channel.

Vegetation species and their zonation along the valley floors of these ephemeral ramblas is highly variable, as exemplified by Table 6.1, which shows a summary of one set of data collected from a series of quadrats along the Rambla Torrealvilla that were positioned to take account of spatial variability in terms of morphology within each reach. Highlighted is the variation within these quadrats both in terms of coverage and species which are dependent upon their position within a reach, within a rambla and between ramblas. These differences are driven by a range of environmental factors. At the scale of individual ramblas, the bedrock can be seen as an important factor. For example, *Retama sphaerocarpa, Juncus bulbosus,* and *Artemisia barberia* were consistently recorded in the Rambla de Nogalte on a schist bedrock, while in the marl-dominated Rambla Salada, *Inula viscosa, Auriculae ursifolium* and *Thymelaea hirsuta* were

Table 6.1   Example of vegetation quadrat data collection for one site in the Rambla Torrealvilla

Rambla Torrealvilla (marl and gravel bedrock)		
Site details i. approx % vegetation cover ii. sediment type iii. mean size of coarser fraction (cm)	**Position of quadrat within channel**	**Dominant vegetation types within quadrats**
Aqueduct site (mid reach) *Quadrat 1* i. 30 ii. cobble/sand iii. 10.25	Piedmont	*Tamarix parviflora* *Artemesia barberia* *Lygeum spartum*
Quadrat 2 i. 40 ii. gravel/sand iii. 1.68	Channel – centre	*Thymelaea hirsuta* *Inula crithmoides* *Lygeum spartum* *Moricandia arvensis*
Quadrat 3 i. 15 ii. gravel/sand iii. 3.78	Bar top	*Artemesia barberia* *Inula crithmoides* *Moricandia arvensis* *Lygeum spartum* *Spergularia* (sp.)
Quadrat 4 i. 20 ii. sand iii. <0.2	Channel – centre	*Artemesia barberia* *Moricandia arvensis* *Lygeum spartum*
Quadrat 5 i. 95 ii. sand iii. < 0.2	Channel – side	*Tamarix parviflora* *Artemesia barberia* *Lygeum spartum* *Spergularia* (sp.)

noted as more prevalent. Salada means salty or brackish, and so some adaptation to saline conditions is apparent. However, even within each rambla there are marked zones of vegetation. Some of the more species-rich parts may indicate areas that have been less affected by flood events in recent years, or have been affected by the impact of land use. Conversely, other areas along the channels are dominated by swards of specific types of vegetation. This is especially so within marl catchments which appear more sensitive to smaller flow events. Here, densely vegetated zones comprise primarily either *Tamarix parviflora* and an associated succulent type plant *Arthrocnemum* (sp.), *Lygeum spartum* grass, or *Saccharum ravennae*. Such species appear more resistant to moderate flow events and thus may be good indicators of where such events have occurred, or are likely to occur. However, coverage may also be influenced by groundwater levels.

6.3 CONCEPTUAL MODEL OF CHANGE

In order to understand the characteristics of ephemeral channels, their evolution over time, and their stability or otherwise in time-scales of decades, it is useful to have a conceptual model of the interactions of form and process and how these operate in episodically flowing systems. However, in a holistic approach to the channel system the vegetation must be considered as a key component through its influence on process and, therefore, on morphology through the effects of roughness on flow hydraulics and of resistance on sediment erodibility. Channel morphology may be described through its characteristics and behaviour on the reach scale – that is, for lengths of channel with reasonably homogeneous characteristics; for example, between tributaries or major structural controls or major changes in morphology. At this scale, and over a time-scale of a few years to decades, the major inputs to the system can be regarded as the water and sediment which arrive in discrete and infrequent events of various magnitudes. The characteristics and occurrence of these events are influenced by both climate and catchment characteristics, which are themselves prone to change over time. Between flood events, climate conditions prevail which affect the growth of vegetation and rates of evapotranspiration. The other major external influence upon the channel, both directly and indirectly, is human activity (Figure 6.5). Because the Mediterranean region has a long history of settlement and land use, this influence is profound and continuing. Current direct activities include construction of dams, water diversion and water harvesting, flood protection embankments and structures, waste disposal, and gravel-digging, while major indirect influences include land use changes through crop choices, grazing and practices such as ploughing and irrigation; water abstraction and groundwater depletion; settlement and urbanisation; fire; mining and quarrying.

All of these external influences and driving forces act on the morpho-sedimento-ecological channel system, whose state can be characterised by a range of variables. Key characteristics of the channel are its width and depth, cross-sectional shape, pattern, and gradient and the dimensions of the floodplain or valley floor. These influence the hydraulics of the flow and the distribution of velocity and shear stress as well as the extent and depth of inundation and water flow (Figure 6.6). The size and spatial distribution of the surface sediment influences the erodibility of the surface. The resistance of the surface to erosion is also influenced by vegetation cover, type and density. In addition, the vegetation also affects the hydraulics through its influence upon roughness.

Changes in channel form take place by erosion and deposition produced by spatially varying sediment transport during flow events. Detachment occurs when the force of water flow exceeds the threshold for entrainment of that particular material. If sediment is readily available then it

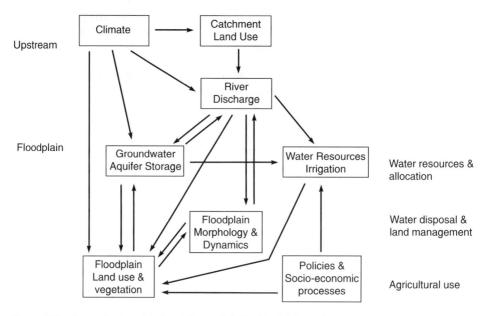

Figure 6.5   Conceptual model of variations within the floodplain system

is assumed that the flow erodes material until the load capacity of the flow is satisfied. Hyperconcentrated flows may even occur, altering the transport capabilities of the flow (Thornes, 1994). Likewise, as the energy and forces of flow are reduced (either spatially by change in channel morphology and roughness downstream or, temporally as flow recedes) then capacity is reduced, and the coarsest material is deposited first and in the zones of lowest shear stress. Because of the rapid rise in stage at the beginning of hydrographs in these channels, there is considerable evidence to suggest that, in alluvial sections, much of the bed becomes mobilised very quickly. Scour tends to take place at these early peak flows but much fill may occur on the recession limb of the hydrograph. The net result in terms of change in channel morphology very much depends on the sediment supply and the relative duration of the water and sediment flows. Not only may morphology be changed by a flow event, but, the distribution and calibre of sediment in a reach may be altered and vegetation may be damaged or destroyed. All of these modifications will then affect the response of the reach to a subsequent flow.

The generally accepted conceptual model of ephemeral channels is that of Wolman and Gerson (1978) in which changes take place during the limited and infrequent flow events and little adjustment takes place in between events. The morphological response is simply to that event and thus affects features for considerable lengths of time. No equilibrium form is identifiable, and instead the morphology reflects the most recent flow or possibly a complex mosaic of the effects of flow of different magnitude (Schick, 1974; Pickup and Reiger, 1976). However, some of the ephemeral channels of the Mediterranean might be considered to occupy an intermediary position between the humid and arid models. This is particularly the case where there is significant vegetation and several flow events take place each year, as experienced in many of these Mediterranean channels (e.g. Conesa-García, 1995). These ephemeral channels also differ from both humid and arid channels in often having significant non-aquatic vegetation within the channels themselves, not just on the banks or floodplains where vegetation cover may even be sparser. The vegetation can therefore significantly affect the processes

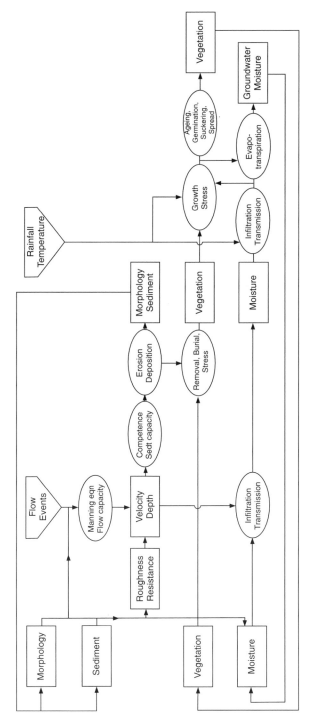

Figure 6.6 Flow chart of process links in the computer simulation model of channel changes

and impacts of a flow event. Adjustment and alteration between high magnitude events may therefore take place. Nevertheless, a model of continuous change rather than equilibrium and morphological stability is deemed more appropriate.

6.4 EVIDENCE AND TECHNIQUES OF ANALYSIS OF CHANGE

Various types of evidence and approaches may be taken to the measurement and assessment of channel changes over the time-scale of years to decades, most of which are widely applied geomorphological techniques and discussed extensively in the literature. Here, the particular issues of applying such techniques to ephemeral channels will be addressed.

Direct monitoring of channel changes can take place using conventional techniques of mapping by ground survey, photogrammetry or other remote-sensing imagery, cross-sectional surveys and instrumentation for flows. One of the major advantages of this type of environment is, of course, that these surveys can be undertaken at times of no flow such that the whole bed topography can be mapped. The major problem is that of spatial sampling and selection of monitoring sites such that events and changes do occur within a reasonable time span. A few sites have been established to measure processes in detail and are providing invaluable information. Far fewer sites are being monitored for their morphological, sedimentary and vegetational changes. Because of the sporadic nature of the flow in the Mediterranean it is advocated that there is great merit in the establishment of numerous monitoring sites at which relatively low technology and low-cost techniques can be applied, and a network of such sites would increase enormously our knowledge of river behaviour and response to events. If such sites could also be maintained at relatively low effort for many years, then a much better picture would be obtained and any patterns or trends in responses to climate change or land use change are more likely to be detected. A hierarchical approach to monitoring is also advocated in order to detect the details of processes and mechanisms of change and the subtle effects of small flows as well as large-scale flood effects.

Table 6.2 sets out a system of monitoring that has been implemented at several sites in the Guadalentín basin in southeast Spain and which, over a period of four years on those channels, has proved to work very well and has provided much valuable data. At the reach scale, topographic mapping is essential to detect three-dimensional changes in form. This may be done by ground survey and GPS is increasingly feasible for rapid acquisition of such data though care still has to be taken, even with the most recent and sophisticated equipment, on the scale and resolution of changes that can actually be measured. Considerable care also has to be taken in the definition of form and aerial coverage. The ability to return to exact points, which the most modern differential GPS systems allow, is enormously advantageous. In the monitoring undertaken from 1996 to 1999 at these sites, the surveying was by total station. This gave very high accuracy of individual points and reasonably high resolution in a number of points with rapid automatic recording of points, but it was found useful also to undertake surveys of cross-sections in order to ensure exact replication. Technology now allows Digital Terrain/Elevation Models to be constructed from these various types of topographic mapping and repeat surveys thus allowing changes to be plotted and analysed in various ways. Thus, amounts and spatial distribution of erosion and deposition are very easily exhibited and statistical analyses of changes can produce depths, areas, and volumes of change for reaches or selected parts. This provides the facility not only to produce detailed quantitative data of extremes, averages and variability of morphological changes but also to calculate sediment fluxes.

Table 6.2   Methods of field measurement

Flow events and rainfall	
Rainfall:	At least one continuous recording rain gauge in each catchment.
Flow:	Crest stage recorder at each site. Contains strip of water-sensitive tape which changes colour on contact with water. Continuous pressure stage recorder on Torrealvilla at aqueduct site.
Morphology	
Method:	Detailed topographic survey using geodimeter (100 m length reach).
Products:	Maps, DEMs, cross-sections.
Derived data:	Morphological zones, floodplain and channel elevations and relief, channel width, depth, slope, pattern.
Resurveys:	Change in morphology, erosion, deposition, volumes of sediment moved.
Vegetation	
Method:	(i) Quadrat surveys (3 m square), 3–5 quadrats at each site; (ii) mapping of valley floors.
Products:	Maps and spreadsheets of cover and species for each quadrat.
Derived data:	% cover, species, average height, state.
Resurveys:	Growth, death, germination, flood impacts.
Sediment	
Method:	(i) Mapping of zones; (ii) quadrats (0.5 m square); (iii) sediment samples.
Products:	Maps, size classifications, surface state.
Derived data:	Average and maximum sizes.
Resurveys:	Change in size, movement of particles, erosion, deposition.

The technique of using quadrats to monitor vegetation, sediment and minor morphological changes has been found invaluable. The composition, state and coverage of vegetation in the plots have been measured at various time intervals and show seasonal changes as well as the effects of flows. The use of sediment quadrats to measure changes in the size of surface sediments has proved very effective, allowing both overall changes such as coarsening or fining to be detected and also the movement of individual particles out of or into the quadrat (see Figure 6.8 below). The spatial relations of vegetation cover, sediment and morphology can also be examined from these quadrats.

Other techniques that have been used in ephemeral channels for direct monitoring include the use of scour chains, famously employed by Leopold et al. (1966) and later by Hassan et al. (1999) among others, and the use of painted pebbles and magnetic tracers (e.g. Schick et al., 1987; Hassan, 1990). Supplementary, rapid appraisal techniques can also prove invaluable. Simple ground photography has been shown to be very effective by Graf (1987) in his repetition of historical photography to demonstrate major changes over a time-scale of decades. More routine photography and systematic archiving of such sources would be very valuable to future research. The insertion of simple instrumentation such as cheap crest-stage recorders can also help to provide wider spatial coverage of flows than is possible with more expensive instrumentation. Again, such techniques are complementary to more intensive instrumentation and measurement.

For more wide-scale monitoring, repeat aerial photography is useful. Some historical aerial photography is available in regions of the Mediterranean but, as digital photography or possibly other remote-sensing systems such as LIDAR become more readily available, systematic programmes of survey, at least after major events, need to be carefully considered. The problem with both historical photography and historical maps is that they only provide snapshots in time. However, given the relatively few large flows that occur in these channels, it may be possible from rainfall records to identify major events and examine relational changes. While this may not allow the detection of the impact of the individual event, it does allow some perspective on the changes. Such an approach is being applied to the analysis of changes in vegetation in the channels of the Guadalentín in southeast Spain using air photographs over the period 1956–1999 (Mant, 1999) and has been used by Conesa-García (1986). Changes in planform of channels have been analysed from historical maps, for example, of the Po (Braga and Gervasoni, 1989), and several of these types of analyses have shown large-scale transformations of pattern, particularly metamorphosis from braided to meandering channels, mainly as a result of river regulation (e.g. Bravard et al., 1997).

Other historical information may also supplement the above approaches. The occurrence and effects of very large events are usually reported in newspapers, which helps to establish occurrence and usually a relative magnitude. However, most newspaper reports understandably focus on the human impact and damage of such events, and the amount of information on any physical impact to be gleaned from such sources is usually meagre. The extent of systematic archiving of such sources is also very varied. In major events affecting the main areas of settlement and infrastructure, official reports may be produced but are often not readily accessible. In major cities, or at specific sites prone to extreme flooding, flood marks may be recorded. Combinations of these various types of historical information have been used successfully to compile records of major events in various places – López-Bermúdez (1973), La Barbera et al. (1992), Arnaud-Fassetta et al. (1993) and López-Áviles et al. (1998). For example, Benito et al. (1996) used a combination of various official and ecclesiastical documents and chronicles together with previously published compilations to produce a synthesis of historic floods in Spain.

The potential for analysis of sedimentary evidence within Mediterranean channels remains largely unexploited, particularly at the historical time-scale, in spite of significant work more generally on ephemeral channel deposits, though much of it on ancient sediments (Frostick and Reid, 1987). Various authors, such as Lewin et al. (1991) and Mather et al. (1995), have demonstrated the enormous value of analysis of terraces and alluvial deposits in the longer-term, and the classic work of Vita-Finzi (1969) still remains a landmark in analysis of river channel changes spanning the historical period. The analysis of slack water deposits has begun with the work of Benito et al. (1998) but merits much wider attention within the Mediterranean region. The use of GPR for detecting recent changes and sedimentary structures, as well as longer term deposits, remains to be more fully explored but may help in detection of mobilisation and scour depths in channels.

Botanical evidence relating to channel change and flood impacts has mainly been the preserve of American literature. This research has concentrated upon riparian vegetation with an emphasis on woody riparian plants such as Tamarisk, Cottonwood and Willows. For example, in terms of channel change, Graf (1978) has noted that Tamarisk becomes increasingly stable, eventually restricting channels. The impact of floods has also been observed, with Horton et al. (1960) and Warner and Turner (1975) observing that, although these types of vegetation are generally recognised as growing above high-water stages, they are capable of surviving periods of inundation for up to six weeks. Everitt (1980) also indicated that Tamarisk could survive inundation and that where deposition occurs during a flood, root layering may occur and vegetation may regrow after such an event. Equally, Simon and Hupp (1987) have also noted similar evidence, adding that exhuming buried stems to the depth of the original germination point, can provide an estimate of sediment-accretion rates. Conversely, if low-water conditions prevail, Tamarisk seedlings will mature, creating stabilisation of the channel and, ultimately, new channel morphology. The extent of flood impacts is also discussed by

Stromberg (1997) who noted that where extensive scour occurs it may prevent herbaceous species from colonising very rapidly. This in turn enables colonisation by woody riparian vegetation to occur. While this type of botanical evidence is to a limited extent available for parts of the USA, similar observations within the Mediterranean ramblas are still being analysed.

Archaeological evidence has, of course, been of widespread use in the Mediterranean region given its long history of settlement (e.g. Vita-Finzi, 1969) and the use of modern artefacts continues to be applicable. The enormous advances in geochemical and isotopic techniques will also help to date changes in Mediterranean valleys, though the time-scale of decades still remains problematical. The use of fingerprinting of sediment sources could prove one of the most valuable techniques in showing the amounts and sources of various sediment contributions and, therefore, effects of land use change and incidence of storm events.

6.5 IMPACTS OF FLOOD EVENTS

The number of case studies of physical impacts of flood events in the Mediterranean region is still limited, so that the extent to which systematic relations and general patterns occur is difficult to assess. Much reliance still tends to be placed on evidence of the behaviour of ephemeral channels in other semi-arid regions of the world. The global literature on these channels has been reviewed by Thornes (1994) and Reid and Frostick (1992), therefore the few articles and newer literature from the Mediterranean, together with field evidence from current studies by the authors, will be discussed here – in particular, the detailed evidence provided by the occurrence of a moderate-sized flood at monitored sites in the Guadalentín basin, southeast Spain.

The depths of net scour cited for Mediterranean rivers include 0.5 m (Segura, 1983), 3 m (Pardo 1991) and 4 m in the Ouvèze flood (Flageollet et al., 1993). Within nine sites monitored in the Guadalentín basin and affected by a significant flood event in September 1997, the maximum depth of scour actually measured was 1.05 m. Observations elsewhere on the same channels indicated that scour could have reached 2 m in some locations. Maximum erosion appeared to take place downstream of confluences, i.e. where flow energy increased markedly, or downstream of check dams and track crossings, where sediment transport was impeded thus increasing entrainment capacity (flow may also have surged). However, analysis of the maximum scour at each monitored site against calculated peak discharge, from cross-sections and flood marks or crest stage recorder, for an event gives a strong relationship (R^2 of 76%) (Figure 6.7). This is somewhat surprising given that lithology and morphology vary between these sites. Net channel incision along lengths of channel as opposed to scour holes averaged 0.47 m over a length of 120 m at one site and 0.76 m over 110 m at another, as measured from surveyed long profiles. Calculations of net erosion from volume changes computed using DEMs compiled from detailed field surveys, produce values of 14 cm and 39 cm net lowering over the whole site at the same locations (Hooke and Mant, 2000). The maximum erosion depths measured on these Spanish streams are much more than those predicted by Leopold et al.'s (1966) equation for mean scour depths (as derived from scour chains) but comparable with those measured by Hassan et al. (1999).

In relation to depositional amounts, patterns or volumes from single events, very few figures are available for Mediterranean streams though Ergenzinger (1988) does provide some figures of volumes resulting from massive landslide inputs to streams in southern Italy. Harvey (1984) states that up to 0.4 m was deposited as dunes in lee sites in the event on the Rambla Honda in southeast Spain. Conesa-García (1995) measured up to 25 cm deposition in a period of five

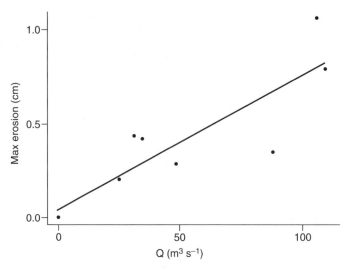

Figure 6.7 Plot of maximum erosion against peak discharge

years. In the September 1997 flood event in southeast Spain the resurveys of the monitored channel reaches reveal three main types or patterns of deposition:

1. Deposition, mainly of fine material, tends to take place overbank, on the floodplain with erosion occurring in the channel and only rarely any scouring of floodplain deposits in this size event.
2. Deposition takes place in the channel, downstream of major scour holes and as a lobe of sediment, presumably derived from the scour hole and deposited almost immediately.
3. Some sections or reaches of channel show net aggradation. These are not widespread and, in spite of systematic mapping along several kilometres of channel, it is difficult to detect a systematic pattern or relationship, e.g. to sediment source and supply or to channel morphology and flood hydraulics. Longer-term monitoring is needed to detect such relationships.

Vegetation is significant in encouraging sedimentation, both at the very local level of the wake of bushes and on floodplains and at the reach level. However, sedimentation is not necessarily high in all or throughout vegetated reaches because, within a short distance, the supply is lost due to the lack of sediment entrainment where there is a vegetation cover. Actual amounts of deposition measured from the September 1997 event reached a maximum depth of 0.64 m. Again, analyses of a maximum depth of deposition at the monitored sites, in relation to various controlling factors, reveal strong relationships with stream power: 93% of maximum deposition can be explained by a combination of discharge, channel width and slope.

A significant feature that does develop during individual flood events is that of headcuts in the channel bed. These are often associated with scour holes but are differentiated here as features with a near vertical backwall. They may occupy the whole channel width, usually in the largest headcuts, or more often are small headcuts within the channel. In the September flood of 1997, a number of these features, about 40–60 cm in height, developed. In some cases, the causes of the initiation were obvious such as a morphological, sedimentary or vegetational change. Others may have been related to slight disturbances such as gravel digging on a small scale. Some may represent the continued headward development of previously formed head-cuts, for which the initiation point should be sought downstream and often relates to check dams and tracks or other such structures.

Channel widening has been reported in several case studies though much of the broader geomorphological literature implies a lack of importance of bank erosion in ephemeral channels. The lack of high or well-defined banks makes bank erosion less obvious but if high floodplains and terrace banks are present they can exhibit similar features of collapsing to those reported in perennial streams (Hooke, 1979). Harvey (1984) reported on the significant role of bank erosion in the Rambla Honda flood and Sala (1983) has discussed its importance in subhumid channels. Fifteen metres of bank erosion in one day is reported by Benchetrit (1972). In the September 1997 flood in southeast Spain the maximum channel widening measured at the sites was 1.28 m, representing a 36% increase in inner channel width at a section where flood width was 30.7 m. The averaged change in channel width for the seven sites experiencing flow showed the strongest relationship to be with flood depth (R^2 70%), while that with original channel width only provided 57% explanation. A combination of flood depth and sediment types produced an 88% level of explanation while flood depth, sediment type and vegetation type produced 96% explanation. Change in width from 21 cross-sections at the seven sites experiencing flow showed little relation to any of the site variables.

Channel switching and change in planform is obviously much greater in larger events and on braided streams. Again, the number of specifically documented cases is small though the wider issue of channel pattern change and channel metamorphosis on the longer time-scale is identified (see next section). The lower the banks and the more ill-defined the channel, the more likely it is to change position within an event. Some of the mechanisms and extent of this have been discussed in Chapter 5. In terms of net change from an event, the non-cohesive gravel-bed channels show much greater propensity to change and commonly the form and position of inner channels are recreated in each event. Minor avulsions and temporary bar formations are very common within the overall active channel. Large-scale avulsions, chute-cutoffs and channel migration are much less common in the single-thread channels, partly because of their cohesion and partly because of the lack of well-defined meandering patterns and the tendency towards irregular and low-sinuosity patterns. In the September 1997 event in southeast Spain, the one braided gravel-bed site only experienced a relatively low flow and the two minor inner channels present in the pattern were not modified significantly. On the Torrealvilla channel, with a marl and gravel catchment, some chute flows and avulsions did take place at high flow but these appear not to have altered the inner channels in most cases and were only possible in the widest parts of the valley.

Some discussion of the nature of flood deposits in ephemeral channels can be found in the literature (Reid and Frostick, 1992) but is mostly concerned with facies and sedimentary structures or types of location. Particular types of depositional and sedimentary features such as boulder berms have been identified, though again very little in the Mediterranean environment. Changes in size of sediment or distribution of sediment zones have rarely been described in relation to specific events, an exception being the analysis by Thornes (1976) of the Pulio Rambla in southern Spain. Much discussion and interest focuses on the possible presence of sediment slugs and waves and the mechanisms of translation downstream. The system of monitoring at the Spanish sites has allowed both the overall changes in reaches and the more detailed changes of surface characteristics to be detected. Results show a general tendency to coarsening of the surface in the 1997 event within the channels, and additional fine deposition on the floodplains. The degree of sorting of deposits and the patterns of deposition vary, some coarse deposits occurring as distinct lobes, some as riffle type features, some as near plane beds, and others as distinct lobate bars. Considerable evidence points towards the high capacity and competence of these flows. At one site in the $90 \, m^3 \, s^{-1}$ event of September 1997, a masonry block 2 m by 1 m by 1 m was found 1 m above the bed of the channel and the only obvious source of masonry was a track crossing 200 m upstream. Massive deposits of cobbles of 10–20 cm in size are not uncommon in these channels. Even a 40-cm stage flow at one site was able to rework the bed and deposit a 17 cm depth of gravel material. Harvey (1984), in one of the few analyses of the geomorphic effects of a major event, which occurred on Rambla Honda in 1980, describes how scour was widespread during the flood in the main channel and widening occurred through bank collapse. On the recession limb, dunes of up to 0.4 m were deposited in lee sites on the channel floor. Beyond the channel, headcut trenches formed in the farmland. There were distinct differences in behaviour and hydraulic geometry between confined and unconfined sites though scour predominated in the upper sites.

The behaviour of these channels is also controlled by both the impact of vegetation upon flood events and its sensitivity to an event, or series of events. Both the type and density of vegetation vary within different reaches thus making quantification of the exact impact of vegetation difficult to assess. Thus, although it has been acknowledged that the role of vegetation in semi-arid channels cannot be underestimated as a potential factor in influencing the sediment load of a flood (Smith et al., 1993), to date its precise role remains an unresolved issue. It should be noted that some research into the role of channel vegetation in semi-arid areas has been completed, but the focus of this has centred upon more arid areas of the USA (Osterkamp and Costa, 1987; Wolman and Gerson, 1978; Stromberg, 1997). The implication of ephemeral channel vegetation upon the channel development processes within the Mediterranean basin remains virtually non-existent. The primary influences of a plant operate as a combination of individual plants' resistance to damage or erosion during and after an event, and as a bed surface roughness component. Research that directly addresses these issues is primarily related to temperate environments and at the reach scale (Cowen, 1956; Hey, 1986; Prosser and Slade, 1992; and Ferguson, 1994), or laboratory experiments to consider vegetation flexibility (Fathi-Maghadam and Kouwen, 1997) or the impact of flow velocities on floodplain plants (Li and Shen, 1973; Klassen and Zwaard, 1974). Understanding the dynamics of vegetation and its importance in terms of affecting channel geometry is not easy. For example, while the spatial and temporal variability of the flood events are themselves important, the seasonal impacts of vegetation must also be acknowledged. Furthermore, the relationship between the substrate, vegetation geometry, assemblages and the surface roots, particularly in terms of their strength factor, are important issues that add to the complexity. Thus different species or assemblages of vegetation behave differently for a given event and recovery rates are not only dependent on the size of an event, but also on the different types and extent of the vegetation. Field observations have highlighted such variations. For example, some vegetation types such as *Lygeum spartum* are more conducive to the trapping of fine sediment within the vegetation itself, while others of a similar but different structure (for example, *Inula viscosa*) are more likely to cause deposition of large sediment upstream and wakes of fines downstream. At the larger scale, and where vegetation is subject to larger flow events, similar patterns can be noted. While reaches associated with mature *Tamarix parviflora* may cause wakes of sediment to occur, and some damage to the vegetation in terms of bending or even partial burial may be apparent, sections of channel dominated by *Saccharum ravennae* may indicate complete removal of vegetation after an event. Similarly, the response of different types of vegetation within these channels is extremely variable. Recovery is not only driven by a combination of moisture and seasonal factors, but is also related to the sensitivity of a species, to a flood or series of flood events, and its age.

The largest floods in ephemeral channels in the Mediterranean region have killed hundreds of people. Spain is particularly vulnerable and Benito et al. (1998) cite deaths in recent floods. Buildings and portions of towns and villages have been destroyed, as have bridges; for example, an iron bridge on the Alamanzora was swept downstream by an 18 m high flood wave in the October 1973 floods in southeast Spain (Mairoto et al., 1998). Track crossings can be taken out by even quite low flows, and crops and tree groves, e.g. almond, olives and citrus, are frequently damaged or destroyed. The temporary earth embankments built in many channels to divert water into fields and agricultural terraces are obviously destroyed by the higher flows but also serve their purpose of diversion in the more frequent low flows. Overall, several authors have stated that in ephemeral channels of semi-arid environments, it is the sediment that causes the greatest problems rather than the force of water flow or the depth of inundation (Poesen and Hooke, 1997).

6.6 EFFECTS OF SEQUENCES OF FLOODS

The large flood events tend to receive almost all the attention because it is thought that these are the channel-forming and effective discharges. The number of smaller flows in these ephemeral channels of the Mediterranean, and their subsequent geomorphological effects, have perhaps directly and indirectly been rather neglected, at least in the international literature, even if well known to the inhabitants of the areas. Table 6.3 gives the number, stage and calculated discharge of flows measured at the nine monitoring stations in the Guadalentín over a period of three years. This shows not only the high variability of occurrence of flows, but that, on some reaches, even in this relatively dry part of the Mediterranean, and in years that have had below long-term average rainfall, several events a year have occurred at some of the sites.

The greatest number of flows have occurred at the monitored site Salada 3 which is just downstream of a particularly active tributary. In the three events exceeding the 1-m stage between January 1997 and 30 September 1997, some changes in the calibre of the sediment cover took place (Figure 6.8) but modifications to the channel morphology were mostly slight. The amount of fine material deposited in those comparable size events did, however, differ. In the event of 30 September, which nearly reached the 2-m stage, the channel at the upstream end was eroded and much coarser material was introduced. In the middle, straight narrow part, a 60-cm deep scour hole was formed. Downstream, a deep narrow part below a marked steepening of the channel was completely infilled with 20-cm diameter cobbles, probably derived from erosion of the valley wall immediately upstream. In another event in February 1999, which nearly reached the 2-m stage, at the upstream end the thalweg was eroded a further 20 cm but the scour hole in the middle part was almost completely infilled with very fine material. Downstream, a veneer of fine material was deposited on top of the coarse, and the slope profile was further smoothed.

This example illustrates that the responses to similar size floods are not always the same. Analysis of several sites indicates three main factors influencing the sequence of behaviour at any point:

1. The channel morphology prior to the event and, in some locations, also the state of the vegetation.
2. The calibre and amount of sediment load in the event and the loading relative to the runoff magnitude and timing.
3. The passage of sediment waves down through the channel and the supply of material immediately upstream.

The effects of morphology can also be examined by calculating the hydraulics of the channel before and after modification. At one monitored site on the Rambla Torrealvilla in southeast

Table 6.3　Maximum flow stage readings at the nine field sites in the Guadalentín basin

Site	Date	Max stage (cm)	Site	Date	Max stage (cm)	Site	Date	Max stage (cm)
Salada 1	***8/4/97***	Installed	**Torreal villa 1**	***6/4/97***	Installed	**Nogalte 1**	***6/4/97***	Installed
	9/4/97	10						
	−16/9/97	99		−17/9/97	Recorder gone >2 m			
	−22/9/97	39					−18/9/97	0
	−26/10/97	138		−30/10/97	10		−31/10/97	15
	−1/4/98	11.7					−2/4/98	0
	−17/9/98	11.5		−14/9/98			−15/9/98	2
				16/9/98	rebuilt			
	−16/2/99	12		−13/2/99	0		−15/2/99	0
	−4/99	135		−4/99	4		−4/99	0
	−8/6/99	0		−11/6/99	14.5		−12/6/99	0
			Torreal villa 2	***20/1/97***	Installed			
Salada 2	***9/4/97***	Installed		***7/4/97***	11	**Nogalte 2**	***6/4/97***	Installed
		9		***18/6/97***	198			
	−16/9/97	132		***20/9/97***	55			
	−22/9/97	123					−17/9/97	2
	−27/10/97	eroded, rebuilt		−30/10/97	248		−31/10/97	5
	−1/4/98	3.5		−30/3/98	0		−2/4/98	0
	−17/9/98	0		−12/9/98	55		−15/9/98	0
							−15/2/99	0
	−4/99	eroded					−4/99	0
	14/6/99	new gauge					−12/6/99	0
Salada 3	***23/1/97***	Installed	**Torreal villa 3**		Installed	**Nogalte 3**	***19/1/97***	Installed
	2/4/97	123		***6/4/97***			−4/4/97	0
	9/4/97	23						
	−16/9/97	109		−17/9/97	90		−17/9/97	4
	−22/9/97	103		−22/9/97	13 eroded			
	−26/10/97	>161		−30/10/97	c 150 rebuilt		−31/10/97	59
							−2/4/98	0
	−31/3/98	0		−29/3/98	0			
	−11/9/98	0		−12/9/98	vandalised rebuilt		−15/9/98	3.5
	−10/2/99	12		−11/2/99	0		−14/2/99	0
	−20/4/99	190		−4/99	0		−4/99	0
	−9/6/99	20		−10/6/99	0		−12/6/99	0

Key:
23/1/97 = Actual date
−16/9/97 = Reading in period to date

Spain, as a result of a flood event with a peak discharge of 107 m³ s⁻¹ on 30 September 1997, the former shallow channel was incised by about 0.6 m (Figure 6.9) creating a new well-defined rectangular channel. If the bankfull capacities for two cross-sections are calculated for before and after this change, then the capacity has increased from 1.52 to 11.98 m³ s⁻¹ on one section and from 6.53 to 24.40 m³ s⁻¹ on another. Mean velocities changed from 0.95 to 2.80 m s⁻¹ and 1.80 to 3.16 m s⁻¹ respectively, and shear stresses at bankfull from 16.1 to 80.7 N m⁻² and 41.9 to 96.5 N m⁻². Thus, particularly at bankfull discharges, the incision of the channel is likely to have a positive feedback effect, making it more likely that the channel will erode further (though moderate flow stages will be lower for a given volume of flow). A similar type of effect

Figure 6.8 Sediment change in one quadrat resulting from flow events in the period 1996–1999

was found on the Gila River, Arizona, by Hooke (1996). Although some changes in morphology would tend to have a positive feedback effect, with partially incised channels becoming deeper, this may not happen if the next event brings a large influx of sediment.

It emerges from this analysis that the larger events are generally erosional, though they may not necessarily be erosional down-channel or cross-channel. The smaller flows are largely depositional and tend to produce a veneer of fine material over the bars and to fill in the scour holes. The fine material, certainly in the marl catchments, can then act as a seal on the surfaces, becoming hard and resistant and with low infiltration capacity. This field evidence from other channels in the Guadalentín basin further confirms Conesa-García's (1995) proposals on the role of different size events.

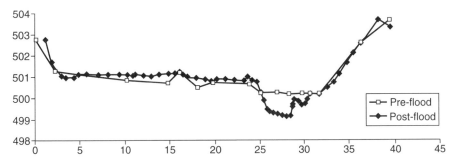

Figure 6.9 Cross-sectional change at Oliva site, Rambla Torrealvilla

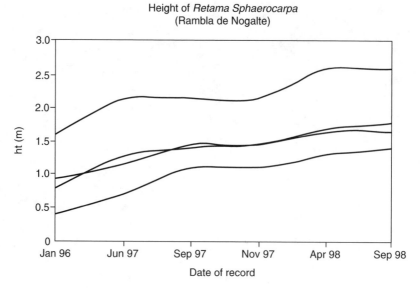

Figure 6.10   Example of individual plant growth rates of vegetation for one species

The importance of the size and sequencing of flood events in these channels also has a strong influence upon the types of plants that are likely to flourish, and where. Mediterranean vegetation growth is primarily restricted to the main periods of moisture availability in early spring and late autumn. Different species favour different periods of the year for their primary growth period, as indicated in Table 6.4 and highlighted in Figure 6.10, which have been compiled from field data. Furthermore, although these species are generally tolerant of high temperatures, many are more susceptible to frost and temperatures below 3 °C (Blamey and Grey-Wilson, 1993).

This vegetation seasonality allows for the identification of the dominant species along the valley floors and floodplains on a yearly cycle. However, the diversity of species found varies through time and space, which may be directly attributable to flood frequency, event size, season of occurrence, or local land use. Thus, for example, should a period of only minor events occur, vegetation more usually associated with the floodplain may start to invade the channel. Such vegetation, however, is often more sensitive to moderate flow events and will, therefore, not necessarily recover from such events as readily as species that are more adaptable

Table 6.4   Growth periods for plants compiled from a combination of field observations and results from Blamey and Grey-Wilson (1993)

Vegetation	Growth period
Tamarix parviflora	April–June
Inula viscosa	July–November
Juncus bulbosus	March–September
Lygeum spartum	April–July
Retama spaerocarpa	March–May

to variation in flow events. Such an episode was observed following an extended period of drought conditions at the Oliva reach in the upper section of the Rambla Torrealvilla – a section of the channel that was noted as being floristically rich. Species such as *Reichardia picroides*, *Phlomis fructicosa* and *Dianthus vultaria* were identified as plants that are more commonly associated with cultivated land and very dry areas. Also, at the Serrata field site, vegetation more commonly associated with dry rocky hillslopes and floodplains including *Thymelaea hirsuta* and *Artemisia barberia* were present in the channels. However, following a series of moderate sized events, most of these shrubs have not survived. In particular, at the Serrata reach there has been a return to a domination of species associated with land degradation, and saline conditions such as *Inula viscosa* and *Lygeum spartum*. Such observations have begun to indicate that the role of vegetation in modifying the effects of floods, and the movement of sediment within them, is dependent upon more than just the growth cycle of specific species. Instead the importance of spacing and sequencing of the flood events themselves, rather than just their statistical frequency, has a direct impact upon the diversity of species present. Therefore although the effects upon vegetation depend upon the regime of a given sequence of flood events, equally the change in geometry of a channel is determined in part by the state and density of vegetation present at a particular time. This, in turn, is intrinsically tied to past flow regimes.

6.7 CHANGES ON DECADAL TIME-SCALES

On the time-scale of a few decades, one would expect the signal of climate changes and land use changes affecting river channels to be detectable. Unfortunately, the lack of suitable documentation, recording and measurement, incidental or organised, means that it is difficult to provide evidence for, or to infer, changes. If one looks at the longer term of hundreds of years then the accumulation of archaeological and sedimentary evidence indicates major changes and a high sensitivity of these fluvial systems to changes in the forcing functions of runoff and sediment supply (Lewin et al., 1995).

Overall, the evidence from Vita-Finzi (1969), papers in Lewin et al. (1995), Benito et al. (1998) and elsewhere is that many of the streams of the Mediterranean are undergoing an incision or have recently undergone an incision into the medieval fill. Barker (1995), from intensive archaeological work, found that a phase of sedimentation and a more recent incision had taken place in the Biferno valley of Italy in the twentieth century. Over the longer-term, phases of sedimentation were closely related to periods of intensification of settlement and agriculture. Maas et al. (1998), working in a small catchment in Crete, found that renewed valley sedimentation occurred in the latter part of Little Ice Age, continuing until the 1930s. More recently incision has taken place in the upper parts and deposition downstream as a result of two major floods. Likewise, López-Áviles et al. (1998) recorded a phase of alluviation in a subcatchment of the Ebro basin in medieval times, related to both climate and anthropogenic activity. By a comparison of historic records with recent flood measurements, they show the influence of major flood events. Conacher and Sala (1998) indicate that streams of southern Italy, a tectonically active area, are presently degrading though they can aggrade in major storms, mainly due to sediment supply from landslides. In the Valencia area of Spain, recent changes are quite well documented and show that gravel abstraction and dam construction have caused incision in many reaches (Conacher and Sala, 1998).

In analyses of channel changes, mainly using historical maps of southern France over timescales of about a century, Bravard et al. (1997) have shown the widespread transformation of many streams from very braided at the end of the last century to much simpler or single-thread patterns now. Braga and Gervasoni (1989) have shown that most of the changes on the Po River since the eighteenth century are anthropogenic, induced due to river regulation but that in the historic period there may have been some neotectonic influences.

One of the best documented studies is that by Rozin and Schick (1996) of changes in channel form of a stream in the Mediterranean/semi-arid transition of Israel since 1917. Channel morphology remained similar in the period 1917–1945 when little land use change took place, but after 1945 the braided channel narrowed, became single-thread, was incised and then stabilised by vegetation. This change is attributed to a major increase in water and sediment storage within the catchment and to a major decrease in the volume and frequency of flow events. Bank erosion also decreased due to a reduction in grazing. The study shows a very high sensitivity and rapid response of channels to changes in land use and thus of runoff and sediment supply, with response evident within a few years.

Compilations of various records of flow events and storms demonstrate that within the Mediterranean region there are distinct periods of occurrence of storms and floods and droughts (e.g. Maheras, 1988; Perry, 1997; López-Áviles et al., 1998; Diez-Herrero et al., 1998; Pavese et al., 1992) with phases persisting for several years. From the examples available for the Mediterranean region, it is evident that the response to these phases and the sequence of events is not simple and depends very much on actual locations and attributes in a basin. The evidence of recent channel responses tends to support the model produced in other semi-arid regions (e.g. Hereford, 1984; Burkham, 1972) of floodplain construction and channel narrowing during periods of lower flood magnitude and frequency. However, contrasting phases are to be expected and should be allowed for in channel management, in the same way as Warner (1987) has indicated for flood-dominated and drought-dominated phases in Australia.

6.8 MODELLING CHANNEL CHANGES

The lack of records of channel changes and responses to changed conditions means that the empirical basis for prediction of impacts of future changes in forcing functions or basin attributes is very limited. Modelling may therefore offer a basis for trying to answer questions of 'what if?', assuming that there is sufficient knowledge of the processes to provide the components of the model. A simulation model of changes in a channel reach has been developed under the EU-funded MEDALUS III project (Brookes et al., 2000; Hooke, 1999).

The model is based on the conceptual framework outlined earlier in this chapter and shown in Figure 6.6. The aim of the model is to simulate the outcome of flow events and of climatic conditions and other activities in a reach of channel over the time-scale of years to decades. The aim is not to produce a prediction of exact amounts and locations of erosion and deposition but to provide some guidance on likely channel behaviour given certain attributes and conditions of a reach. Using such a tool it should then be possible to run the model on various sites and under various climatic and land use scenarios to identify most vulnerable reaches and most critical conditions. The approach that has been taken is 'top-down' and holistic – that is, to try to simulate outcomes and the complex combinations and interactions that occur.

The model comprises a program written in C++ with links to GIS for input and output and is run on a PC platform. The inputs are morphology of the reach, sediment cover and vegetation cover. Morphology is derived as a DEM from the survey input then converted to elevation data per pixel. Sediment data are maps of zones of sediment of different class sizes. Vegetation data are maps of zones of different types of vegetation, classified into herbs, shrubs and phreatophytes, and assigned an age, state (degree of stress), and density of cover (Brookes et al., 2000). The model has typically been run for reaches of about 100 m with pixels representing 30 cm, X and Y. All the data are converted to raster format so values are assigned for each attribute to each cell. The processes outlined in Figure 6.6 are then operated. Flow capacity and distribution of flows and velocity are calculated for each row for a given input of peak discharge and a linear recession through an event. Sediment capacity is calculated for a cross-section using the Bagnold (1966) equation at present. Erosion takes place if there is excess capacity and the velocity exceeds thresholds set by the Hjulstrom (1935) equation. Both roughness and resistance are affected by sediment and vegetation characteristics in each cell, so the model is sensitive to local variations. Within an event, erosion and deposition are calculated in relation to capacity and competence as the flow recedes. The net outcome produces changes in the morphology. Sediment cover may also change, depending on where erosion has taken place, the size of material entrained and already in the specified sedigraph, and the rate of recession. Vegetation may be battered or destroyed in an event, and this has a significant effect upon the erodibility of the sediment as well as affecting velocity through its roughness effects.

Examples of input and output of the changes produced in a simulation by the model are shown in Figure 6.11. The model can be run for any length of period from a month (basic iteration unit) to 30 years. Flow occurrences are input as discrete events with a peak flow and duration and a sediment load, calibre and duration. In between flow events, climate conditions control vegetation growth and the moisture processes, and different combinations and sequences of flow events can be tested. Various other human impacts can be input but have not yet been operationalised. In results so far, the model has been found to replicate conditions of clearwater flow reasonably well but is highly sensitive to the sediment loadings. The model has been validated with the results of the events at the monitored sites, but detailed sediment data are lacking for these or, indeed, many sites at all. If some validation can be obtained then the model would at least allow comparison of the effects of, for example, an increase in supply of fine sediment due to changes in land use, but the deposition of fine material has been simulated from high sediment load events. Further tests with the sediment component and the specification of the sediment processes are to be undertaken. However, a simulation model has now been constructed which is capable of running over time-scales of decades and incorporates the effects of vegetation and the feedback effects of changes in site characteristics. Even without detailed sediment data, it may be possible to use the field observations of actual outcomes to test conditions in the model which produce such results.

6.9 IMPLICATIONS FOR MANAGEMENT

It is evident from the Mediterranean region and other semi-arid areas, and has long been accepted, that ephemeral streams tend to be highly variable in their characteristics spatially and are highly variable in behaviour over time. They tend to respond readily to each flow event and may exhibit very large changes in the highest magnitude floods and even those occurring about every 10 years. Several authors have identified many of the major problems and established that damage in such events is due to the sediment processes and erosion and sedimentation occurrence. Major choices need to be made in management strategy between appropriation of resources put into catchment management and into channel management. That requires much greater understanding of the sources and dynamics of sediment supply and transport which produce channel erosion and sedimentation. Within channel management, a major distinction is usually made between structural and non-structural approaches. There is now, generally, much movement in river management against structural controls and towards either softer engineering or non-structural approaches, to the extent that in some cases structural controls have even been removed from rivers (Brookes and Shields, 1996). Ephemeral

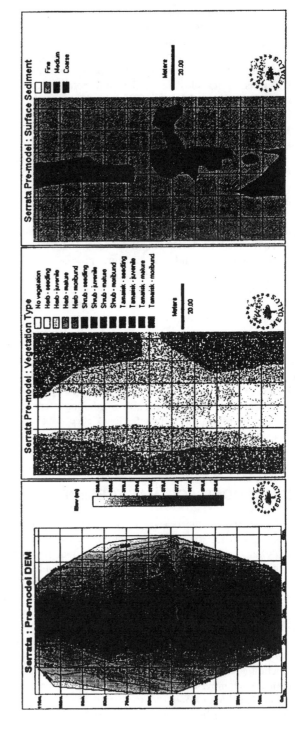

Figure 6.11   Example of the input and output from CHANGISM, the channel change simulation model

Plots demonstrating starting conditions showing, from left to right, digital elevation model, vegetation type and surface sediment

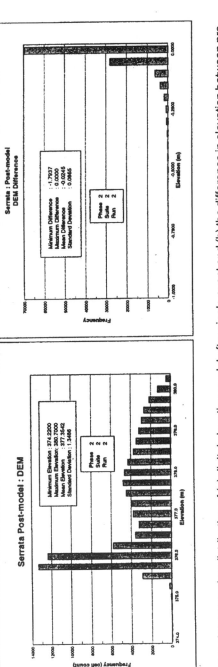

Bar charts demonstrating distribution of (a) digital elevation model after single event and (b) the difference in elevation between pre- and post-event for clearwater flow

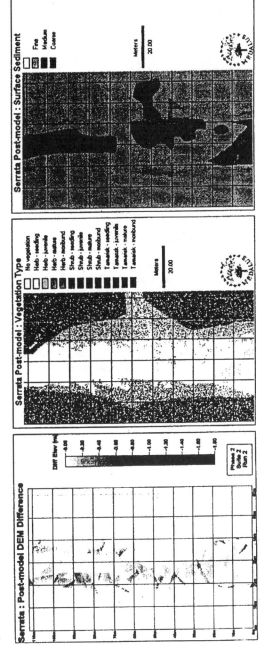

Plots demonstrating end-of-run conditions showing, from left to right, difference in digital elevation models, vegetation type and surface sediment

channels, because of their variability and instability, are particularly unsuitable for structural approaches. The lessons are being learned to some extent with considerable doubts being raised, particularly as a result of destructive floods in the Italian Piedmont and elsewhere, on the wisdom of river regulation and channelisation for such streams (Masson, 1993). The erratic behaviour of these channels must be recognised and land use and activities in valley floors planned to adjust and adapt to these characteristics.

The other lesson that must be learned from the examination of changes over the decadal time-scale is that there are marked phases of behaviour, both in climate and river response, over lengths of periods of years. Any management strategy must recognise the natural occurrence of such phases and allow for these extremes. It must also be acknowledged that there are great dangers in basing management decisions on data for restricted periods which may coincide with a particular climatic phase, such as that experienced during the 1980s, when prolonged arid conditions prevailed compared with the preceding and succeeding periods. This has been shown to occur elsewhere in semi-arid areas (Hooke, 1994; Warner, 1987) and the acknowledgement of these climatic phases is of enormous importance for water resources as well as for catchment and channel planning.

6.10 CONCLUSIONS

Much more work is needed to understand the behaviour of these channels at both the time-scale of decades and the response to individual events. If examined statistically, or solely on either a spatial or temporal basis, a lack of systematic patterns becomes apparent. This may be inherent within these systems or it may be due to their complexity, both in terms of their form and the difficulty in predicting the probability of a specific reach being affected by a flood event. Therefore, it is important that spatio-temporal behaviour is documented to detect the effects of sediment waves and the variability of the sediment supply, particularly since the few studies that are available indicate that channel erosion is a major source of sediment to the channel system. On an individual event basis, it is the large flood events which produce significant channel changes and whose effects may persist for many years. However, other events, including the quite minor ones, move sediment and could ultimately contribute to longer phase trends of channel form, particularly in terms of narrowing and stabilisation by vegetation during periods of limited high flow events. Changes over periods of years to decades must also be considered. These are primarily influenced by the magnitude and frequency of flow events, their spacing and sequencing, the interaction of sediment and vegetation and the consequential morphological feedback mechanisms. Therefore, as exemplified within this chapter, for both single events and over the longer time-scale, greater documentation is needed to elucidate the extent to which channel changes within the Mediterranean are synchronous, and thus show similar and widespread responses to climatic and land use changes. Efforts should be made to implement more routine monitoring but, in the meantime, modelling can provide some useful indicators of vulnerability and possible outcomes.

ACKNOWLEDGEMENTS

Much of the work in the Guadalentín basin was undertaken under Contract ENV4-LT95-0118 of the European Commission funded MEDALUS III Project 4. We are grateful for that funding and to all who cooperated in that project from King's College London, University of Leeds (UK), K. University of Leuven (Belgium) and University of Murcia (Spain) and particularly to Robert Perry and Bill Duane (University of Portsmouth) for field survey and data-processing assistance. Christopher Brookes and Bill Duane created and developed the channel change simulation model. We are grateful to the Department of Geography, University of Portsmouth, for facilities and to Margaret Fairhead for word processing.

REFERENCES

Alexander, D. 1980. *Observations on the regularity of ephemeral channel form*. Occasional Paper No.36. University College, London, 41pp.

Arnaud-Fassetta, G., Ballais, J.L., Begnin, E., Jorda, M., Meffre, C., Provansal, M., Roclitis, J. and Suanez, S. 1993. La crue de l'Ouvèze à Vaison-la-Romaine (22-9-92). Ses effets morphodynamiques, sa place dans le fonctionnement d'un géosystème anthropisé. *Revue de Géomorphologie Dynamique*, XLII, 34–48.

Bagnold, R.A. 1966. An approach to the sediment transport problem from general physics. Physiographic and hydraulic studies of rivers. *Geological Survey Professional Paper, 422-I*, 1–20.

Barker, G. (Ed.) 1995. *The Biferno Valley Survey. The Archaeological and Geomorphological Record*. Leicester University Press.

Benchetrit, M. 1972. *L'Erosion Actuelle et ses consequences sur l'amengement de l'Algerie*. Publi. Univ. Poitiers, Vol. 11. Presses Universitaires de France, Paris.

Benito, G., Machado, M.G., Passmore, D.G., Brewer, P.A., Lewin, J., Branson, J. and Wintle, A.G., 1996. Climate change and flood sensitivity in Spain. In A.J. Brown and K.J. Gregory (Eds) *Global Continental Changes: the Context of Palaeohydrology*. Geological Society Special Publication, 115. Geological Society, London, 85–98.

Benito, G., Machado, M.J., Perez-Gonzalez, A. and Sopena, A. 1998. Palaeoflood hydrology of the Tagus River, central Spain. In G. Benito, V.R. Baker and K.J. Gregory (Eds) *Palaeohydrology and Environmental Change*. John Wiley & Sons Ltd, Chichester, 317–334.

Blamey, M. and Grey-Wilson, C. 1993. *Mediterranean Wild Flowers*. Harper Collins, St Helier, Jersey, 560pp.

Braga, G. and Gervasoni, S. 1989. Evolution of the Po River: an example of the application of historic maps. In G.E. Petts (Ed.) *Historical Change of Large Alluvial Rivers: Western Europe*. John Wiley & Sons Ltd, Chichester, 113–126.

Bravard, J.-P., Amoros, C., Pautou, G., Bornette, G., Bournaud, M., Creuz des Chtelliers, M., Gibert, J., Peiry, J.-L., Perrin, J-F. and Tachet, H. 1997. River incision in South-East France: morphological phenomena and ecological effects. *Regulated Rivers: Research and Management*, 13, 75–90.

Brookes, A. and Shields, F.D. 1996. *River Channel Restoration*. John Wiley & Sons Ltd, Chichester, 433pp.

Brookes, C., Hooke, J.M. and Mant, J. 2000. Modelling vegetation interactions with channel flow in river valleys of the Mediterranean region. *Catena*, 40, 93–118.

Burkham, D.E. 1972. Channel changes of the Gila River in Safford Valley, Arizona 1946–70. *US Geological Survey Professional Paper*, 655-G.

Conacher, A.J. and Sala, M. (Eds) 1998. *Land Degradation in Mediterranean Environments of the World*. John Wiley & Sons Ltd, Chichester, 491 pp.

Conesa-García, C. 1986. Movilidad de las barras de grava en lechos de rambla del sureste peninsular (Espana). In F. López-Bermúdez and J.B. Thornes (Eds) *Estudios Sobre Geomorfologia del Sur de Espana*. Murcia, 55–58.

Conesa-García, C. 1990. Soil erosion and fluvial sedimentation in the 'Ramblas' of south-east Spain. *Mediterranée*, 71, 63–74.

Conesa-García, C. 1995. Torrential flow frequency and morphological adjustment of ephemeral channels in south-east Spain. In E.J. Hickin (Ed.) *River Geomorphology*. John Wiley & Sons Ltd.

Cowen, W.L. 1956. Estimating hydraulic roughness coefficients. *Agricultural Engineering*, 37, 473–475.

Diez-Herrero, A., Benito, G. and Lain-Huerta, L. 1998. Regional palaeoflood databases applied to flood hazards and palaeoclimate analysis. In G. Benito, V.R. Baker and K.J. Gregory (Eds) *Palaeohydrology and Environmental Change*. John Wiley & Sons Ltd, Chichester, 335–348.

Ergenzinger, P. 1988. Regional erosion: reaches and scale problems in the Buaonamico Basin, Calabria. *Catena Supplement*, 13, 97–108.

Everitt, B.L. 1980. Ecology of saltcedar – a plea for research. *Environmental Geology*, 3, 77–84.

Fathi-Maghadam, M. and Kouwen, N. 1997. Nonrigid, nonsubmerged, vegetative roughness on floodplains. *Journal of Hydraulic Engineering*, 123(1), 51–57.

Ferguson, R.I. 1994. Critical discharge for entrainment of poorly sorted gravel. *Earth Surface Processes and Landforms*, 19, 179–186.

Flageollet, J., Fraipont, P., Gourbesville, P., Tholey, N., Trautmann, J. 1993. La crue de l'Ouvèze de Septembre 1992: Origines, effets, enseignements. *Revue de Géomorphologie Dynamique*, 42, 57–72.

Frostick, L. and Reid, I. (Eds) 1987. *Desert Sediments: Ancient and Modern*. The Geological Society Special Publication 35. Blackwell, Oxford.

Graf, W.L. 1978 Fluvial adjustments to the spread of tamarisk in the Colorado Plateau region. *Geological Society of America Bulletin*, 89, 1491–1501.

Graf, W. L. 1987. Late Holocene sediment storage in canyons of the Colorado Plateau. *Geological Society of America Bulletin*, 99, 261–271.

Harvey, A. 1984. Geomorphological response to an extreme flood: a case from southeast Spain. *Earth Surface Processes and Landforms*, 9, 267–279.

Hassan, M.A. 1990. Scour, fill and burial depth of coarse material in gravel bed streams. *Earth Surface Processes and Landforms*, 15, 341–356.

Hassan, M.A., Schick, A.O. and Shaw, P.A. 1999. The transport of gravel in an ephemeral sandbed river. *Earth Surface Processes and Landforms*, 24, 623–640.

Hereford, R. 1984. Climate and ephemeral-stream processes: twentieth century geomorphology and stratigraphy of the Little Colorado River. Arizona. *Geological Society of America Bulletin*, 95, 654–668.

Hey, R.D. 1986. River mechanics. *Journal of the Institution of Water Engineers and Scientists*, 40, 139–158.

Hjulstrom, F. 1935. Studies of the morphological activity of rivers as illustrated by the River Fyris. *Bulletin of the Geological Institute of the University of Uppsala*, 25, 221–527.

Hooke, J.M. 1979. An analysis of the processes of river bank erosion. *Journal of Hydrology*, 42, 39–62.

Hooke, J.M. 1994. Hydrological analysis of flow variation of the Gila River in Safford Valley, Southeast Arizona. *Physical Geography*, 15(3), 262–281.

Hooke, J.M. 1996. River responses to decadal-scale changes in discharge regime: the Gila River, S.E. Arizona. In J. Branson, A. Brown and K.J. Gregory (Eds) *Global Palaeohydrology*. The Geological Society Special Publication No. 115, 191–204.

Hooke, J.M. 1999. Modelling response of ephemeral channels and floodplains to flow and to changes in climate and land use. In *Ephemeral Channels and Rivers*. MEDALUS III Final Report, 245–286.

Hooke, J.M. and Mant, J.M. 2000. Geomorphological impacts of a flood event on ephemeral channels in SE Spain. *Geomorphology*, 34, 163–180.

Horton, J.S., Mounts, F.C. and Kraft, J.M. 1960. Seed germination and seeding establishment of phreatophyte species. *US Dept. Agriculture Forest Service Res. Paper*. RM 47, 17pp.

Klassen, G.J. and Zwaard, J.J. 1974. Roughness coefficients of vegetated flood plains. *Journal of Hydrological Research*, 12, 43–63.

La Barbera, P., Lanza, L., Marzano, F., Minciardi, R., Mugnai, A., Paollucci, M. and Siccardi, F. 1992. Multisensor analysis of the flood event of November 23–25th 1987 on the Arno basin. In A.J. Saul (Ed.) *Floods and Flood Management*. Kluwer Academic, London.

Leopold, L.B., Wolman, M.G. and Miller, J.P. 1964. *Fluvial Processes in Geomorphology*. Freeman, San Francisco.

Leopold, L.B., Emmett, W.W. and Myrick, R.M. 1966. Channel and hillslope processes in a semiarid area, New Mexico. *US Geological Survey Professional Paper*, 352-G, 193–253.

Lewin, J., Macklin, M.G. and Woodward, J.C., 1991. Late Quaternary sedimentation in the Voidomatis Basin, Epirus, northwest Greece. *Quaternary Research*, 35, 103–115.

Lewin, J., Macklin, M.G. and Woodward, J.C. (Eds) 1995. *Mediterranean Quaternary River Environments*. Balkema, Rotterdam, 292pp.

Li, R. and Shen, H.W. 1973. Effect of tall vegetation on flow and sediment. *ASCE*, 99, H5, 9748–9813.

Llorens, P. and Gallart, F. 1992. Small basin response in a Mediterranean mountainous abandoned farming area: research design and preliminary results. *Catena*, 19, 309–320.

López-Áviles, A., Ashworth, P.J. and Macklin, M.G. 1998. Floods and Quaternary sedimentation style in a bedrock-controlled reach of the Bergantes River, Ebro basin, northeast Spain. In G. Benito, V.R. Baker and K.J. Gregory (Eds) *Palaeohydrology and Environmental Change*. John Wiley & Sons Ltd, Chichester, 181–198.

López-Bermúdez, F. 1973. *La vega alta del Segura: clima, hidrologia y geomorfologia*. Murcia: Universidad de Murcia.

Maas, G.S., Macklin, M.G. and Kirkby, M.J. 1998. Late Pleistocene and Holocene river development in Mediterranean steepland environments, southwest Crete, Greece. In G. Benito, V.R. Baker and K.J. Gregory (Eds) *Palaeohydrology and Environmental Change*. John Wiley & Sons Ltd, Chichester, 153–166.

Macklin, M.G., Lewin, J. and Woodward, J.C. 1995. Quaternary fluvial systems in the Mediterranean basin. In J. Lewin, M.G. Macklin and J.C. Woodward (Eds) *Mediterranean Quaternary River Environments*. Balkema, Rotterdam, 1–28.

Maheras, P. 1988. Changes in precipitation conditions in the western Mediterranean over the last century. *Journal of Climatology*, 8, 179–189.

Mairoto, P., Thornes, J.B. and Geeson, N. 1998. *Atlas of Mediterranean Environments*. John Wiley & Sons Ltd, Chichester, 205pp.

Mant, J. 1999. *Interactions of vegetation and morphological change in ramblas of SE Spain*. Paper presented at BGRG Postgraduate Symposium, Nottingham.

Masson, M. 1993. Après Vaison-la Romaine. Pour une approche pluridisiplinaire de la prévision et de la planification. *Revue de Géomorphologie Dynamique*, XLII, 73–76.

Mather, A.E, Silva, P.G., Goy, J.L., Harvey, A.M. and Zazo, C. 1995. Tectonics versus climate: an example from late Quaternary aggradational and dissectional sequences of the Mula basin, southeast Spain. In J. Lewin, M.G. Macklin and J.C. Woodward (Eds) *Mediterranean Quaternary River Environments*. Balkema, Rotterdam, 77–88.

Navarro Hervas, F. 1991. *El Sistema Hidrografico del Guadalentin*. Region de Murcia, Murcia, 256.

Osterkamp, W.R. and Costa, J.E. 1987. Changes accompanying an extraordinary flood on a sand-bed stream. In L. Mayer and D. Nash (Eds) *Catastrophic Flooding*. Allen & Unwin, Boston, 201–224.

Pardo, P. 1991. *La erosion antropica en el litoral valenciano*. Conselleria d'Obre Publiques, Urbanisme i transports.

Parsons, A.J. and Wainwright, J. 1995. Response of hillslope hydrology to vegetation change in southern Arizona. Paper presented at the conference *Floods, Slopes and River Beds*, CNRS, Paris, 22–24 March.

Pavese, M.P., Banzon, V., Coldcino, M., Gregori, G.P. and Pasqua, M. 1992. Three historical data series on floods and anomalous climatic events in Italy. In R.S. Bradley and P.D. Jones (Eds) *Climate since AD 1500*. Routledge, London, 155–170.

Pickup, G. and Reiger, W.A. 1976. A conceptual model of the relationship between channel characteristics and discharge. *Earth Surface Processes and Landforms*, 4, 37–42.

Perry, A. 1997. Mediterranean Climate. In R. King, L. Proudfoot and B. Smith (Eds) *The Mediterranean*. Arnold, London, 315pp.

Poesen, J., Vandaele, K. and van Wesemael, B. 1996. *Contribution of gully erosion to sediment production on cultivated land and rangelands*. IAHS Publication 236. IAHS Press, Wallingford.

Poesen, J.W.A. and Hooke, J.M. 1997. Erosion, flooding and channel management in Mediterranean environments of southern Europe. *Progress in Physical Geography*, 21(2), 157–199.

Prosser, I.P. and Slade, C.J. 1992. Gully formation and the role of valley-floor vegetation south eastern Australia. *Geology*, 22, 1127–1130.

Reid, I. and Frostick, L.E. 1992. Channel form, flows and sediments in deserts. In D.S.G. Thomas (Ed.) *Arid Zone Geomorphology*. Belhaven Press, 117–135.

Rozin, U. and Schick, A.P. 1996. Land use change, conservation measures and stream channel response in the Mediterranean/semi-arid transition zone: Nahal Hoga, southern Coastal Plain, Israel. *Proceedings Exeter Symposium, Erosion and Sediment Yield: Global and Regional Perspectives*, IAHS Publication No. 234, 427–444.

Sala, M. 1983. Fluvial and slope processes in the Fuirosos basin, Catalan ranges north east Iberian coast. *Zeitschrift für Geomorphologie*, 27, 393–411.

Schick, A.P., Hassan, M.A. and Leckach, J. 1987. A vertical exchange model for coarse bedload movement numerical considerations. *Catena Supplement*, 10, 73–83.

Schick, A.P. 1974. Formation and obliteration of desert stream terraces – a conceptual analysis. *Zeitschrift für Geomorphologie Supplement*, 21, 88–105.

Schick, A.P. and Lekach, J. 1987. A high magnitude flood in the Sinai Desert. In L. Mayer and D. Nash (Eds) *Catastrophic Flooding*. Allen & Unwin, Boston, 381–410.

Schumm, S.A. 1977. *The Fluvial System*. John Wiley & Sons Ltd, New York.

Schumm, S.A. and Lichty, R.W. 1963. Channel widening and floodplain construction along Cimmarron River in south-western Kansas. *USGS Professional Paper*, 352D, 71–88.

Segura, F. 1983. Procesos fluviales en lechos con materiales gruesos. *Cuadernos de Geografia*, 16, 123–138.

Simon, A. and Hupp, C.R. 1987. Geomorphic and vegetative recovery processes along modified Tennessee streams: an interdisciplinary approach to distributed fluvial systems. In *Forest Hydrology and Watershed Management. Proceedings of the Vancouver Symposium, August 1987*. IAHS-AISH Publ No. 167 Y.

Smith, W.A., Dodd, J.L., Skinner, Q.D. and Rodgers, J.D. 1993. Dynamics of vegetation along and adjacent to an ephemeral channel. *Journal of Range Management*, 46, 56–64.

Stromberg, J.C. 1997. Growth and survivorship of Fremont cottonwood, Goodding willow and cedar seedlings after large floods in central Arizona. *Great Basin Naturalist*, 57(3), 198–208.

Thornes, J.B. 1976. *Semi-arid Erosional Systems*. Occasional Paper 7. London School of Economics, London.

Thornes, J.B. 1980. Structural instability and ephemeral channel behaviour. *Zeitschrift für Geomorphologie*, 36, 233–244.

Thornes, J.B. 1994. Channel processes, evolution and history. In A.D. Abrahams and A.J. Parsons (Eds) *Geomorphology of Desert Environments*. Chapman & Hall, 287–317.

Vidal-Abarca, M.R., Montes, C., Suarez, M.L. and Ramirez-Diaz, L. 1992. An approach to the ecological characterisation of arid and semiarid basins. *Geojournal*, 26, 335–340.

Vita-Finzi, C. 1969. *The Mediterranean Valleys*. Cambridge University Press, Cambridge.

Warner, D.K. and Turner, R.M. 1975. Saltcedar seed production, seedling establishment, and response to inundation. *Arizona, Science Journal*, 10, 131–144.

Warner, R.F. 1987. Spatial adjustments to temporal variations in flood regime in some Australian rivers. In K. Richards (Ed.) *River Channels: Environment and Process*. Blackwell, Oxford, 14–40.

Wolman, M.G. and Gerson, R. 1978. Relative scales of time and effectiveness of climate in watershed geomorphology. *Earth Surface Processes*, 3, 189–208.

7 The Relationships between Alluvial Fans and Fan Channels within Mediterranean Mountain Fluvial Systems

ADRIAN M. HARVEY

Department of Geography, University of Liverpool, UK

7.1 INTRODUCTION: COUPLING OR BUFFERING ROLE OF ALLUVIAL FANS

Alluvial fans function as major sedimentation zones within mountain fluvial systems, storing sediment supplied by mountain catchments (Harvey, 1997). Fans occur where confined mountain streams lose power, typically at mountain-front, tributary junction or intermontane basin situations.

Locating factors are (i) a high sediment supply (controlled by the geology, geomorphology and climate of the mountain source area) and (ii) the topographic setting where reduced confinement causes a rapid downstream loss of transporting power at the mountain front or equivalent location (controlled in part by the geology, tectonics and geomorphology of the mountain-front zone, and in part by the processes on the fan itself).

Alluvial fans have a critical role in mountain fluvial systems (Harvey, 1997). Aggrading fans trap coarse sediment from the mountain source areas, which may be prevented from reaching downstream reaches. Such fans therefore buffer the system. However, if channel continuity is established through the fan environment, and if, during flood events, sediment transport occurs from the mountain source areas to downstream reaches, the fan surfaces become inactive and the system as a whole becomes coupled.

Tectonics, geomorphic history and base level may influence fan location and some aspects of fan morphology, but within this context, the factors that control the processes on fans relate largely to the supply and transport of water and sediment. These are primarily climatic factors and influence: (i) weathering rates, and therefore the availability of sediment; (ii) runoff and erosion rates (partly through the influence of vegetation cover), and therefore the delivery of sediment to the feeder channel; and (iii) the magnitude and frequency of flood flows, and therefore flood power for sediment transport. The style of sediment transport depends on the water : sediment ratio fed to the fan (Wells and Harvey, 1987). Whether erosion or deposition takes place on the fan is, together with the style of deposition, controlled by the critical power relationships (Bull, 1979). Flood power depends on flood discharges, gradient and water depths, which vary with confinement. If flood power falls below the threshold of critical power (Bull, 1979), in other words below that required to transport the sediment supplied, deposition takes place. In conditions of excess power, erosion of the fan will take place.

Dryland Rivers: Hydrology and Geomorphology of Semi-arid Channels. Edited by L.J. Bull and M.J. Kirkby.

Climatic and other changes in the supply and transport of water and sediment, causing temporal changes in critical power relationships, will be expressed by changes in the depositional or erosional regime of the fan. Aggrading fans therefore preserve a sedimentary record of the history of environmental change that, because of the buffering effect, is more directly related to the mountain source areas than are sites further downstream. Fan morphology, too, expresses the history of deposition and erosion on the fan. Especially sensitive is the gradient of the fan surface, which reflects sediment size (Blissenbach, 1952; Bull, 1977), and depositional process (by debris flows or fluvial processes, both by channelised and sheet flows) (Kostaschuk et al., 1986; Wells and Harvey, 1987). If the processes on the fan change, there is likely to be a resulting change in the locations and styles of erosion or deposition, leading to a change in fan morphology.

These may ultimately affect the coupling or buffering status of the fan. A previously buffered aggrading system may become a coupled system by dissection of the former fan surface and establishment of channel continuity. Alternatively, excess sediment may bury the former fan surface including any former channels through the fan, transforming a previously coupled system to a buffered system. The coupling status of a fan will be expressed in the relationships between fan and channel morphologies, and the relationships of the upstream and downstream drainage networks (see Figure 7.1).

This chapter deals with coupling characteristics on fans, through an examination of the relationships between fan and channel properties. These are discussed within the context of changes in fan morphology resulting from late Quaternary environmental changes within the Mediterranean region, with special reference to fan systems in southeast Spain.

7.2 ALLUVIAL FANS IN THE MEDITERRANEAN REGION

Alluvial fans occur in a wide range of climatic contexts (Rachocki and Church, 1990), including arctic (e.g. Leggett et al., 1966; Ritter and Ten Brink, 1986), alpine (e.g. Kostaschuk et al., 1986; Ritter et al., 1995), humid temperate (e.g. Harvey and Renwick, 1987; Brazier et al., 1988; Kochel, 1990), and humid tropical climates (e.g. Kesel and Spicer, 1985; Kesel and Lowe, 1987).

However, fans appear to be particularly well developed in dry or, at least, seasonally dry climates where storm- and flood-dominated systems, involving high rates of erosion (Baker, 1977; Laronne and Reid, 1993), combine with rapid downstream losses of stream power (Reid and Frostick, 1997). There are numerous studies of the classic alluvial fans of the American southwest (e.g. Denny, 1965; Hooke, 1967; Bull, 1977; 1991; Beaty, 1990; Dorn 1994; Blair and McPherson, 1994a), but also many studies related to fans in other dry regions, for example in Australia (e.g. Wasson, 1974, 1979) and in the Middle East (e.g. Al Sarawi, 1988; Al Farraj and Harvey, 2000). Because ongoing tectonism is conducive to creating the gross topography suitable for fan accumulation, many of these studies describe fans in tectonically active mountain regions. Alluvial fans are characteristic landforms of tectonically active mountainous dry-region landscapes.

Three aspects of the Mediterranean region favour the development of Pleistocene to modern alluvial fans. Much of the region comprises young mountains, uplifted during the Pliocene to early Pleistocene and continues to be tectonically active today. The modern climate ranges from near desert to seasonally dry subhumid climates. During Pleistocene global glacials, climates in the Mediterranean basin also tended to be dry, but involved major sediment production from the mountain areas (Amor and Florschutz, 1964; Butzer, 1964; Sabelberg, 1977; Rhodenburg and Sabelberg, 1980).

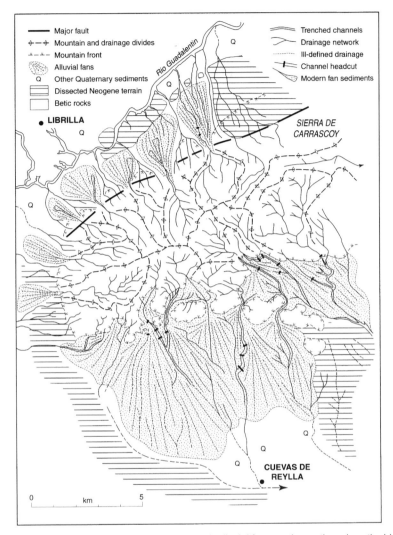

Figure 7.1 Comparison of drainage networks through alluvial fans on the north and south sides of the Sierra de Carrascoy, Murcia, Spain. Note buffering effect of the fans north of the Sierra, compared with the channel continuity through the fans south of the Sierra

Excluding studies related to neighbouring alpine regions (e.g. Marchi et al., 1993; Gomez Villar et al., 1995), studies of fans in Mediterranean regions include studies of fans in Spain (e.g. Harvey, 1984a, 1984b, 1987a, 1990, 1996; Harvey et al., 1999; Somoza et al., 1989; Silva et al., 1992; Delgardo Castilla, 1993; Calvache et al., 1997), Italy (Sorriso Valvo et al., 1998), Greece (e.g. Pope and Van Andel, 1984) including Crete (e.g. Nemec and Postma, 1993), Turkey (Roberts, 1995), Tunisia (White, 1991; White et al., 1996; White and Walden, 1997) and Israel (e.g. Bowman, 1978, 1988; Frostick and Reid, 1989; Gerson et al., 1993; Goldberg, 1994; Amit et al., 1995), though some of those in Tunisia and Israel relate to desert regions rather than Mediterranean regions.

Many of these and other studies demonstrate that while tectonism may be important for determining the location and broad development of alluvial fans in the Mediterranean region (e.g. Harvey, 1988, 1990; Somoza et al., 1989; Silva et al., 1992; Calvache et al., 1997; Gerson et al., 1993; Amit et al., 1995), a major determinant of change within Mediterranean alluvial fan environments has been Quaternary climatic change.

Where they have been dated, many fan sedimentary sequences (e.g. Harvey, 1990; Nemec and Postma, 1993; Roberts, 1995; Harvey et al., 1999) indicate phases of fan aggradation coincident with Pleistocene global glacials, presumably in response to greater sediment production from increased mechanical weathering rates in the mountain source areas. Phases dominated by fan dissection occurred during the intervening interglacials. These include the Holocene, although in some areas this is complicated by human-induced sedimentation (e.g. Pope and Van Andel, 1984). As a result, many Mediterranean fans show a tendency to switch styles between aggradation and dissection, with concurrent variations in the degree of prox-imal-distal coupling within the fan channel systems.

Dissection during periods of reduced sediment supply appears to be enhanced in this region by the development of pedogenic calcrete crusts on abandoned fan surfaces. Such cementation is characteristic of many dry regions (Lattman, 1973), but appears to be particularly well developed in semi-arid Mediterranean regions (Blumel, 1986; Alonso Zarza et al., 1998; Nash and Smith, 1998). Calcrete would tend to increase runoff from the fan surfaces themselves (McDonald et al., 1997), and would also tend to constrain the development of excessive widths in incised fan channels (Van Arsdale, 1982; Harvey, 1987a), thus enhancing stream power. The influence of dissection on the coupling characteristics of the fans in southeast Spain is considered below.

7.3 THE QUATERNARY ALLUVIAL FANS OF SOUTHEAST SPAIN

The study area is coincident with the zone of modern semi-arid climate in the provinces of Alicante, Murcia and Almeria in southeast Spain. The southern part of the region comprises the eastern Betic Cordillera, mountain ranges mostly of metamorphic rocks, block-faulted by a series of left-lateral strike-slip faults (Bousquet, 1979). Between these ranges are Neogene sedimentary basins (Weijermars, 1991). To the north are the Pre-Betic ranges of folded Mesozoic, dominantly carbonate, rocks, with outcrops of Triassic diapirric gypsum-rich marls in the intervening lowlands (Moseley et al., 1981).

In the Betic zone, late Neogene to early Pleistocene fans developed at the margins of the sedimentary basins, which became uplifted and dissected before the accumulation of the modern Pleistocene to Holocene fans (Figure 7.2) at mountain fronts along the basin margins in a variety of tectonic settings (Harvey, 1990, 1996; Silva et al., 1992). In the Pre-Betic zone the modern Pleistocene to Holocene fans accumulated at mountain fronts along the margins of structural depressions (Figure 7.2).

Modern climates are semi-arid Mediterranean, with mean annual rainfall < c. 300 mm (Geiger, 1970; Neumann, 1960), much of which falls in autumn storms. Pleistocene climates apparently were also dry and seasonal, but variable between global glacial and interglacial phases: during glacial phases climates were too cold and dry for extensive tree growth, but the tree cover during interglacials was perhaps more extensive on the higher ground (Amor and Florschutz, 1964; Alonso Zarza et al., 1998; Munuera and Carrion, 1991).

The sedimentary sequences preserved within the modern Pleistocene to Holocene fans suggest a progressive change, over the period from the mid–late Pleistocene, from fan aggradation towards fan progradation (proximal incision with distal aggradation) and dissection (Harvey,

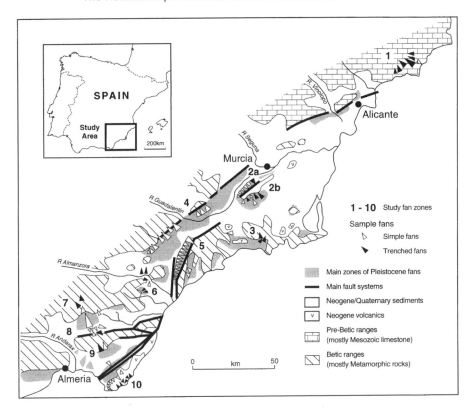

Figure 7.2   Location map of the sampled fan study areas in southeast Spain. Key to the main fan groups: 1, Pre-Betic ranges fans; 2, Sierra de Carrascoy fans (a) north, (b) south (note: this area is shown in more detail in Figure 7.1); 3, Mazarron fans; 4, Guadalentín valley, northern margin fans; 5, Sierra de Almanara fans; 6, Vera basin fans, several minor groups; 7, Sierra de los Filabres, southern margin fans; 8, Tabernas basin, fans fed by catchments on Neogene sedimentary rocks; 9, Sierra de Alhamilla fans; 10, Cabo de Gata range fans

1984b, 1990, 1996). Within this overall context, pulsed periods of aggradation were interspersed with periods dominated by dissection. Where these phases can be dated by correlation with the Quaternary littoral sequence (Harvey, 1978, 1990; Goy and Zazo, 1986, 1989; Goy et al., 1986a, 1986b; Hillaire-Marcel et al., 1986; Harvey et al., 1999), the aggradational phases appear to correlate broadly with Pleistocene global glacials and the dissection phases with interglacials. The sediments themselves range from debris flows to fluvial deposits (as both channelised and sheet bodies), with debris flows more common from small catchments and less common from catchments whose bedrock geologies are dominated by fissile schists (Harvey, 1984a, 1990). In addition, many of the fan sequences show a progressive trend away from debris-flow towards dominantly fluvial deposition (Harvey, 1984a, 1984b, 1990, 1992a). Capping buried palaeo-fan surfaces and forming the exposed surfaces of the older segments of the modern fans are massive pedogenic calcretes. Soil and calcrete development have been used to correlate between fan surfaces and between fan and other depositional surfaces (Harvey et al., 1995; 1999), but the presence of calcrete on fan surfaces in mid-fan and distal zones also affects the dynamics of fan dissection (see below, and Harvey, 1987a).

Table 7.1   Summary of properties of each fan group

	Fan Group (a)	Geology (b)	Tectonics (c)	Terrain (d)	Number of fans Sampled (e)	Number of fans final (f)
1	Pre-Betic	Lst	P	Mt fr, Complex	7	6
2a	Carrascoy N	Met	F	Mt fr	6	5
2b	Carrascoy S	Met	P	Mt fr	6	4
3	Mazarron	Met	Var	Complex	3	–
4	Guadalentín	Met, Sed	F	Mt fr	12	11
5	Almenara	Schist	F	Mt fr	15	14
6	Vera basin	Met, Sed	Var	Complex	9	4
7	Filabres	Schist	P	Complex	8	4
8	Tabernas *	Neogene Sed	F	Mt fr	4	4
9	Alhamilla N	Schist, Met	Var	Complex	4	1
10	Gata	Volc	P	Complex	21	13
				TOTALS	95	67

Notes: (a) For locations see Figure 7.2. (b) Lst – Pre-Betic Cretaceous Limestones; Met – Low and Mod grade Betic basement metamorphic rocks; Sed – Various basement and Mesozoic sedimentary rocks; Schist – mod–high grade Betic basement schists; Neogene Sed – Neogene sedimentary rocks; Volc – Neogene volcanic rocks. (c) P – tectonically stable, passive mountain front; F – active faulted mountain front; Var – various. (d) Mt fr – linear mountain front; Complex – irregular and backfilled mountain fronts. (e) Number of fans sampled in the field. (f) Final number of fans used in regression analyses (equations (7.3) and (7.4)), after rejection of those unsuitable for fan area analysis (see text).
* Tabernas basin, fans from Neogene sedimentary rocks only.

A total of 95 fans have been studied in southeast Spain, within 10 main groups (Table 7.1; Figure 7.2). Detailed descriptions of the fan settings, sequences and morphology for many of these groups are given elsewhere (Harvey, 1978, 1984b, 1987a, 1987b, 1988, 1990; Silva et al., 1992; Harvey et al., 1999). Many of the fans show a modern tendency to dissect, and in extreme cases to couple mountain source–area drainage basins with downstream reaches of the fluvial systems (Harvey, 1996). However, this tendency shows considerable variations between the various fan groups (Table 7.2). The fan channels themselves are ephemeral, carrying flow only

Table 7.2   Numbers of sampled fans of each dissection status, by fan group

	Fan group	Simple aggrading fans	Headcut fans	Dissected headcut fams	Distally-dissected fans (base level)
1	Pre-Betic	1	4*	2	–
2a	Carrascoy N	5	1	–	–
2b	Carrascoy S	3	3	–	–
3	Mazarron	–	1	2	–
4	Guadalentín	12	–	–	–
5	Almenara	15	–	–	–
6	Vera basin	4	2**	1	2
7	Filabres	5	–	–	3
8	Tabernas	4	–	–	–
9	Alhamilla	2	1	–	1
10	Gata	15	–	–	6
Totals		66	12***	5	12

Each asterisk (*) indicates a scoured but non-dissected fan (total 3).

from rare storms (Harvey, 1984c). They are gravel- and cobble-bed channels with morphology ranging from sinuous single-thread channels to wide, shallow-braided channels (Figure 7.3).

7.4 FAN AND FAN CHANNEL MORPHOMETRY: DISSECTING AND AGGRADING FANS

In previous studies fan dissectional status has been summarised by relationships between fan and channel profiles (Harvey, 1987a, 1988, 1996). In those studies, aggrading (or perhaps more strictly, prograding) fans are defined as fans where in proximal locations the fan channel is incised within a fanhead trench, below the fan surface, but has a gentler gradient, and meets the fan surface in mid-fan at an intersection point. There, flow confinement decreases and deposition takes place distally on the active depositional lobe (Blair and McPherson, 1994a, 1994b), as described by Hooke (1967) and Wasson (1974). On the Spanish fans and elsewhere, the gradient downfan from the intersection point normally increases from that of the channel within the fanhead trench. This is the 'normal' fan morphology, described many times in the literature.

Dissecting fans are of two types. In the first type, first described in southeast Spain (Harvey, 1978) but also common in the Valencia lowland further north (Segura, 1986), incision takes place at the intersection point. This produces a headcut or step in the channel profile, which serves to concentrate the flow, and may lead to further headcut development downfan, ultimately perhaps to the total dissection of the distal fan surface. In the second type, the fan is totally dissected by the fan channel. The first type has been interpreted to be the result of excess power at the intersection point, as a result of the gradient increase more than compensating for the reduced confinement of the flow (Harvey, 1987a, 1996). These conditions would be exacerbated by climatically induced reductions in sediment supply or increases in flood runoff, and by the presence of calcrete-cemented fan surfaces at the intersection point (Van Arsdale, 1982; Harvey, 1987a). The second type can be produced either by development of the first type or as a result of base-level induced incision into the distal zones of an alluvial fan.

Of the 95 fans studied in southeast Spain, 66 are simple aggrading or prograding fans, 12 are dissecting or distally trenched fans, where the field evidence suggests that dissection can be related to tectonic or local incisional base-level conditions, and 17 are headcut, intensively scoured, or dissecting fans where dissection appears to be related purely to processes at mid-fan (Table 7.2). All 17 of the last group are calcrete-crusted fans, and most are in the three northern fan groups (Pre-Betic ranges, Carrascoy, Mazarron: Figure 7.2). Many are confined, rather than simple mountain-front fans.

Earlier studies (Harvey, 1987a) suggested that modern mid-fan dissection on calcrete-crusted fans can be related to a threshold (Schumm, 1979) defined by fan and fan-channel morphometric properties: channel widths and gradient characteristics of the fan surfaces and fan channels.

Many previous studies of the morphometric properties of fans show the influence of drainage basin characteristics – commonly, that fan area increases and fan gradients decrease with increasing drainage basin size. For fan area (e.g. Bull, 1962, 1977; Denny, 1965; Hooke, 1968; Hooke and Rohrer, 1977; Lecce, 1991), the relationship for fans in any one region can be expressed by a simple equation:

$$F = pA^q \qquad (7.1)$$

where F and A are the fan and drainage basin areas (km^2) respectively (Harvey, 1997). Fan groups in many regions accord with this general relationship, with the values of the exponent q

(a)

(b)

Figure 7.3 Fan channel types: (a) aggrading fan, active lobe downfan from the intersection point, Villaescusa fan, Murcia; (b) midfan headcut dissection through calcrete crusted older fan surfaces, Torre fans, Mazarron, Murcia

(c)

(d)

Figure 7.3   Fan channel types: (c) single-thread fan channel, Little Honda fan, Almería;. (d) braided fan channel, within fan trench, Ros fan, Murcia

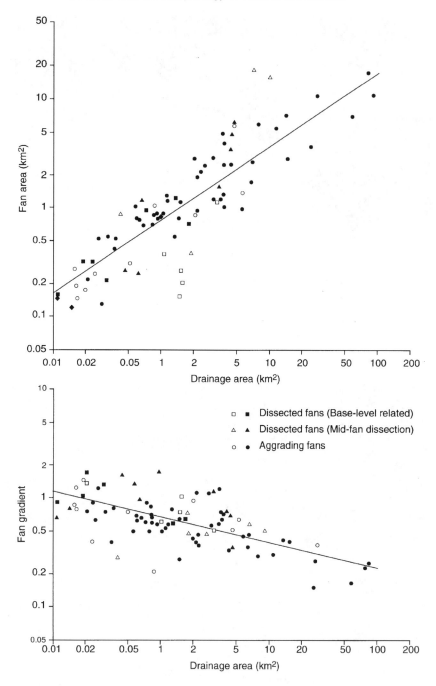

Figure 7.4 Plot of conventional morphometric relationships of fan area and fan gradient data in relation to drainage area, for the southeast Spanish fans. Open symbols relate to data not used in the final regression analysis (see text)

generally ranging between 0.7 and 1.1. The values of p show a wide range, depending on fan age, climate, tectonic setting and bedrock erodibility.

The fan gradient (G), usually measured as the axial fan surface slope of the upper part of the fan, has a simple inverse relationship with drainage area (A) (e.g. Bull, 1962, 1977; Hooke, 1968), which may be expressed by the equation:

$$G = aA^{-b} \tag{7.2}$$

where the exponent b is normally within the range -0.15 to -0.35 (Harvey, 1997). Values of a show a range between c. 0.03 and 0.17, apparently reflecting variations in sedimentary processes and sediment size, but which also may reflect tectonics (Bull, 1962).

Previous studies have shown that the Spanish fans accord with these general relationships for fan area, either as a whole (Harvey, 1997) or for selected fan groups (Silva et al., 1992; Harvey et al., 1999). Similarly, the gradient relationships for the Spanish fans as a whole accord with these general relationships (Harvey, 1987a), but explanation levels can normally be improved by considering different fan groups separately (Harvey, 1987a, 1987b; Silva et al., 1992; Harvey et al., 1999), by distinguishing between debris-flow and fluvially dominant fans (Harvey, 1992b), or by taking into account basin relief characteristics as well as basin size (Harvey, 1987a, 1992b).

From the total of 95 fans studied here (as opposed to 77 in the earlier study in which fan area was not considered; Harvey, 1987a) a subtotal of 67 fans can be identified for which fan areas can be individually related to catchment areas. Conventional regression analyses (Figure 7.4) have been run for these fans, with the following results:

$$F = 0.807A^{0.675} \tag{7.3}$$

$$(n = 67, \text{SE} = 0.208 \log \text{units}, r = 0.908, \ p < 0.0001)$$

$$G = 0.068A^{-0.249} \tag{7.4}$$

$$(n = 67, \text{SE} = 0.173 \log \text{units}, r = 0.693, \ p < 0.0001).$$

In the fan area case no improvement is achieved by using multiple regression and taking into account basin relief or slope characteristics. In the fan gradient case the explanation level is improved if basin slope is also taken into account, yielding the regression equation:

$$G = 0.163A^{-0.084}B^{0.654} \tag{7.5}$$

$$(n = 67, \text{SE} = 0.139 \log \text{units}, r = 0.817, \ p < 0.001, \text{both variables})$$

where B is mean basin slope (basin relief/basin length).

Previous studies have demonstrated that residuals from such regression analyses can be used to identify the different tectonic settings between fan groups (Silva et al., 1992) or, for one fan group (the Cabo de Gata group; Group 10, Figure 7.2), can be used to distinguish between unconfined aggrading, confined and dissecting fans (Harvey et al., 1999).

For the fans from southeast Spain as a whole, however, the residuals from the regressions above do little more than indicate that the dissected fans tend to have relatively steep fan gradients in relation to drainage area.

The channel geometry characteristics – channel gradient and channel width as an inverse measure of flow confinement – are important for influencing transporting power at the intersection point. A similar approach has been used to express the relationships of fan-channel geometry to drainage area as for fan gradients, and to compare fan channels with 'normal' non-fan stream channels in the field area.

Blair and McPherson (1994b) have argued that fluvial processes on fans, dominated by sheetfloods, are fundamentally different from channelised fluvial processes, and are associated with fundamentally different gradient characteristics. They argue for a 'slope gap', equivalent to gradients of c. 0.008–0.025 separating the two regimes. Their concept may be valid in relation to the differences between channelised and sheet flows, expressed here by the increases in gradients which characteristically occur at fan intersection points. However, their concept of a specific slope gap is flawed, as it is in part related to preservation potential and based on comparisons between alluvial fans and *large* rivers. Gradients of both rivers and fans reflect drainage basin size, and for valid comparisons drainage basins of a similar size range should be used.

Within southeast Spain, non-fan ephemeral gravel-bed streams, draining basins of a comparable size range to those feeding the fans, have been examined at 38 sites. Channel widths and gradients have been measured and estimates made of bar sediment sizes from quadrat photo-plots, yielding estimates equivalent to the D_{90} (Harvey, 1987c). Similarly, 52 sites were surveyed on fan channels, in mid-fan environments, within fanhead trenches. In addition the distal gradients were measured on 35 fans downfan from the intersection points.

The gradient measures were related to drainage areas by the following regression equations:

$$S_n = 0.038A^{-0.370} \tag{7.6}$$

$$(n = 38, SE = 0.183 \text{ log units}, r = 0.861, p < 0.0001)$$

$$S_f = 0.042A^{-0.213} \tag{7.7}$$

$$(n = 52, SE = 0.187 \text{ log units}, r = 0.603, p < 0.0001)$$

$$G_d = 0.062A^{-0.214} \tag{7.8}$$

$$(n = 35, SE = 0.187 \text{ log units}, r = 0.605, p < 0.0001)$$

S is channel gradient ('n' on non-fan channels, 'f' on fan channels) and G_d is the distal fan gradient. Both sets of channels plot with similar relationships to drainage area as the Appalachian streams studied by Hack (1957). In neither of the cases involving channel gradients did inclusion of basin slope or sediment data improve the explanation level. In fact, for neither dataset did the sediment data show any correlation with drainage area nor was there any correlation between sediment size and channel gradients or channel widths. Testing the Spanish non-fan channels against the Hack (1957) relationship of channel gradients on the ratio of sediment size to drainage area does yield a significant correlation, but it is weaker than that for drainage area alone. No direct comparison can be made with Hack's (1957) results, because of differences in the sediment measures used.

When plotted (Figure 7.5), the gradient regressions for the Spanish data illustrate that all gradients decrease with increasing drainage area, and for drainage areas less than 10 km^2 there is little difference in channel gradient between non-fan and fan channels. For larger drainage areas in this dataset non-fan channels tend to be less steep; however, this may simply reflect the sample differences between the two datasets.

When comparing channel gradient regressions with fan gradient regressions the greater fan gradients per drainage area are obvious. The similarity between fan gradient, measured on the upper part of each fan (see equation (7.4)), and the distal fan gradient reflects the general linearity and lack of concavity in the axial fan profiles. These results as a whole suggest the possibility of differences between the gradients of non-fan and fan channels, but they also emphasise the differences between the gradients under channelised conditions within the

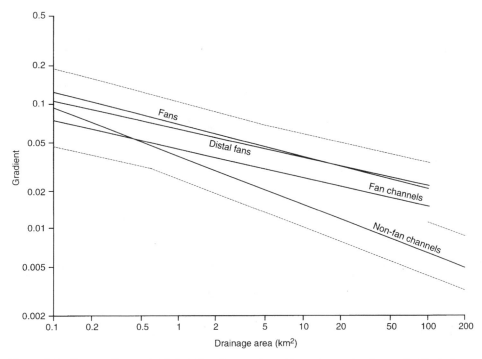

Figure 7.5 Comparative graphs derived from the morphometric analyses, showing gradient to drainage area relationships for non-fan channels, fan channels, distal and main fan surfaces. Error bars are for one SE

fanhead trench and sheetflow conditions affecting the distal fan surfaces. Thus in part they support the Blair and McPherson (1994b) concept of different gradients resulting from deposition from channelised streamflow and sheetfloods. However, the 'slope gap' itself does not exist. Non-fan channels from basins of c. 50–200 km^2 clearly have slopes within the so-called slope gap of gradients in the range c. 0.008–0.025.

The channel width data have been treated in the same way, and can be related to drainage area by the following regressions:

$$W_n = 6.40A^{0.268} \tag{7.9}$$

$$(n = 38, \mathrm{SE} = 0.216 \log \text{units}, r = 0.719, p < 0.0001)$$

$$W_f = 6.68A^{0.319} \tag{7.10}$$

$$(n = 48, \mathrm{SE} = 0.245 \log \text{units}, r = 0.665, p < 0.0001)$$

When plotted (Figure 7.6), these results suggest little difference between non-fan and fan channel widths in situations fed by drainage basins smaller than c. 10 km^2, but perhaps a tendency for wider channels on larger fans. This, together with the tendency for greater slopes noted above, suggests greater bedload transport and a greater likelihood of braiding on fan channels.

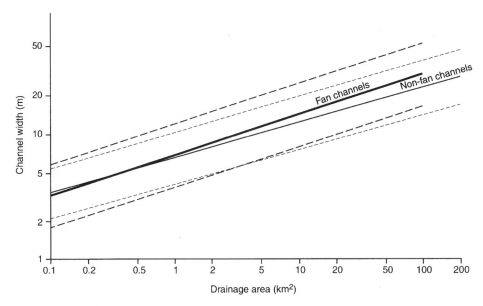

Figure 7.6 Comparative graphs derived from the morphometric analyses, showing channel width to drainage area relationships for non-fan channels and fan channels. Error bars are for one SE

Returning to the question of the differences between channels on aggrading and dissecting fans, this analysis confirms the results of the earlier study (Harvey, 1987a). Residuals from the regressions above, together with those from the regression of fan-channel gradient on fan gradient (Table 7.3) confirm the tendency for dissection of crusted fans to be characterised by: (i) greater than expected fan gradients, (ii) less than expected channel gradients (whether estimated on the basis of drainage area or fan gradient) and (iii) smaller than expected channel widths. In situations involving only limited sediment transport, this would tend to increase unit stream power through the intersection point, and allow scour and incision into the underlying fan sediments.

However, some of the individual cases are anomalous, and these results do not add to the precision with which mid-fan trenching thresholds may be defined. Further consideration needs to be given to individual cases.

Table 7.3 Characteristics of dissecting fans: mean residuals (log units) for dissecting fans from regression analyses for all available data

	Fan gradient regression (see equation (7.4))	Fan channel gradient regression (see equation (7.7))	Channel width regression (see equation 7.10))	Regression of fan channel gradient on fan gradient (see below *)
n (Dissecting fans)	17	15	15	15
Mean residuals for dissecting fans	+0.048	−0.024	−0.074	−0.039

* Regression results: $Sf = 0.418G^{0.871}$ ($n = 44$, SE $= 0.134$ log units, $r = 0.831$, $p < 0.0001$)

7.5 AGGRADATION OR DISSECTION: EXAMPLES OF INDIVIDUAL FANS

Complete fan and channel profiles have been surveyed through 15 of the fans, and channel width and sediment data derived from numerous sections along 11 of these fan channels (some data have been previously reported in Harvey, 1987a; some in Harvey et al., 1999; and the remainder previously unreported). Of the 15 fans, eight are aggrading fans (one of which, Ceporro, Table 7.4, shows minor localised scour at the intersection point, but is essentially an aggrading fan), two are deeply scoured at their intersection points, one is a headcut fan, two are dissected headcut fans, and two are dissected fans, but the dissection is base-level related (Harvey et al., 1999). The fan and channel gradient and channel width data have been summarised for three characteristic zones on each fan: the proximal upper fan zone, the mid-fan zone upfan of the intersection point where present, and the distal zone downfan from the intersection point (Table 7.4). Examples of each type are illustrated on Figure 7.7.

These data summarise downfan trends and express changes in gradient and channel characteristics through the intersection point associated with aggradation, scour or dissection. The fan gradient data show that some fan profiles are more or less rectilinear, sometimes below a steep apex zone, others are concave, especially on backfilled fans, and others show a steepening of profile in mid-fan. There are no obvious differences between aggrading and dissecting fans.

The channel gradient data illustrate the tendency for steepening at the intersection point, and even on markedly concave fans, where there may be no absolute channel steepening (e.g. Little Honda fan), the degree of channel concavity is less than that of the fan profile. On the scoured, headcut and dissected headcut fans, channel gradients in the distal reaches are closer to the fan gradients than further upfan, as though there has been little substantial gradient modification since incision. The generalised figures also obscure localised channel steepenings; for example, on Cayola the channel steepens towards the headcut zone (Harvey, 1978), and on Ros the channel between successive downchannel headcuts comprises a series of locally concave reaches (Harvey, 1988). On the two base-level dissected fans, marine erosion of the fan toes at times of high sea levels caused channel foreshortening and a zone of channel steepening (Harvey et al., 1999). On Cala Carbon fan this coincides with the fan distal zone, but on La Isleta fan this steepening has worked up into mid-fan to produce a local steepening from gradients of 0.045 to 0.054 (Harvey et al., 1999).

The channel width data illustrate how, on most of the aggrading fans, channel widths increase towards the intersection points then rapidly increase downfan before the channels peter out completely. On some of the dissecting fans, widths are restricted throughout, but again local variations of width on Cayola and Ros fans are obscured by these generalised data.

These data amplify the results of the earlier analyses. However, it is clear that although the general association of trenching with transporting power characteristics is evident in most cases, it is difficult to identify specific trenching thresholds associated with particular combinations of gradients and channel widths. On the other hand, scour and trenching appear to be widespread trends on the crusted fans (they are present to some degree on more fans than these data suggest) and are more advanced on some fans than on others. Response of the non-crusted Filabres fans to a major flood in 1980 (Harvey, 1984c) and to subsequent floods shows no sign of vertical incision. However, field observations suggest that recent floods affecting the Carrascoy, and especially the Pre-Betic fan groups, caused localised incisions into the channel floors. Incision appears to be a progressive trend with differing degrees of development in evidence on the range of examples shown here. There is no sign of incision on the Filabres fans (Honda, Little Honda, Mezquita: Table 7.4) and little or none on the non-dissected fans of the Cabo de Gata group (Salina and Michelin). Villaescusa shows the first signs of incision with

Table 7.4 Proximal, mid-fan and distal fan characteristics of selected fans

Fans and group*	Mean fan gradients			Mean channel gradients			Mean channel width (m)		
	Prox.	Mid-fan	Distal	Prox.	Mid-fan	Distal	Prox.	Mid-fan	Distal
Aggrading fans									
Villaescusa (5)	0.067	0.056	0.044	0.043	0.036	0.040	7.3	5.9	11.2†
Honda (7)	0.036	0.031	0.027	0.019	0.020	0.029	9.7	30.6	56.4†
Little Honda (7)	0.176	0.126	0.090	0.136	0.047	0.040	4.7	6.2	7.1†
Mezquita (7)	0.111	0.081	0.052	0.062	0.050	0.052	4.4	8.7	14.7†
Ceporro (8)(x)	0.085	0.061	0.058	0.061	0.044	0.051	2.4	2.2	4.2†
Feos (9)	0.255	0.198	0.149	0.163	0.095	0.063	2.0	1.9	1.3
Salina (10)	0.070	0.080	0.067	0.060	0.074	0.067	–	–	–†
Michelin (10)	0.029	0.032	0.024	0.032	0.028	0.024	–	–	–†
Scoured fans									
La Nucia (1)	0.182	0.176	0.132	0.149	0.134	0.142	7.1	5.0	5.2
Lisbona (6)	0.092	0.087	0.067	0.091	0.053	0.056	2.6	5.5	9.1
Headcut fan									
Cayola (1)	0.123	0.100	0.070	0.121	0.074	0.057	4.5	13.2	27.4†
Headcut dissecting fans									
Corachos (2b)	0.023	0.026	0.018	0.023	0.014	0.018	9.0	7.2	4.9
Ros (2b)	0.080	0.070	0.055	0.056	0.052	0.048	5.6	13.0	22.7
Base-level induced dissecting fans									
Cala Carbon (10)	0.047	0.060	0.045	0.051	0.040	0.060	–	–	–
La Isleta (10)	0.082	0.050	0.046	0.057	0.054‡	0.029	–	–	–

* For groups see Table 7.1; for locations of groups, see Figure 7.2; (x) shows minor local scour.

† Indicates that channels continue to increase in width, then disappear altogether downfan.

‡ Channel gradient decreases to 0.045 then increases downfan to 0.054.

Note: No width data are available for Cabo de Gata group fans.

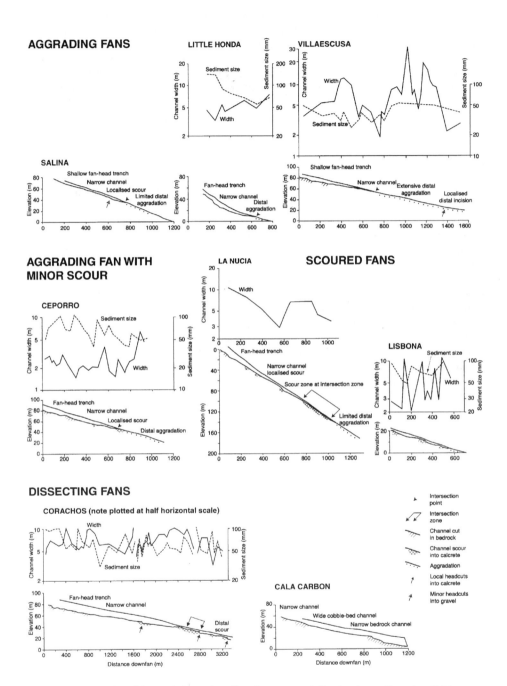

Figure 7.7 Examples of fan and channel profiles, for representative fan types. Based on field survey. Also shown are plots of channel width, where data are available

shallow headcuts developed into unconsolidated sediments of the distal fan zone. There is localised minor scour on Ceporro, and major intersection point scour on La Nucia. At Lisbona, scour and minor incision have created several secondary intersection points, between which there are wide aggrading channels. At Cayola scour is expressed as one major marked headcut, followed downfan by a secondary aggrading intersection point. At Ros the channel as a whole is trenched below the fan surface, but two major headcuts separate wider locally aggrading (?) reaches. At Corachos there is a continuously trenched channel, within which there is major headcut development, especially in the vicinity of the former intersection point.

7.6 DISCUSSION

The alluvial fans of southeast Spain exhibit varying degrees of modern incision, and in a few cases the incision can be directly related to local base-level conditions. The most obvious examples are the Cabo de Gata coastal fans (Harvey et al., 1999), where modern incision has followed coastal erosion related to high Holocene sea levels. On other coastal fans, including examples in western Almeria and in Murcia, which have not been studied here, modern incision may have the same causes. Additionally, tectonically induced river incision may have led to fan dissection on, for example, the Alamanzora valley fans (in the Vera group 6, Figure 7.2) and the Rambla Sierra on the northern margin of the Sierra de Alhamilla (Figure 7.2). Locally, base-level incision of a tributary fan may be caused by the (non-base-level related) incision of a main fan channel (some of the smaller fans in the Filabres group are of this type).

However, the major cause of incision of the modern channels into the older fan surfaces involves critical power relationships and is controlled by the supply of water and sediment to the fan environments. The dry cold phases of the late Pleistocene appear to have been periods of excess sediment supply to the alluvial fan systems of southeast Spain, causing fan aggradation (Harvey, 1990; Harvey et al., 1999). During the Holocene either a reduction of sediment supply as a result of reduced sediment availability or an increase in stream power must be postulated to have been the cause of incision in fan proximal zones. Apart from minor sedimentation, possibly related to human activity, the Holocene appears to have been associated with progressive incision, not only in proximal zones but particularly in mid-fan zones. This incision has been spatially variable, with the non-crusted Filabres fans showing only fanhead dissection, but other zones showing variable degrees of mid-fan scour and dissection. The greatest degree of dissection is shown by the strongly calcreted fans of the Pre-Betic and Carrascoy ranges. As dissection takes place there is progressive integration of the drainage networks, from the mountain catchments to the axial river systems. The fan environments cease to be buffering zones, and as coupling increases the fan's surfaces become inactive and the channels become simply the middle reaches of continuous drainage networks.

To return to the themes introduced earlier, it is relevant to ask whether this pattern is typical of Mediterranean fans as a whole, and whether, as a result, there is a particular Mediterranean type of alluvial fan.

Pleistocene and modern tectonism are characteristic of most Mediterranean regions, and although it may have been important for the creation of topography favouring the development of alluvial fans, it appears to have had at most a minor role in influencing processes on Mediterranean fans from the late Pleistocene through the Holocene.

The modern climate is undoubtedly important. Throughout the range of Mediterranean climates, intense rainstorms and high runoff rates appear to be important for fan processes (Harvey, 1984c; Frostick and Reid, 1989). Past climates also appear to have had similar influences on fans in many Mediterranean regions and there is evidence that many, but not all,

Mediterranean areas had pulsed periods of fan aggradation during the late Pleistocene, followed by progradation and dissection during the Holocene (e.g. Spain: Harvey, 1990; Crete: Nemec and Postma, 1993; Turkey: Roberts, 1995; Tunisia: White et al., 1996). A major factor appears to have been increased availability of coarse sediment from the mountain source areas during late Pleistocene cold, dry climates.

Influence from human activity is another common characteristic of Mediterranean environments. However, most studies of Mediterranean fans identify a relatively modest role for human influence in comparison with the climatic influence (e.g. Pope and Van Andel, 1984).

An important characteristic of the fan surfaces in Spain that makes them prone to dissection is the presence of calcrete, indurating the distal fan surfaces. While this might initially protect the surfaces from erosion, it also appears to focus the flow and, once the calcrete is eroded, the flow width is restricted, maintaining relatively high unit stream power. Calcrete crusts, which appear to be important in other relatively dry Mediterranean regions, have also been observed on fans in Crete (Nemec and Postma, 1993) and Tunisia (White et al., 1996), and are undoubtedly present in other areas of the Mediterranean region and the Middle East.

The interaction between the calcrete-crusted surfaces and high runoff power is the main cause of modern dissection on the Spanish fans and is a tendency that would be expected in other Mediterranean regions. Even without current climatic trends, which suggest greater storminess, we could expect the present dissectional trend to continue.

ACKNOWLEDGEMENTS

I am grateful to a number of sources for funding field work in Spain over a number of years, particularly to the University of Liverpool Research and Development Fund and to the British Council/Acciones Integradas programme for joint work with Spanish scientists. I am grateful to Sandra Mather of the University of Liverpool, Department of Geography, Graphics Section, for producing the diagrams and to Lindy Walsh of Cortijo Urra for providing accommodation in the field in Spain.

REFERENCES

Al Farraj, A. and Harvey, A.M. 2000. Desert pavement characteristics on wadi terrace and alluvial fan surfaces: Wadi Al Bih, UAE and Oman. *Geomorphology*, 35, 279–297.

Al Sarawi, A.M. 1988. Morphology and facies of alluvial fans in Kadhmah Bay, Kuwait. *Journal of Sedimentary Petrology*, 58, 902–907.

Alonso-Zarza, A.M., Silva, P.G., Goy, J.L. and Zazo, C. 1998. Fan-surface dynamics and biogenic calcrete development: interactions during ultimate phases of fan evolution in the semiarid SE Spain (Murcia). *Geomorphology*, 24, 147–167.

Amit, R., Harrison, J.B.J. and Enzel, Y. 1995. Use of soils and colluvial deposits in analyzing tectonic events – The southern Arava Rift, Israel. *Geomorphology*, 12, 91–107.

Amor, J.M. and Florschutz, F. 1964. Results of the preliminary palynological investigation of samples from a 50 m boring in southern Spain. *Boletin de la Real Sociedad Espanola de Historia Natural (Geologica)*, 62, 251–255.

Baker, V.R. 1977. Stream channel response to floods, with examples from central Texas. *Geological Society of America, Bulletin*, 88, 1057–1071.

Beaty, C.B. 1990. Anatomy of a White Mountains debris flow – the making of an alluvial fan. In A.H. Rachocki and M. Church (Eds) *Alluvial Fans: A Field Approach*, Wiley, Chichester, 69–89.

Blair, T.C. and McPherson, J.G. 1994a. Alluvial fan processes and forms. In A.D. Abrahams and A.J. Parsons (Eds) *Geomorphology of Desert Environments*. Chapman & Hall, London, 354–402.

Blair, T.C. and McPherson, J.G. 1994b. Alluvial fans and their natural distinction from rivers based on morphology, hydraulic processes, sedimentary processes and facies assemblages. *Journal of Sedimentary Research*, A64, 450–489.

Blissenbach, E. 1952. Relation of surface angle distribution to particle size distribution on alluvial fans. *Journal of Sedimentary Petrology*, 22, 25–28.

Blumel, W.D. 1986. Calcretes in southeast Spain – genesis and geomorphic position. In F. López-Bermúdez and J. Thornes (Eds) *Estudios sobre geomorphologia del sur de Espana*. Universidad de Murcia, Murcia, 23–26.

Bousquet, J.C. 1979. Quaternary strike-slip faults in southeastern Spain. *Tectonophysics*, 52, 277–286.

Bowman, D. 1978. Determination of intersection points within a telescopic alluvial fan complex. *Earth Surface Processes and Landforms*, 3, 265–276.

Bowman, D. 1988. The declining but non-rejuvenating base-level – the Lisan Lake, the Dead Sea, Israel. *Earth Surface Processes and Landforms*, 13, 239–249.

Brazier, V., Whittington, G. and Ballantyne, C.K. 1988. Holocene debris cone evolution in Glen Etive, Western Grampian Highlands, Scotland. *Earth Surface Processes and Landforms*, 13, 525–531.

Bull, W.B. 1962. Relations of alluvial fan size and slope to drainage basin size and lithology, in western Fresno County, California. *United States Geological Survey Professional Paper*, 430B, 51–53.

Bull, W.B. 1977. The alluvial fan environment. *Progress in Physical Geography*, 1, 222–270.

Bull, W.B. 1979. Threshold of critical power in streams. *Geological Society of America, Bulletin*, 90, 453–464.

Bull, W.B. 1991. *Geomorphic Responses to Climatic Change*. OUP, Oxford.

Butzer, K.W. 1964. Climatic-geomorphic interpretation of Pleistocene sediments in the Eur-African sub tropics. In F.C. Howell and F. Bouliere (Eds) *African Ecology and Evolution*, Methuen, London, 1–25.

Calvache, M., Viseras, C. and Fernandez, J. 1997. Controls on alluvial fan development – evidence from fan morphometry and sedimentology; Sierra Nevada, SE Spain. *Geomorphology*, 21, 69–84.

Delgardo Castilla, L. 1993. Estudio sedimentologico de los cuerpos sedimentarios Pleistocenos en la Rambla Honda, al N. de Tabernas, provincia de Almería (SE de España). *Cuaternario y Geomorfologia*, 7, 91–100.

Denny, C.S. 1965. Alluvial fans in Death Valley region, California and Nevada. *United States Geological Survey Professional Paper*, 466, 59pp.

Dorn, R.I. 1994. The role of climatic change in alluvial fan development. In A.D. Abrahams and A.J. Parsons (Eds) *Geomorphology of Desert Environments*, Chapman & Hall, London, 593–615.

Frostick, L.E. and Reid, I. 1989. Climatic versus tectonic controls of fan sequences: lessons from the Dead Sea, Israel. *Journal of the Geological Society, London*, 146, 527–538.

Geiger, F. 1970. Die ariditat in sudostspanian. *Stuttgarten Geographische Studien*, Band 77, 173pp.

Gerson, R., Grossman, S., Amit, R. and Greenbaum, N. 1993. Indicators of faulting events and periods of quiescence in desert alluvial fans. *Earth Surface Processes and Landforms*, 18, 181–202.

Goldberg, P. 1994. Interpreting late Quaternary continental sequences in Israel. In O. Bar-Yosef and R.S. Kra (Eds) *Late Quaternary Chronology and Palaeoclimates of the Eastern Mediterranean*, Radiocarbon, Special Publication 1994, 89–102.

Gomez Villar, A., Montserrat, G., Ortigosa, L.M. and García-Ruíz, J.M., 1995. Colonizacion vegetal y actividadgeomorfologica en abanicos aluviales del Pirineo espanol. *Cuaternario y Geomorfologia*, 8(3–4), 53–63.

Goy, J.L. and Zazo, C. 1986. Synthesis of the Quaternary in the Almeria littoral, neotectonic activity and its morphologic features, Western Betics Spain. *Tectonophysics*, 130, 259–270.

Goy, J.L. and Zazo, C. 1989. The role of neotectonics in the morphologic distribution of the Quaternary marine and continental deposits of the Elche Basin (SE Spain). *Tectonophysics*, 163, 219–225.

Goy, J.L., Zazo, C., Bardaji, T. and Somoza, L. 1986a. Las terrazas marinas del cuaternario recente en los litorales de Murcia y Almeria (Espana): El control de la neotectonica en la disposicion y numero de las mismas. *Estudios Geologicos*, 42, 439–443.

Goy, J.L., Zazo, C., Hillaire-Marcel, C. and Causse, C. 1986b. Stratigraphie et chronologie (U/Th) du Tyrrhenian du Sud-Est de l'Espagne. *Zeitschrift für Geomorphologie, Supplement Band*, 62, 71–82.

Hack, J.T. 1957. Studies of longitudinal stream profiles in Virginia and Maryland. *United States Geological Survey Professional Paper*, 294, 45–97.

Harvey, A.M. 1978. Dissected alluvial fans in southeast Spain. *Catena*, 5, 177–211.

Harvey, A.M. 1984a. Debris flows and fluvial deposits in Spanish Quaternary alluvial fans: implications for fan morphology. In E.H. Koster and R. Steel (Eds) *Sedimentology of Gravels and Conglomerates*. Canadian Society of Petroleum Geologists, Memoir 10, 123–132.

Harvey, A.M. 1984b. Aggradation and dissection sequences on Spanish alluvial fans: influence on morphological development. *Catena*, 11, 289–304.

Harvey, A.M. 1984c. Geomorphological response to an extreme flood: a case from southeast Spain. *Earth Surface Processes and Landforms*, 9, 267–279.

Harvey, A.M. 1987a. Alluvial fan dissection: relationships between morphology and sedimentation. In L. Frostick and I. Reid (Eds) *Desert Sediments, Ancient and Modern*. Geological Society of London, Special Publication 35. Blackwell, Oxford, pp. 87–103.

Harvey, A.M. 1987b. Patterns of Quaternary aggradational and dissectional landform development in the Almeria region, southeast Spain: a dry-region tectonically-active landscape. *Die Erde*, 118, 193–215.

Harvey, A.M. 1987c. Discussion: of Church, M.A., McClean, D.G and Wolcott, J.F., River bed gravels: Sampling and analysis. In C.R. Thorne, J.C. Bathurst and R.D. Hey (Eds) *Sediment Transport in Gravel-Bed Rivers*. Wiley, Chichester, 639.

Harvey, A.M. 1988. Controls of alluvial fan development: the alluvial fans of the Sierra de Carrascoy, Murcia, Spain. In A.M. Harvey and M. Sala (Eds) Geomorphic Processes in Environments with Strong Seasonal Contrasts – Volume II: Geomorphic Systems. Catena Supplement, 13, 123–137.

Harvey, A.M. 1990. Factors influencing Quaternary alluvial fan development in southeast Spain. In A.H. Rachocki and M. Church (Eds) *Alluvial Fans: A Field Approach*, Wiley, Chichester, 247–269.

Harvey, A.M. 1992a. Controls on sedimentary style on alluvial fans. In P. Billi, R.D. Hey, C.R. Thorne and P. Tacconi (Eds) *Dynamics of Gravel-Bed Rivers*, Wiley, Chichester, 519–535.

Harvey, A.M. 1992b. The influence of sedimentary style on the morphology and development of alluvial fans. *Israel Journal of Earth Sciences*, 41, 123–137.

Harvey, A.M. 1996. The role of alluvial fans in the mountain fluvial systems of southeast Spain: implications of climatic change. *Earth Surface Processes and Landforms*, 21, 543–553.

Harvey, A.M. 1997. The role of alluvial fans in arid zone fluvial systems. In D.S.G. Thomas (Ed) *Arid Zone Geomorphology: Process, Form and Change in Drylands*, 2nd Edn. Wiley, Chichester, 231–259.

Harvey, A.M. and Renwick, W.H. 1987. Holocene alluvial fan and terrace formation in the Bowland Fells, northwest England. *Earth Surface Processes and Landforms*, 12, 249–257.

Harvey, A.M., Miller, S.Y. and Wells, S.G., 1995. Quaternary soil and river terrace sequences in the Aguas/Feos river systems: Sorbas basin, southeast Spain. In J. Lewin, M.G. Macklin and J.C. Woodward (Eds) *Mediterranean Quaternary River Environments*. Balkema, Rotterdam, 263–281.

Harvey, A.M., Silva, P.G., Mather, A.E., Goy, J.L., Stokes, M. and Zazo, C. 1999. The impact of Quaternary sea-level and climatic change on coastal alluvial fans in the Cabo de Gata ranges, southeast Spain. *Geomorphology*, 28, 1–22.

Hillaire-Marcel, C., Carro, O., Causse, C., Goy, J.L. and Zazo, C. 1986. Th/U dating of *Strombus bubonius*-bearing marine terraces in southeastern Spain. *Geology*, 14, 613–616.

Hooke, R. le B. 1967. Processes on arid region alluvial fans. *Journal of Geology*, 75, 438–460.

Hooke, R. le B. 1968. Steady state relationships on arid-region alluvial fans in closed basins. *American Journal of Science*, 266, 609–629.

Hooke, R. le B. and Rohrer, W.L. 1977. Relative erodibility of source area rock types from second order variations in alluvial fan size. *Geological Society of America, Bulletin*, 88, 1177–1182.

Kesel, R.H. and Lowe, D.R. 1987. Geomorphology and sedimentology of the Toro Amarillo alluvial fan in a humid tropical environment, Costa Rica. *Geografiska Annaler*, 69A, 85–99.

Kesel, R.H. and Spicer, B.E. 1985. Geomorphic relationships and ages of soils on alluvial fans in the Rio General valley, Costa Rica. *Catena*, 12, 149–166.

Kochel, R.C. 1990. Humid fans of the Appalachian Mountains. In A.H. Rachocki and M. Church (Eds) *Alluvial Fans: A Field Approach*. Wiley, Chichester, 109–129.

Kostaschuk, R.A., MacDonald, G.M. and Putnam, P.E. 1986. Depositional processes and alluvial fan – drainage basin morphometric relationships near Banff, Alberta, Canada. *Earth Surface Processes and Landforms*, 11, 471–484.

Laronne, J.B. and Reid, I. 1993. Very high rates of bedload sediment transport by ephemeral desert rivers. *Nature*, 366, 148–150.

Lattman, L.H. 1973. Calcium carbonate cementation of alluvial fans in southern Nevada. *Geological Society of America, Bulletin*, 84, 3013–3028.

Lecce, S.A. 1991. Influence of lithologic erodibility on alluvial fan area, western White Mountains, California and Nevada. *Earth Surface Processes and Landforms*, 16, 11–18.

Leggett, R.F. Brown, R.J.E. and Johnston, G.H. 1966. Alluvial fan formation near Aklavik, Northwest Territories, Canada. *Geological Society of America, Bulletin*, 77, 15–30.

Marchi, L., Pasuto, A. and Tecca, P.R. 1993. Flow processes on alluvial fans in the eastern Italian Alps. *Zeitschrift für Geomorphologie*, 37, 447–458.

McDonald, E.V., Pierson, F.B., Flerchinger, G.N. and McFadden, L.D. 1997. Application of a process-based soil-water balance model to evaluate the influence of Late Quaternary climate change on soil-water movement. *Geoderma*, 74, 167–192.

Moseley, F., Cuttell, J.C., Lange, E.W., Stevens, D. and Warbrick, J.R. 1981. Alpine tectonics and diapiric structures in the Pre-Betic zone of south-east Spain. *Journal of Structural Geology*, 3, 237–251.

Munuera, M. and Carrion, S.J. 1991. Palinologia de un deposito arqueologico en el suresteiberico semiar-ido; Cuerva del Algarobo (Mazarron, Murcia). *Cuarternario y Geomorfologia*, 5, 107–118.

Nash, D.J. and Smith, R.F. 1998. Multiple calcrete profiles in the Tabernas basin, southeast Spain: their origins and geomorphic implications. *Earth Surface Processes and Landforms*, 23, 1009–1029.

Nemec, W. and Postma, G. 1993. Quaternary alluvial fans in southwest Crete: sedimentation processes and geomorphic evolution. In M. Marzo and C. Puigdefàbregas (Eds) *Alluvial sedimentation*. International Association of Sedimentologists, Special Publication 17, 235–276.

Neumann, H. 1960. El Clima del sudeste de Espana. *Estudios Geograficos*, 21, 171–209.

Pope, K.O. and Van Andel, T.H. 1984. Late Quaternary alluviation and soil formation in the southern Argolid: Its history, causes and archaeological implications. *Journal of Archaeological Science*, 11, 281–306.

Rachocki, A.H. and Church, M. (Eds) 1990. *Alluvial Fans: A Field Approach, Wiley*. Chichester, 391 pp.

Reid, I. and Frostick, L.E. 1997. Channel form, flows and sediments in deserts. In D.S.G. Thomas (Ed) *Arid Zone Geomorphology: Process, Form and Change in Drylands*, 2nd Edn. Wiley, Chichester, 205–229.

Rhodenburg, H. and Sabelberg, U. 1980. Northwest Sahara Margin: terrestrial stratigraphy of the Upper Quaternary and some palaeoclimatic implications. In E.M. Van Sinderen Bakker, Sr. and J.A. Coetsee, (Eds) *Palaeoecology of Africa and the Surrounding Islands*, 12, 267–276.

Ritter, D.F. and Ten Brink, N.W. 1986. Alluvial fan development and the glacial-glaciofluvial cycle, Nenana Valley, Alaska. *Journal of Geology*, 94, 613–625.

Ritter, J.B., Miller, J.R., Enzel, Y. and Wells, S.G. 1995. Reconciling the roles of tectonism and climate in Quaternary alluvial fan evolution. *Geology*, 23, 245–248.

Roberts, N. 1995. Climatic forcing of alluvial fan regimes during the Late Quaternary in Konya basin, south central Turkey. In J. Lewin, M.G. Macklin and J. Woodward (Eds) *Mediterranean Quaternary River Environments*. Balkema, Rotterdam, 205–217.

Sabelberg, U. 1977. The stratigraphic record of late Quaternary accumulation series in southwest Morocco and its consequences concerning the pluvial hypothesis. *Catena*, 4, 209–214.

Segura, F.S. 1986. La Rambla de Cervera: some aspects concerning the sedimentology, the hydrology and the geomorphology. In M. Sala, F. Gallert and N. Clotet (Eds) *Excursion Guide Book*. COMTAG Symposium, University of Barcelona, 88–103.

Schumm, S.A. 1979. Geomorphic thresholds: the concept and its application. *Institute of British Geographers, Transactions*, New Ser. 4, 485–515.

Silva, P.G., Harvey, A.M., Zazo, C. and Goy, J.L. 1992. Geomorphology, depositional style and morpho-metric relationships of Quaternary alluvial fans in the Guadalentin depression (Murcia, southeast Spain). *Zeitschrift für Geomorphologie*, 36, 325–341.

Somoza, L. Zazo, C., Goy, J.L. and Morner, N.A. 1989. Estudio geomorfologico de seguencias de abanicos aluviales cuaternarios (Alicante-Murcia, Espana). *Cuaternario y Geomorfologia*, 3, 73–82.

Sorriso-Valvo, M., Antronico, L. and Le Pera, E. 1998. Controls on modern fan morphology in Calabria, Southern Italy. *Geomorphology*, 24, 169–187.

Van Arsdale, R. 1982. Influence of calcrete on the geometry of arroyos near Buckeye, Arizona. *Geological Society of America, Bulletin*, 93, 20–26.

Wasson, R.J. 1974. Intersection point deposition on alluvial fans: an Australian example. *Geografiska Annaler*, 56A, 83–92.

Wasson, R.J. 1979. Sedimentation history of the Mundi Mundi alluvial fans, western New South Wales. *Sedimentary Geology*, 22, 21–51.

Weijermars, R. 1991. Geology and tectonics of the Betic Zone, SE Spain. *Earth Science Reviews*, 31, 153–236.

Wells, S.G. and Harvey, A.M. 1987. Sedimentologic and geomorphic variations in storm generated alluvial fans, Howgill Fells, northwest England. *Geological Society of America, Bulletin*, 98, 182–198.

White, K.H. 1991. Geomorphological analysis of piedmont landforms in the Tunisian Southern Atlas using ground data and satellite imagery. *The Geographical Journal*, 157, 279–294.

White, K. and Walden, J., 1997. The rate of iron oxide enrichment in arid zone alluvial fan soils, Tunisian Southern Atlas, measured by mineral magnetic techniques. *Catena*, 30, 215–227.

White, K., Drake, N., Millington, A. and Stokes, S. 1996, Constraining the timing of alluvial fan response to Late Quaternary climatic changes, southern Tunisia. *Geomorphology*, 17, 295–304.

Part III Channel Network Expansion

8 Gully Erosion in Dryland Environments

J. POESEN, L. VANDEKERCKHOVE, J. NACHTERGAELE,
D. OOSTWOUD WIJDENES, G. VERSTRAETEN
AND B. VAN WESEMAEL
Laboratory for Experimental Geomorphology, K.U. Leuven, Belgium

8.1 INTRODUCTION

Soil erosion by water is considered to be one of the most important land degradation processes in the Mediterranean. Over the last decades, most studies dealing with water erosion in this environment have mainly focused on interrill and rill erosion (e.g. Poesen and Hooke, 1997). This is seen in the use of many runoff plots throughout the Mediterranean when assessing water erosion rates (e.g. Kosmas et al., 1997), but also in the use of both empirical (e.g. the RUSLE, i.e. Revised Universal Soil Loss Equation; Renard et al., 1997) and process-based (e.g. EUROSEM, i.e. European Soil Erosion Model; Morgan et al., 1998) erosion models quantifying only interrill and rill erosion. There are indications, however, that gully erosion is a significant sediment source in Mediterranean upland areas. In addition, given that gullies act as effective links between upland areas and channels, transferring both overland flow and sediment relatively rapidly to lower areas in the landscape, gully erosion aggravates flooding problems and silting up of reservoirs. This justifies an analysis of gully erosion in Mediterranean environments in order to better quantify sediment production by these erosion features, and to assess their controlling factors. Such information is crucial for predicting the effects of anticipated environmental changes (climatic and land use changes) on gully erosion rates.

Gully erosion is defined as the erosion process whereby runoff water accumulates and often recurs in narrow channels and, over short periods, removes the soil from this narrow area to considerable depths. Gullies are often defined for agricultural land in terms of channels that are too deep to easily ameliorate with ordinary farm tillage equipment, typically ranging from 0.5 m to as much as 25 to 30 m (Soil Science Society of America, 1996). In the 1980s, the term 'ephemeral gully erosion' was introduced to include concentrated flow erosion larger than rill erosion but less than classical gully erosion, as a consequence of the growing concern that this sediment source was once ignored in traditional soil erosion assessments (Foster, 1986; Grissinger, 1996a, 1996b). According to the Soil Science Society of America (1996), *ephemeral gullies* are small channels eroded by concentrated overland flow that can easily be filled by normal tillage, only to reform again in the same location by additional runoff events. Poesen (1993) observed the formation of ephemeral gullies in concentrated flow zones, located not only in natural drainage lines (thalwegs of zero-order basins or hollows) but also along (or in) linear landscape elements (e.g. drill lines, dead furrows, headlands, parcel borders, access

Dryland Rivers: Hydrology and Geomorphology of Semi-arid Channels. Edited by L.J. Bull and M.J. Kirkby.
© 2002 John Wiley & Sons, Ltd.

roads, etc.). Channel incisions in linear landscape elements are usually classified as rills accord-ing to the traditional definitions which associate rill formation with the micro-relief generated by tillage or landforming operations (Haan et al., 1994). However, as such incisions may also become very large, this classification seems unsuited. In order to account for any type of concentrated flow channels that would never develop in a conventional runoff plot used to quantify interrill and rill erosion, Poesen (1993) distinguishes rills from gullies by a critical cross-sectional area of 929 cm^2 (square foot criterion). Hauge (1977) first used this criterion. Other criteria include a minimum width of 0.3 m and a minimum depth of about 0.6 m (Brice, 1966), or a minimum depth of 0.5 m (Imeson and Kwaad, 1980). Nevertheless, it must be acknowledged that the transition from rill erosion to ephemeral gully erosion to classical gully erosion represents a continuum, and any classification of hydraulically related erosion forms into separate classes (microrills, rills, megarills, ephemeral gullies, gullies) is, to some extent, subjective (Grissinger, 1996a, 1996b). As to the upper limit of gullies, no clear-cut definition exists. In other words, the boundary between a large gully and a(n) (ephemeral) river channel is very vague.

This chapter addresses a number of issues that are considered to be important when dis-cussing gully erosion in dryland environments. First, the contribution of gully erosion to sediment production in a range of Mediterranean dryland environments is reviewed. Then, after providing a classification of gullies and an overview of the main processes shaping gullies, topographic thresholds for incipient gullying in drylands are determined before discussing ways by which sediment production through gully erosion can be predicted. Finally, this chapter concludes with a discussion of research needs.

8.2 GULLIES AS IMPORTANT SEDIMENT SOURCES IN DRYLANDS

Despite the spectacular nature of gully erosion, relatively few studies have taken their contribu-tion into account when assessing soil losses in upland areas or when quantifying sediment production at the catchment scale in dryland environments. Therefore, this section attempts to quantify the sediment volumes produced by these erosion features at various temporal and spatial scales in typical drylands using different approaches.

Table 8.1 lists eroded soil volumes by ephemeral gully erosion for three intensively cultivated areas in Portugal and Spain. Gully volumes were assessed either in the field by mapping ephemeral gully length, width and depth or by extracting total ephemeral gully length from aerial photos and multiplying this length with a representative gully cross-section measured in the field. These gully volumes were compared with volumes of topsoil eroded by interrill and rill erosion on similar upland areas.

For relatively wet winters with intense rains, ephemeral gully erosion may be held respon-sible for 47% and 51% of total sediment produced by water erosion over a six-month period in wheat fields in northeast Portugal and in almond groves in southeast Spain (Table 8.1) respec-tively. Over a 3- to 20-year period, the contribution of ephemeral gully erosion rate to total sediment production amounts to 80% in southeast Portugal (Table 8.1).

On abandoned agricultural land (southeast Spain) the contribution of permanent gullies to mean sediment production over a 10-year period equals 83% of total sediment produced by water erosion (Table 8.1). This figure was obtained by surveying a 200-m wide and 500-m long hillslope section at the footslopes of the Sierra de Gata (southeast Spain), currently being used as rangeland (Poesen et al., 1996). On the hillslope subsections where only rills developed over

Table 8.1 Contribution of discontinuous (ephemeral) gully erosion to total sediment production by water erosion in various Mediterranean environments

Study area	Obs. period (years)	Method	Gully (ton ha^{-1} yr^{-1})	Interrill & rill (ton ha^{-1} yr^{-1})	% gully*	Source
Bragança, (northeast Portugal)	< 1 (wet winter)	Field mapping (March 1996)	16.1 (a)	18 (a, b)	47	(1)
Guadalentín (southeast Spain)	< 1 (wet winter)	Field mapping (Dec. 1996)	37.6 (c)	36.6 (b, c)	51	(2)
Alentejo (southeast Portugal)	3–20	Aerial photo and runoff plots	3.2 (d)	0.8 (e)	80	(3, 4)
Almeria (southeast Spain)	10	Field mapping	9.7 (f)	2.0 (b, f)	83	(5)

* % gully equals the percentage of total sediment production in uplands due to (ephemeral) gully erosion.

Notes

(a) Figures for interrill and rill as well as for ephemeral gully erosion are average values for 4 study sites (cultivated fields for wheat production) ranging between 0.5 and 4.1 ha.

(b) Soil loss due to interrill erosion was assumed to be 10% of total soil lost by rill and gully erosion. This figure is based on data reported in the literature for comparable conditions (Govers and Poesen, 1988).

(c) Figures for interrill and rill as well as for ephemeral gully erosion were measured in a 0.24-ha zero-order catchment (intensively cultivated almond groves).

(d) Eroded volumes by ephemeral gullies were calculated using two sets of channel cross-sectional data (depth x width) as measured in the cultivated fields (for wheat production): i.e. 0.15 × 0.50 m and 0.25 × 1.50 m.

(e) Figures for interrill and rill erosion rates are means for 13 runoff plots with different crop rotations (wheat–leguminosa–wheat, wheat–leguminosa, fallow–wheat).

(f) Figures for interrill and rill as well as for ephemeral gully erosion were measured in a 10-ha hillslope section (abandoned agricultural land, rangeland), straight in plan form.

References

(1) Vandekerckhove et al. (1998)
(2) Poesen et al. (1999b)
(3) Vandaele et al. (1997)
(4) Tomas (1992)
(5) Poesen et al. (1996)

the last decade, mean rill erosion rates equalled 1–2 m^3 ha^{-1} yr^{-1}. However, for hillslope sections where gullies developed in the same period, corresponding mean gully erosion rates ranged between 5 and 13 m^3 ha^{-1} yr^{-1} (Poesen et al., 1996).

An indication of the importance of sediment production by gullies in drylands can also be found when comparing mean sediment deposition rates in Spanish reservoirs (see Figure 8.1) with sediment production rates by interrill and rill erosion measured on runoff plots. Mean sediment deposition rate measured over a period of 5–101 years (Avendaño Salas et al., 1997) in Spanish reservoirs with corresponding catchments ranging between 31 km^2 and 16 952 km^2, equals 4.3 ton ha^{-1} yr^{-1} and can even go up to 10 ton ha^{-1} yr^{-1} or more (Avendaño Salas et al., 1997, López-Bermúdez, 1990; Romero-Díaz et al., 1992; Sanz Montero et al., 1996). These figures are significantly higher than reported short- to medium-term mean rates of interrill and

Figure 8.1   Silted reservoir (Valdinfierno) near Zarcilla de Ramos (southeast Spain, December 1995). Mean rate of sediment deposition in this reservoir equals 4.8 ton ha^{-1} yr^{-1} (Sanz Montero et al., 1996). Gully erosion in the catchment plays an important role in sediment production

rill erosion measured on runoff plots (Castillo et al., 1997; Kosmas et al., 1997; Andreu et al., 1998; Puigdefàbregas et al., 1999; Romero-Díaz et al., 1999): i.e. 0.1 and less ton ha^{-1} yr^{-1} for shrubland (matorral, $n = 95$ plot-years) and olive groves ($n = 3$), 0.2 ton ha^{-1} yr^{-1} for wheat ($n = 65$) and Eucalyptus plantations ($n = 12$), and 1.4 ton ha^{-1} yr^{-1} for vines ($n = 9$) (Kosmas et al., 1997). There are various possible reasons to explain the discrepancy between the reported sediment production rates at the catchment scale and at the runoff plot scale. One of these is that, at the catchment scale, sediment production processes other than interrill and rill erosion also operate, such as gully and channel erosion. Also, sediment produced by interrill and rill erosion in uplands is often deposited at the foot of hillslopes or in depressions within the landscape and therefore does not reach the river channel. Hence, other sediment-generating processes in catchments must play an important role in the production of sediments which are transported by (ephemeral) rivers and which cause reservoir infilling. Gully and channel erosion are likely to be involved in this process which is confirmed by observations reported by Plata Bedmar et al. (1997) who studied Cs^{137} content of sediments deposited in the Puentes reservoir (southeast Spain). These authors reported that only 40% of the sediment deposited between 1970 and 1994 in the Puentes reservoir originated in the 10-cm thick topsoil from the catchment (which was assumed to accumulate most of the Cs^{137} fallout). Hence, 60% of the sediment accumulated in the reservoir originated in subsurface soil horizons, which contained no Cs^{137}. It is hypothesised that gully and river channel processes could be held responsible for the erosion and transport of this sediment volume to the reservoir.

A recent survey in the catchments of 22 Spanish reservoirs clearly indicates that specific sediment yield increases when the frequency of gullies increases in the catchment (Figure 8.2; Koninckx, 2000). For catchments in which no gullies were observed, mean specific sediment

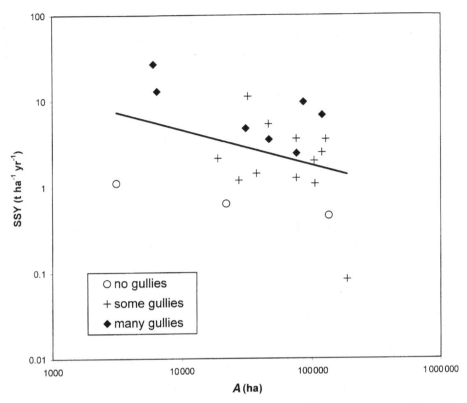

Figure 8.2 Impact of the presence of active gullies in 22 selected Spanish catchments (draining to a reservoir), on area-specific sediment yield (SSY). Specific sediment yield was calculated based on published reservoir sedimentation data (Avendaño Salas et al., 1995; Sanz Montero et al., 1996). Presence of active and permanent gullies in an area of within 10 km from the reservoir was recorded during field surveys in 1999 (Koninckx, 2000). A is catchment area

yield was 0.74 ton ha^{-1} yr^{-1} ($n = 3$). For catchments where numerous gullies could be observed, however, mean specific sediment yield was one order of magnitude larger, i.e. 9.61 ton ha^{-1} yr^{-1} ($n = 7$). Catchments that had some gullies had an intermediate mean specific sediment yield of 2.97 ton ha^{-1} yr^{-1} ($n = 12$). In other words, the presence of (active) gullies in a Mediterranean catchment seems to be an important indicator for the magnitude of sediment production within these catchments.

Other field observations in drylands confirm that gullies are an important or even a dominant sediment source: e.g. in Algeria (e.g. GTZ, 1996), in southern France (Olivry and Hoorelbeck, 1989–90; Muxart et al., 1990; Bufalo and Nahon, 1992; Lhénaff et al., 1993; Wainwright, 1996; Lecompte et al., 1997), Israel (e.g. Seginer, 1966; Nir and Klein, 1974; Perath and Almagor, 2000), northern Morocco (e.g. Heusch, 1970; Laouina et al., 1993), Romania (Radoane et al., 1995), Spain (Donker and Damen, 1984; Thornes, 1984; Ternan et al., 1994; Faulkner, 1995; Martínez-Casasnovas, 1998; Casali et al., 1999; Meyer and Martínez-Casasnovas, 1999; Nogueras et al., 2000) and Tunisia (e.g. De Ploey, 1974). Published data from other semi-arid or arid regions indicate that the proportion of total sediment production generated by gully erosion (% gully) is considerable: i.e. 53% for a semi-arid catchment in Kenya (Oostwoud Wijdenes and Bryan, 1994), 58% for an arid environment in Argentina (Coronato and del Valle, 1993), 60–81% for semi-arid rangelands in Arizona, USA (Osborn and Simanton, 1989), whereas it amounts to 80% for the semi-arid part of Niger (Heusch, 1980).

In conclusion, the data discussed above clearly indicate that gully erosion represents an important sediment source in dryland environments, contributing on average 50 to 80% of overall sediment production. Despite the fact that gully erosion is far from negligible and that gully erosion is often the dominant source of sediment aggravating flooding and silting of reservoirs, too little attention has been given to this soil degradation process in the past. Hence more research related to the rates and controlling factors of gully erosion is needed for this environment.

8.3 CLASSIFICATION OF GULLIES AND PROCESSES OF GULLY EROSION

Gullies have been classified according to different criteria: e.g. plan form of gully (Ireland et al., 1939; De Ploey, 1974), position in the landscape (e.g. Brice, 1966; Poesen et al., 1996), shape of gully cross-section and soil material in which a gully developed (Imeson and Kwaad, 1980), ephemeral or permanent nature of the gully. Ireland et al. (1939) recognised six characteristic gully forms which were produced by physical and land use factors influencing drainage: i.e. linear, bulbous, dendritic, trellis, parallel and compound gullies. De Ploey (1974) also used plan form of the gully as a criterion and found three main gully types in Tunisia: i.e. axial gullying with a single headcut, digitate gullying involving the development of several headcuts, and frontal gullying (with pedimentation) which generally starts from river banks or vertically stabilised gully walls. If one considers position in the landscape as a criterion, gullies can be classified into valley-floor, valley-side and valley-head (Brice, 1966), and each type can be discontinuous or continuous. Continuous gullies form part of a drainage network, whereas discontinuous gullies are isolated from the rest of the drainage network. Valley-side and valley-head gullies are basically the same, and reflect an expansion of the drainage network. Valley-floor gullies re-establish a drainage channel in the alluvium (Schumm et al., 1984). Poesen (1993) also recognised bank gullies in Mediterranean landscapes. This gully type always develops in association with a bank (e.g. river, gully or terrace bank). Imeson and Kwaad (1980) recognised V- and U-shaped gullies in Morocco and subdivided each gully type further according to the type of material in which the gully is developed. Land use has a strong control on the life span of a gully: ephemeral gullies typically develop in intensively cultivated areas whereas permanent gullies develop on abandoned fields or rangelands. Practical considerations led Poesen (1993) to subdivide ephemeral gullies further according to their width/depth ratio (w/d). Wide ephemeral gullies with a $w/d \gg 1$ cause important crop damage. In addition, a high percentage of total soil lost through this gully type consists of fertile topsoil with relatively high organic matter and fertiliser content. On the other hand, wide and shallow gullies are, however, easily erased by conventional tillage. Narrow and deep ephemeral gullies with a $w/d = 1$ or even <1 cause relatively little crop damage and the percentage of total soil loss consisting of fertile topsoil is limited but heavy equipment is required to reshape the areas where these gullies form. The infilling of both gully types by tillage or land-levelling operations often leaves topographic depressions in which new (ephemeral) gullies will develop subsequently.

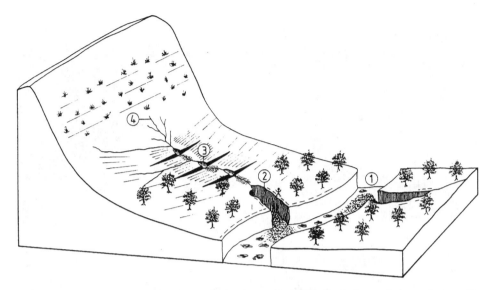

Figure 8.3 Sketch of a typical Mediterranean landscape illustrating the various gully types discussed in this chapter. 1, river channel; 2, bank gully which developed in a river bank; 3, bank gully which developed in a terrace bank; and 4, ephemeral gully in cultivated land or permanent gully in rangeland

In this chapter, a distinction is made between ephemeral gullies, permanent gullies and bank gullies. The square foot criterion is used to identify gullies. Figure 8.3 illustrates these gully types and their typical position in Mediterranean landscapes. Ephemeral gullies are found in cultivated fields and will be erased by the next deep tillage operation (Figure 8.4). Permanent gullies are typically found in abandoned agricultural fields, rangelands or shrubland (Figure 8.5). Once formed, they can be filled by natural processes (such as splash, creep, mass wasting of gully walls and sediment deposition triggered by vegetation growing on the gully floor) in the long term or by anthropogenic processes such as land-levelling operations (e.g. Clarke and Rendell, 2000) or, in contrast, they may enlarge by further incision. Bank gullies (base-level controlled) form where a wash-line, a rill or an ephemeral gully crosses an earth bank (Poesen, 1993; Poesen and Hooke, 1997). In the Mediterranean, as in other semi-arid and arid areas, bank gullies typically occur as tributary channels initiated at the bank of an ephemeral river (Figure 8.6) or at a terrace bank (Figure 8.7) from where gully heads retreat into low-angled pediments, river or agricultural terraces.

Various erosion processes are involved in gully initiation and development. For gullies to initiate by runoff, concentrated surface or subsurface flow is a prerequisite. Concentrated flow intensity (expressed by, for instance, mean flow shear stress, τ) must be large enough and must last long enough (1) to overcome the resistance to detachment and transport of the topsoil (or of the subsoil in the case of macro-pore flow) by concentrated flow and (2) to scour a channel with a cross-section equal to or exceeding the square foot criterion. Whereas rills in loamy cultivated topsoil develop once $\tau > 1$ Pa (Govers, 1985), ephemeral gullies develop at somewhat higher flow intensity values, i.e. $\tau > 4$ Pa, as calculated from field measurements (Figure 8.8). In saturated zones, seepage (ex-filtration) may catalyse or enhance gully development by lowering the erosion resistance of the topsoil (Gabbard et al., 1998). In the Mediterranean, such saturated zones develop frequently where shallow soils (leptosols) occur.

Figure 8.4   Ephemeral gully in wheat field near Bragança (Portugal, March 1996)

Figure 8.5   Permanent gully developed in colluvial deposits of valley bottom near Jerusalem (Israel, May 1995). Land use is (degraded) rangeland

Figure 8.6   Bank gully in ephemeral channel bank located in the catchment of the Valdinfierno reservoir (see also Figure 8.1) near Zarcilla de Ramos (southeast Spain, September 1997)

Figure 8.7   Bank gully developed in agricultural terrace bank near Lorca (southeast Spain, November 1997)

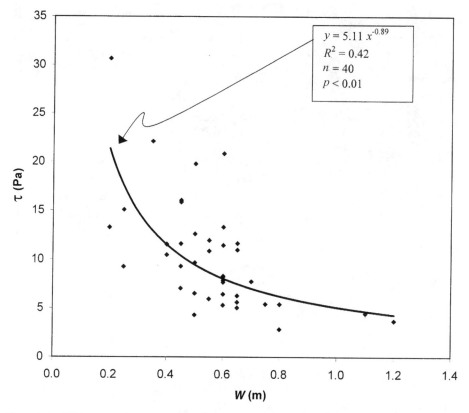

Figure 8.8 Critical mean flow shear stresses (τ) as a function of channel width (W) for ephemeral gully development (i.e. square foot criterion). Data are based on field measurements of ephemeral gully characteristics in the Alentejo (Portugal). Ephemeral gullies developed in sandy loam and silt loam wheat seedbeds on 6 November 1997, during a rain event in which 74 mm was recorded in 24 hours. τ was calculated using peak flow discharge (calculated with the Rational formula), mean flow velocity (calculated with the Manning formula), measured channel slope, and W

Once a gully channel is formed, several processes and process combinations lead to channel expansion: i.e. piping, headcut migration, undercutting by plunge pool erosion, tension cracking, mass failure, fluting and channel bifurcation.

Piping has an important role in the initiation and, to a lesser extent, development of some gully systems such as bank gullies and gullies forming badland areas in the Mediterranean. Several authors discuss the factors controlling piping and tunnelling and its relation to gully head development (e.g. Harvey, 1982; Crouch, 1983; López-Bermúdez and Romero-Díaz, 1989; Martín-Penela, 1994; García-Ruíz et al., 1997; Gutiérrez et al., 1997; Torri and Bryan, 1997). According to Harvey (1982), the occurrence of piping is mainly controlled by the nature of the soil materials at depth – particularly the presence of differential porosity, solubility and strength – together with surface features allowing concentrated penetration of overland flow into deep tension cracks or desiccation cracks, and surface crusting over less consolidated layers. According to Martín-Penela (1994), the most decisive factors for piping are the presence of poorly indurated silty-clayey materials containing cracks, fractures or other discontinuities, such as (gypsum-filled) joints and faults.

A *headcut* is a natural, nearly vertical drop in channel-bed elevation (Stein and Julien, 1993; Robinson and Hanson, 1996; Bennett et al., 2000). The dissipation of kinetic flow energy of the flowing water at the drop causes excessive erosion and results in headcut upstream migration, which deepens and tends to widen the channel. Headcuts propagating in small channels, such as rills and ephemeral gullies, contribute significantly to total upland soil losses during erosive storms.

Plunge pools are formed by falling water at the base of vertical overfalls such as the headcuts of gullies (van der Poel and Schwab, 1988). Plunge pool erosion is essentially controlled by flow erosivity (which in turn depends on water fall height and unit flow discharge) and soil erodibility. The development of plunge pools in gullies decreases the stability of gully walls.

Gully-head and gully-wall collapse are a composite and cyclical process resulting from downslope creep, *tension crack development* (Figure 8.9), crack saturation by overland flow, head or wall collapse followed by debris erosion which facilitates the next failure (Collison, 1996).

There are two common types of *mass failure* of homogeneous, cohesive gully banks. One is progressive, continuous failure by creep movement over long periods of time. The other is catastrophic shear failure of the bank (Alonso and Combs, 1990). The latter is the most frequent mode of failure in cohesive banks. Rapid movement usually occurs when the shear strength along a slip surface is exceeded, either because of a reduction in the shear strength of the bank material (caused by an increase in pore water pressure) or an increase in the stress due to saturation or human activities. In contrast to non-cohesive gully banks maintained at the

Figure 8.9 Tension crack development in bank gully wall (marl) near Fuensanta (southeast Spain, March 1998)

natural angle of repose, where stability is independent of height, the stability of cohesive banks is strongly dependent on both the bank slope angle and height. Most often failure occurs by a deep-seated slip, although shallow slips also occur. The types of failure mechanisms most frequently associated with gully banks are rotational slip, plane slip (in association with tension cracking) and cantilever failure (Alonso and Combs, 1990).

In areas where dispersible soil exists, *fluting* can cause pronounced gully wall retreat (Figure 8.10). Flutes can be described as vertically elongated grooves, generally tapering towards the top that furrows into the wall of the gully. These flutes result predominantly from the action of running water (Veness, 1980). In a study of 55 active bank gullies in southeast Spain, Vandekerckhove et al. (2000a) observed fluting to occur in about 33% of the studied gullies.

Gully channel *bifurcation* may arise either at an actively extending gully head or along a gully channel (Thornes, 1984; Bull and Kirkby, 1997). At the channel head, a division of the drainage area into two tributary parts – resulting, for instance, from vegetation or other surface roughness elements splitting the concentrated flow – is likely to lead to some delay in further extension, as each subdivision must separately exceed its own threshold for gullying. At the channel

Figure 8.10   Fluting on gully wall in dispersible loams near Zarcilla de Ramos (southeast Spain, May 2000)

margin, near-critical side-slope areas may exceed their threshold without interfering with the main channel growth, and it is therefore argued that lateral budding is the more frequent mode of gully branching (Bull and Kirkby, 1997).

The gully initiation processes may be different for each gully type, but most gully types expand by headcut migration. Gully heads are therefore important links between hillslopes and channels and function as sediment sources, sediment stores or sediment conveyors depending on the environmental conditions. Kirkby (1988) and Kirkby and Bull (2000) recognised two types of gully heads: gradual gully heads and sharp headcuts. They observed gradual gully heads to be most common where these channels developed in coarse gravel or boulders, whereas sharply incised gully heads were found in fine-grained sediments (such as marls). On the basis of the morphology of gully head profiles on the Iberian peninsula, Oostwoud Wijdenes et al. (1999) distinguished four gully head types: gradual, transitional (with a short inclined section), rilled-abrupt and abrupt (Figure 8.11). Gradual types start as small (micro) rills and gradually deepen and widen into gullies (i.e. channel cross-sections exceeding the square foot criterion). Transitional gully head types showed an inclined channel section whereas the abrupt types included vertical headwalls (Figure 8.12). When a rill extended upslope from the headcut, it was classified as rilled-abrupt. Gradual and abrupt types represent the two extreme situations, whereas transitional and rilled-abrupt are intermediate types. Oostwoud Wijdenes et al. (1999) did not find a strong relation between gully head type and properties of the material in which the gully head developed. They therefore suggested that the various gully types reflect various evolutionary stages of the corresponding gullies. In intensively cultivated areas, all gully heads are of the gradual type (Figure 8.13). When the land is abandoned, gradual type gully heads evolve over time towards more abrupt gully head types, as indicated by Figure 8.13. Field observations indicated that abrupt gully heads developed from secondary headcuts (knickpoints) in the channel, which migrated upstream. The abrupt gully heads were always formed in more than one soil layer, of which one was a resistant (stony) layer. Field data also indicated that gradual type headcuts were essentially controlled by fluvial processes, whereas abrupt headcuts were controlled by a combination of fluvial and mass-wasting processes. Gradual types usually occurred more downslope along catenas than the abrupt ones, suggesting that the incisions started by fluvial processes and migrated upslope when knickpoints developed in the channel. The rilled-abrupt types are still actively retreating until the knickpoint reaches the most upstream point of incision. Thus, the abrupt types correspond to slower retreat rates, possibly due to a declining catchment. Abrupt gully heads may deteriorate into transitional types when plunge-pool erosion becomes less effective. Transitional types may thus represent a terminal stage or a transitional stage towards abrupt types. The observations reported by Oostwoud Wijdenes et al. (1999) point to an effect of land use change (for instance, from intensive cultivation to land abandonment and to rangeland) on gully head morphology.

8.4 TOPOGRAPHICAL THRESHOLDS FOR INCIPIENT GULLYING IN MEDITERRANEAN ENVIRONMENTS

For given environmental conditions gully heads develop in the landscape where a certain topographic threshold is exceeded. This topographic threshold can be represented by an inverse relation between two geomorphic attributes: i.e. drainage-basin area (A, proportional to surface or subsurface runoff discharge) and critical soil surface slope (S) for incision (e.g. Patton and Schumm, 1975; Dietrich and Dunne, 1993; Montgomery and Dietrich, 1994; Poesen et al., 1998). The topographic threshold is based on the assumption that in a given landscape with a

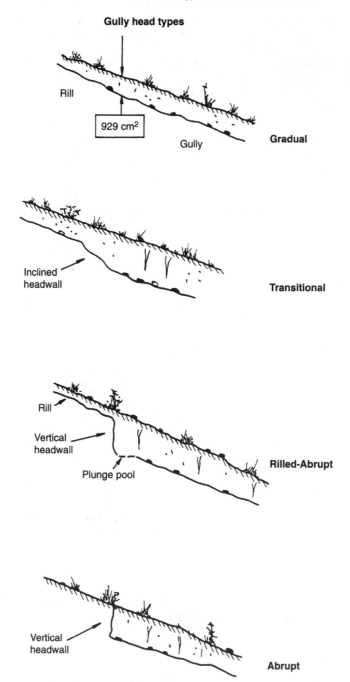

Figure 8.11   Classification of gully head types according to their longitudinal profile (after Oostwoud Wijdenes et al., 1999)

Figure 8.12   Gully with a vertical headwall (abrupt type), Sierra de Gata (southeast Spain, September 1996)

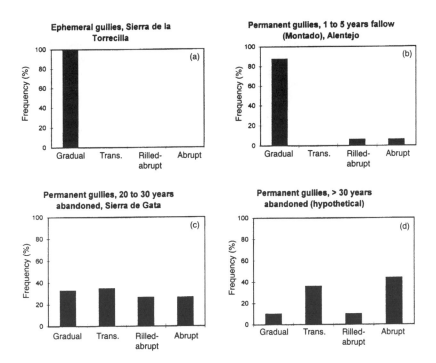

Figure 8.13   Frequency distribution of gully head types in different Mediterranean environments: (a) ephemeral gullies in intensively cultivated almond groves (Sierra de la Torrecilla, Spain); (b) permanent gullies in fallow land (1–5 years), Alentejo, Portugal; (c) permanent gullies in degraded rangelands (20–30 years after abandonment), Sierra de Gata, Spain; and (d) permanent gullies in degraded rangelands (>30 years after abandonment (partly based on field observations and aerial photo-analysis) (after Oostwoud Wijdenes et al., 1999)

given climate, soils and land use, there exists for a given drainage area a critical soil surface slope, necessary for gully incision. As drainage area increases, this critical slope decreases and vice versa. For different environmental conditions and different gully initiating processes (e.g. Horton overland flow, saturation overland flow, shallow small-scale landsliding) different topographic thresholds apply (Montgomery and Dietrich, 1994; Kirkby, 1994; Tucker and Bras, 1998). Several studies have attempted to define this topographic threshold with the aim of understanding (1) where gully heads are located within the valley network, and (2) how the position of gully heads may change under environmental change (climate and land use change) scenarios. In addition, such a topographic threshold helps to identify zones that are prone to gullying and where preventive measures can most economically and successfully be undertaken (Patton and Schumm, 1975; Desmet et al., 1999).

8.4.1 Ephemeral Gullies and Permanent Gullies

Poesen et al. (1998) summarised and compared 10 published datasets for ephemeral gullies and permanent gullies in a range of environments and found that, in addition to the environmental characteristics, the methodology used to assess A and S also affects the topographic threshold for incipient gullying. For instance, local S derived from topographic maps usually under-estimates local S measured in the field, and A values derived from topographic maps can be very different from A values measured in the field, taking into account the effects of small linear landscape elements deviating and concentrating overland flow. This summary also revealed that few topographic threshold data exist for Mediterranean environments. Therefore, new datasets on critical S and A values have been recently collected for a range of Mediterranean conditions in southern Europe using the same field methodology to quantify S and A at ephemeral and permanent gully heads (i.e. where the square foot criterion applies) (Vandekerckhove et al., 1998). Figure 8.14 clearly illustrates that for a given slope gradient, gully channels are found for corresponding drainage areas, which are at least one order of magnitude larger than those for rill channels.

In an attempt to quantify the impact of climate (annual and daily rain) and land use on the topographical threshold for gully head development, Vandekerckhove et al. (2000b) analysed field data collected in six different Mediterranean study areas: five datasets dealt with ephemeral gullies in intensively cultivated fields and three datasets corresponded to permanent gullies in rangelands. Integrating all datasets (Figure 8.15) resulted in a widely scattered data cluster in which an overall negative trend can be discerned. A negative power relationship of the form

$$S = aA^{-b} \tag{8.1}$$

where S is the representative slope gradient of the soil surface at the gully head or gully initiation point (m m^{-1}), A is the drainage basin area (ha) and a and b are environment-specific coefficients, was fitted through all datasets and defined as the mean topographic threshold for gullying in the respective areas (Figure 8.15). Compared to theoretical relationships for channel initiation by overland flow ($0.50 < b < 0.86$; Montgomery and Dietrich, 1994), relatively low values for b ($0.10 < b < 0.41$) were obtained for the studied European Mediterranean environments, suggesting a dominance of overland flow and an influence of subsurface flow, which has been observed on top of the bedrock (or petrocalcic horizon) at shallow depth in the soil profile (Figure 8.16). Subsurface flow processes involve a positive relationship between A and S, as shown by the wetness index $\ln(A/S)$ (Moore et al., 1988) and by Montgomery and Dietrich (1994). Consequently, the influence of subsurface flow weakens the negative trend implied by

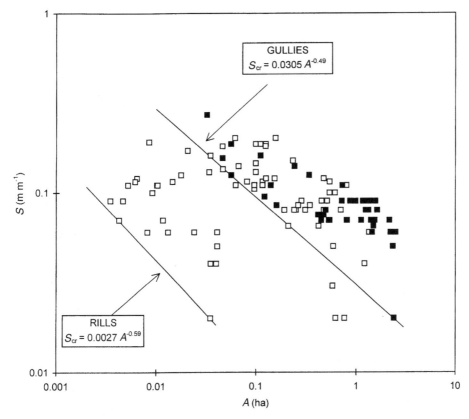

Figure 8.14   Plot of interrill soil surface slope gradient ($S = \tan a$, a = slope angle) at a rill or gully cross-section versus drainage area (A). Rills and gullies developed over a 10-year period on a 10-ha degraded rangeland on the footslopes of the Sierra de Gata (southern Spain; Berael, 1994). Rill cross-sections ($n = 37$) are indicated by open squares whereas gully cross-sections ($n = 37$) are represented by closed squares. Threshold lines indicate critical slope (S_{cr}) of soil surface for rill and for gully development. Rills are defined as concentrated flow channels having a depth of at least 1 cm and a cross-section <929 cm^2, whereas gullies have a cross-section equal to or larger than 929 cm^2 (square foot criterion)

Hortonian overland flow and may be underlying the trends obtained from the European Mediterranean datasets.

It can also be seen in Figure 8.15 that the datasets for gully channel initiation by Hortonian overland flow in the Humboldt Range, Nevada, and in the Stanford Hills, California (Montgomery and Dietrich, 1994), in Gungoandra Creek, Australia (Prosser and Abernethy, 1996) and in southeast Spain (headcut development and trenching of alluvial fans; Harvey, 1987) have steeper slopes than the individual trends for the datasets collected by Vandekerckhove et al. (2000b). Moreover, the negative trend in the integrated dataset of Montgomery and Dietrich (1994) is more pronounced in the lower slope range ($S < 0.40$) and tends to flatten out in the higher slope range where drainage-basin areas decrease more rapidly with slope. According to the models of Montgomery and Dietrich (1994), the process of shallow landsliding becomes more important on steep slopes (i.e. > 0.45), whereas incision by

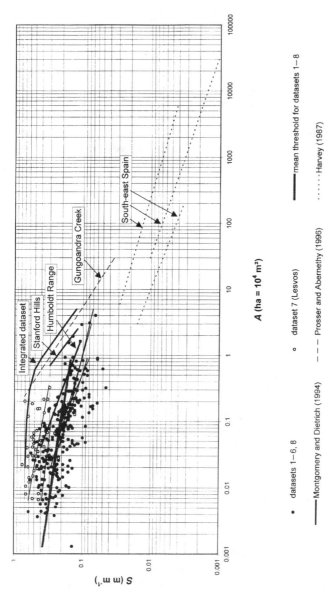

Figure 8.15 Comparison of S–A datasets for gully initiation in the European Mediterranean with published S–A relationships from literature for other environments (after Vandekerckhove et al. 2000b). A is drainage basin area, S = local slope gradient of the soil surface at the channel head. Datasets 1 to 6 and 8 correspond to ephemeral gully heads in cultivated lands and permanent gullies in intensively grazed and sparsely vegetated rangelands (Portugal and Spain). Dataset 7 was collected for permanent gullies in well-vegetated rangelands (with a dense cover of *Sarcopoterium spinosum* bushes) on Lesvos island (Greece).

Trendline through the integrated dataset of Montgomery and Dietrich (1994) for gully heads which developed in various landscapes in western USA as a consequence of both concentrated overland flow and shallow landsliding is also shown. Two other lines were fitted for the Humboldt range (open oak woodland and grasslands in Nevada, USA) and for the Stanford Hills (coastal prairie, California, USA) in which overland flow is dominant (Montgomery and Dietrich, 1994). A threshold line for channel heads initiated by overland flow in the Gungoandra catchment (pastures, Australia) was taken from Prosser and Abernethy (1996). Also plots from Harvey (1987) were added, showing empirical relationships between drainage area and fanhead channel slopes for headcut development and trenching of alluvial fans in southeast Spain

Figure 8.16 Exfiltration at the top of bedrock outcropping in gully channel, which developed in fallow land near Mertola (Alentejo, south Portugal, March 1998)

Hortonian overland flow is dominant in more gentle areas. This trend can also be discerned in the data for the European Mediterranean (Figure 8.15; Vandekerckhove et al., 2000b) although the flattening of the overall negative trend between S and A starts at $S > 0.30$. This flattening is mainly due to the Lesvos dataset, which also includes gullies initiated by landsliding, as indicated by their typical morphology (Figure 8.17).

From a comparison of the datasets corresponding to the European Mediterranean, Vandekerckhove et al. (2000b) also found that vegetation type and cover were far more important in explaining topographical thresholds for different areas than climatic conditions. In cultivated fields, soil structure and soil moisture conditions, as determined by the antecedent rainfall distribution, are critical factors affecting the S–A relationships, rather than daily rainfall events of the gully initiating events. For rangelands, vegetation cover and type at the time of gully head formation appears to be the most important factor differentiating between topographical thresholds, overruling the effect of annual rainfall. This reinforces the conclusion

Figure 8.17   Gully development by shallow landsliding in volcanic soils near Antissa, Lesvos (Greece, April 1998). Land use is rangeland and the small bushes are *Sarcopoterium spinosum*

made by Poesen et al. (1998), after comparing 10 published S–A datasets for various regions in the world, that for a given drainage area (A), critical slopes for gully initiation (S_{cr}) are lowest for ephemeral gullies in intensively cultivated and sparsely vegetated fields, whereas S_{cr} is significantly larger for permanent gullies in uncultivated and well-vegetated land such as grasslands, open woodland and reed-covered valley floors. The importance of vegetation biomass in concentrated flow zones for reducing gully initiation risk in semi-arid environments was also stressed by Graf (1979) and Nogueras et al. (2000). These observations are also in line with the following conclusions drawn from various studies in Australia by Prosser (in press):

> 'Natural vegetated surfaces in humid environments are highly resistant to scour by concentrated overland flow and consequently are only sensitive to gully erosion from extreme events or climate change experienced at 1000 y or longer timescales. Once vegetation cover is degraded, however, these systems become more sensitive to climate change and decadal scale changes can contribute to gully initiation. Many of these degraded hollow or valley bottoms would ultimately scour from large events, regardless of climate changing toward more intense runoff. Particularly areas of high intensity cropland have periods of low resistance to concentrated flow erosion, which make them quite sensitive to relatively small storms and changes to the intensity of rainfall and runoff.'

From these conclusions it becomes clear that any land use change implying a vegetation biomass decrease in the landscape, and particularly in concentrated flow zones, will decrease the threshold for incipient gullying. This implies that for a given slope gradient (S), the critical drainage area (A) for gully head development will decrease, and therefore gully density will increase, as pointed out by Kirkby (1988).

8.4.2 Bank Gullies along Ephemeral River Channels

By definition, bank gullies develop wherever concentrated flow crosses an earth bank. Given that the local slope gradient of the soil surface at the bank riser is very steep (i.e. subvertical to vertical), bank gullies can readily develop (e.g. due to mass movement) with very small drainage basin areas (A) (Poesen and Govers, 1990). This is in agreement with the relations reported in Figure 8.15. Once initiated, bank gullies retreat by headcut migration into the more gentle sloping soil surface of the bank shoulder and further into low-angled pediments, river or agricultural terraces.

Vandekerckhove et al. (2000a) studied the topographic position and characteristics of 55 active bank gully heads which developed into pediments of two study areas (Guadalentín and Guadix) in southern Spain and found a significant negative relation ($R^2 = 0.56$; $p = 0.0001$) between the slope of the soil surface at the gully headcut and the corresponding drainage area. The good correlation for the entire dataset is mainly based on the significant correlation for the Guadalentín study area and also on the difference between both study areas, i.e. significantly smaller catchment areas and steeper slopes in the Guadalentín compared to the Guadix area. The negative slope–catchment area relationship at the bank gully heads results from the more pronounced topography of the Guadalentín. In contrast, it is less evident in the relatively flat Guadix basin, where there is little variation in slope with distance from the drainage divide. Nevertheless, this inverse $S–A$ relationship illustrates the average topographical position of bank gully heads in Mediterranean landscapes.

8.5 PREDICTING SOIL LOSSES AND SEDIMENT PRODUCTION DUE TO GULLYING

8.5.1 Ephemeral Gullies

At present, only a few models claim to be capable of predicting ephemeral gully erosion rates (Poesen et al., 1998): i.e. CREAMS (Chemicals, Runoff, and Erosion from Agricultural Management Systems; Knisel, 1980); GLEAMS (Groundwater Loading Effects of Agricultural Management Systems; Knisel, 1993); EGEM (Ephemeral Gully Erosion Model; Merkel et al., 1988; Woodward, 1999); and WEPP, watershed model (Water Erosion Prediction Project; Flanagan and Nearing, 1995). The channel erosion routines from both EGEM and WEPP watershed models are slightly modified procedures from the CREAMS channel erosion routines (Lane and Foster, 1980). In these models, concentrated flow detachment rate is proportional to the difference between (1) flow shear stress exerted on the bed material and the critical shear stress, and (2) the transport capacity of the flow and the sediment load. Net detachment occurs when flow shear stress exceeds the critical shear stress of the soil or gully-bed material and when sediment load is less than transport capacity. Net deposition occurs when sediment load is greater than transport capacity.

Although these models claim to have a great potential in predicting soil losses by ephemeral gully erosion, they have never been thoroughly tested for this erosion process. Recently, the suitability of the EGEM model for predicting ephemeral gully erosion rates in Mediterranean environments was evaluated by Nachtergaele et al. (2001a). Using a detailed dataset for 86 ephemeral gullies in southeast Spain (Guadalentín study area) and in southeast Portugal (Alentejo study area), Nachtergaele et al. (2001a) found a very good relationship between predicted and measured ephemeral gully volumes ($R^2 = 0.88$). However, as ephemeral gully length is an EGEM input parameter, both predicted and measured ephemeral gully volumes have to be divided by this ephemeral gully length in order to test the predictive capability of EGEM. The resulting relationship between predicted and measured ephemeral gully cross-sections is rather weak ($R^2 = 0.27$). Therefore, Nachtergaele et al. (in press) concluded that EGEM is not capable of predicting ephemeral gully erosion properly for the studied Mediterranean environments. One of the reasons which explains the low predictive capability of EGEM in Mediterranean environments is the way in which EGEM accounts for channel erodibility and critical flow shear stress, which are auto-generated by the model and may therefore not be representative for the stony soils in the Mediterranean. For instance, rock fragment content and moisture content of the fine earth affect resistance of the topsoil to ephemeral gully development, as shown by Poesen et al. (1999a). From the study conducted by Nachtergaele et al. (2001a) it becomes clear that ephemeral gully length is a key parameter in determining the ephemeral gully volume, as illustrated by Figure 8.18.

All (ephemeral) gully erosion models listed above require knowledge of the location of gullies in order to assess total gully length (Poesen et al., 1998). This information is also important for land managers and for predicting the impact of environmental change on the spatial distribution and frequency of gullies. In other words, the main question here is: Where do (ephemeral) gullies start and where do they end in the landscape?

Where do ephemeral gullies start? A possible approach to predict locations in the landscape where gully heads might develop is to apply the topographic threshold concept, as explained

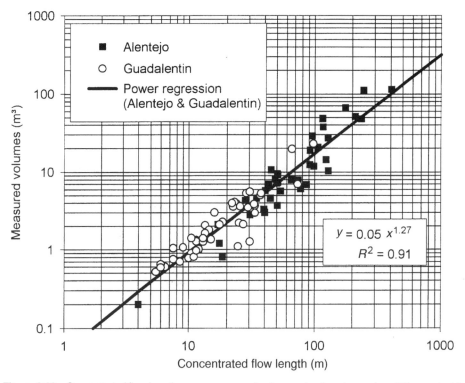

Figure 8.18 Concentrated flow length versus measured ephemeral gully volumes ($n = 86$) on a double logarithmic scale for two Mediterranean areas: Alentejo (southeast Portugal) and Guadalentín (southeast Spain) (after Nachtergaele et al., 2001a)

above and illustrated in Figures 8.15 and 8.19. For each pixel in the landscape A and S must be calculated, and using equation (8.1) and the appropriate a and b values for that environment, one can then assess the risk of having a gully head developing in this pixel.

Where do ephemeral gullies end? Ephemeral gullies end in a downslope direction where massive sediment deposition and fan-building occurs. This is where either surface roughness increases suddenly (e.g. where a different land use begins) or where local slope gradient decreases (Beuselinck et al., 2000). Here transport capacity of the concentrated flow will drop, leading to sediment deposition. Very few S–A relationships for sediment deposition exist. For Mediterranean conditions, Nachtergaele et al. (2001a) published two datasets indicating that the topographic threshold (S–A relationship) for sediment deposition at the bottom end of ephemeral gullies was smaller than the corresponding S–A relationship for incipient ephemeral gullying (Figure 8.19). From this figure it can be seen that ephemeral gullies in the Guadalentín form sedimentation fans at much steeper slopes than is the case in the Alentejo. This is partly attributed to smaller drainage areas (and hence smaller concentrated flow discharges) as well as to the coarser sediment load (rock fragments) transported in the ephemeral gullies of the Guadalentín compared to those of the Alentejo.

De Ploey (1984) proposed a sediment deposition (i.e. colluviation) model for topographically induced sediment deposition:

$$S_{sed} = A_t C^{-0.8} q^{0.5} \tag{8.2}$$

where S_{sed} = critical slope angle (degrees) below which sediment deposition occurs in overland flow

A_t = an empirical coefficient depending on the median sediment size,

C = sediment concentration ($g\,l^{-1}$) in overland flow,

q = unit flow discharge ($cm^3\,cm^{-1}\,s^{-1}$)

Field measurements in different environments of Europe reveal that topographically induced sediment deposition at the downslope end of ephemeral gullies, which developed in non-stony loamy soils, usually occurs in a narrow range of local slopes along catenas with the same land use: i.e. 2–4%. However, topographically induced sediment deposition occurs on steeper Mediterranean slopes when rock fragment content of the topsoil increases, as illustrated by Figure 8.20. These few datasets allow one to locate the initiation point and the sediment deposition point of an ephemeral gully, based on topographic attributes and rock fragment content. Consequently, ephemeral gully length can be derived by routeing concentrated flow from the gully head towards the fan at the gully end.

Desmet et al. (1999) investigated the possibility of predicting the location of ephemeral gullies using an inverse relationship between local slope gradient (S) and upslope contributing area per unit length of contour (A_s). Predicted locations of ephemeral gullies were confronted with the locations recorded in three intensively cultivated catchments over a five-year observation period. The optimal relative area (A_s) exponent (relative to the slope exponent) ranged from 0.7 to 1.5. A striking discrepancy was found between the high relative area exponent required to predict optimally the entire trajectory of the ephemeral gullies and the low relative area exponent (0.2) required to identify the spots in the landscape where ephemeral gullies begin. This indicates that zones in the landscape where ephemeral gullies start are more controlled by slope gradient, while the presence of concavities control the trajectory of the gullies until the slope gradient is too low and sediment deposition dominates. Such an approach can be improved by incorporating the presence of linear landscape elements, soil surface state, vegetation cover and possibly rain, to the input parameters.

(a)

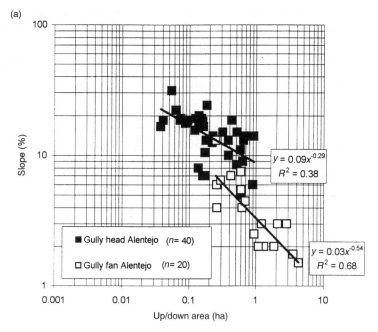

(b)

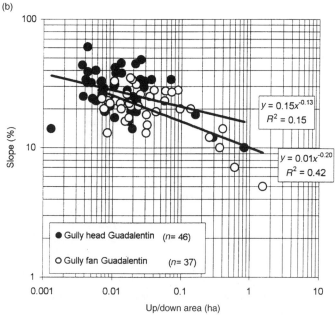

Figure 8.19 Topographical thresholds for ephemeral gully initiation and sediment deposition in both the Alentejo (southeast Portugal, graph (a)) and the Guadalentín (southeast Spain, graph (b)). Up area is the runoff contributing area at the gully head, i.e. the gully initiation point (square foot criterion). Down area is the runoff contributing area at the gully fan, i.e. the sediment deposition point. (After Nachtergaele et al., 2001a)

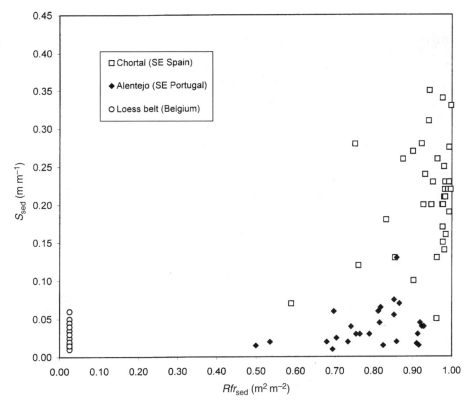

Figure 8.20 Impact of rock fragment content of the topsoil (represented by rock fragment cover at the apex of the sedimentation fan, Rfr_{sed}) on the critical slope (S_{sed}) of the soil surface below which sediment deposition in overland flow occurs and the ephemeral gully channel ends. For comparison, data for silt loam soils having low rock fragment contents, are also depicted ($n = 102$, based on datasets published by Nachtergaele et al., 2001a, 2001b and Vandekerckhove et al., 2000b)

8.5.2 Bank Gullies

Once initiated, bank gullies essentially expand by gully headcut retreat and, to a lesser extent, by gully wall retreat. Whether a bank gully retreats by a single headcut or by multiple headcuts is controlled by factors such as topography and material type, and the processes involved have been discussed above.

Figure 8.21 illustrates an actively retreating bank gully headcut in a typical Mediterranean environment.

Oostwoud Wijdenes et al. (2000) investigated the factors that control the spatial distribution of the activity of 458 bank gully headcuts along a 3.6-km reach of a typical Mediterranean ephemeral channel in southeast Spain. They found that land use has a significant impact on bank gully head activity indicated by features such as sharp headcut edges, presence of plunge pools, tension cracks, recent deposited sediments and flow marks. Recent land use changes involving the extension of almond cultivation appeared to intensify bank gully head activity.

Figure 8.21   Large bank gully near Darro (southeast Spain, May 2000) with two retreating headcuts (in the front and on the right) into Quaternary loams and sandy loams. The gully headcut in the front was caused by an overflowing irrigation canal. Because of the retreating gully headcut at the right, the unmetalled road crossing this infilled valley had to be shifted towards the right over the last century (based on information provided by a local farmer). Aerial photo analysis indicates that between 1956 and 1994, mean linear retreat rate of the headcut at the right equalled 0.82 m yr^{-1}

Also lithology had a clear impact on bank gully headcut activity: for the same land use type, headcuts in marls, sandy loams and loams were significantly more active compared to headcuts which developed in gravels and conglomerates. Similar observations were reported for Romania by Radoane et al. (1995). These authors found that mean rate of gully head cutting was over 1.5 m yr^{-1} for gullies developing in sandy deposits and under 1 m yr^{-1} for gullies cut in marls and clays.

Several studies have attempted to quantify gully headcut retreat in a range of environments, including linear measurements (e.g. Thompson, 1964; Seginer, 1966; Soil Conservation Service, 1966; De Ploey, 1989; Burkard and Kostaschuk, 1995, 1997; Radoane et al., 1995; Vandekerckhove et al., 2001), area measures (e.g. Beer and Johnson, 1963; Burkard and Kostaschuk, 1995, 1997), volumetric measures (e.g. Stocking, 1980; Sneddon et al., 1988; Vandekerckhove et al., 2001) and weight measures (e.g. Piest and Spomer, 1968). According to Stocking (1980), volumetric measures are the best compromise avoiding difficult considerations of bulk density of soils no longer in situ. Few of these studies investigated gully head advancement in Mediterranean type environments (i.e. Seginer, 1966; Radoane et al., 1995; and Vandekerckhove et al., 2001).

Seginer (1966) studied the advancement of gully headcuts in southern Israel and proposed the following equation:

$$R = aA^{0.50} \tag{8.3}$$

where R = average medium-term (15 years) retreat rate (m yr^{-1}) of gully head
         A = area of drainage basin (km^2)
         a = coefficient ranging between 2.1 and 6.0 depending on the studied catchment.

Radoane et al. (1995) reported on an extensive study of gully head advancement in Moldavia (Romania) and proposed the following regression models:

$$R = aA^b L^c E^d P^e \quad \text{(for gullies cut in marls and clays)} \tag{8.4}$$

and

$$R = a + bA + cE + dL + eP \quad \text{(for gullies cut in sandy rocks)} \tag{8.5}$$

where R = medium-term (14 years) retreat rate (m yr^{-1}) of gully head
         A = drainage basin area upstream of gully head (ha)
         L = gully length (m)
         E = relief energy of drainage basin (m)
         P = drainage basin inclination (m (100 m)$^{-1}$)

and a, b, c, d and e are empirical coefficients or exponents.

By far the most important independent variable controlling gully recession rate was drainage-basin area, explaining 54% of the variance in the case of marls-clay rocks and 68% in the case of sandy rocks.

Vandekerckhove et al. (2001) monitored 46 active bank gullies in southern Spain and found that the present drainage basin area (A_p, in m^2) was the most important factor explaining mean short-term (2 years) gully headcut retreat rate, in terms of annual eroded volume (V_e, in m^3 yr^{-1}):

$$V_e = 0.04 \ A_p^{0.38} \quad (R^2 = 0.39) \tag{8.6}$$

This relation was compared with the relation between original drainage basin area (A_o, in m^2) and total eroded bank gully volume (V, in m^3, representing long-term gully head retreat) for the same bank gullies:

$$V = 1.71 \ A_o^{0.60} \quad (R^2 = 0.65) \tag{8.7}$$

Comparing equations (8.6) and (8.7) as a function of drainage basin area for both short-term and long-term gully head erosion shows the importance of runoff production. The smaller exponent (b) and smaller coefficient of determination (R^2) for the short-term relationship ($b = 0.38$ and $R^2 = 0.39$) compared to the long-term relationship ($b = 0.60$ and $R^2 = 0.65$) suggests a greater variability of the annual gully erosion rates measured over two years (V_e) which is smoothed out over a gully's lifespan (V). Several processes explain the smaller weight given to drainage basin area in predicting gully head retreat in the short term, compared to the long term. (1) The influence of tension crack development on gully head retreat rates is an important process interfering with the relationship between V_e and A_p, because tension crack development (and related soil fall or topple) is responsible for a discontinuity of the gully head erosion process. Tension cracks promote collapse of the gully head wall by decreasing the length of the failure plane over which cohesive resistance is mobilised, and by exerting a hydrostatic pressure when filled with runoff water (Selby, 1993; Collison, 1996). (2) Spatial rainfall variability may be responsible for important variations in short-term headcut retreat within the study areas. (3) In the long term, one may expect an increased contribution of extreme rainfall events to gully head retreat whereby the role of drainage basin area becomes more pronounced, as runoff is produced from a larger proportion of the entire catchment during such events and runoff transmission losses are much lower than at low-intensity rain events.

Applying stepwise multiple regression analysis to the short-term gully head retreat data ($n = 46$) for southern Spain, yielded the following result (Vandekerckhove et al., in press b):

$$\log V_e = -2.844 + 0.258 \ (\log A_p) + 0.173 \ H_{hc} + 0.017 \ CN \qquad (8.8)$$

where H_{hc} is the height of the headcut (m), and CN is the runoff curve number (dimensionless; Soil Conservation Service, 1989).

Again, drainage basin area explained the largest part of the variation in short-term headcut retreat rate ($R^2 = 0.37$), followed by height of the headcut ($R^2 = 0.17$) and runoff curve number ($R^2 = 0.048$). Drainage basin area (A_p) and runoff curve number (CN) reflect runoff production whereas the height of the headcut (H_{hc}) indicates the role of runoff energy transfers and under-cutting of the headcut.

Spatial extrapolation of the mean measured volumetric headcut retreat rate of 4.0 m^3 yr^{-1} for an active gully (Vandekerckhove et al., 2001) to a 12 760-ha study area (representing 12% of the total catchment of the Puentes Reservoir, Spain) resulted in 6% of the annual sediment volume filling up the reservoir (Oostwoud Wijdenes et al., 2000). From theoretical considera-tions, Vandekerckhove (2000) calculated the mean proportion of headcut erosion from total bank gully erosion (including gully sidewall erosion and deepening of the gullies) for 32 selected active bank gullies in southeast Spain to be only 25% (assuming a linear gully retreat between 1% and 10% of the original bank gully length). Based on this result, an average total gully erosion rate of 16.0 m^3 yr^{-1} was calculated from the average headcut retreat rate of 4.0 m^3 yr^{-1}. This implies that the contribution of total bank gully erosion to total sediment production in the catchment of the Puentes Reservoir increases to 24%. These simplified calculations again emphasise that bank gully expansion in southeast Spain is a major source of sediment and, therefore, an important process of land degradation.

8.6 CONCLUSIONS

Data from dryland environments compiled in this chapter clearly indicate that whereas mean medium-term interrill and rill erosion rates are in the order of 1–2 m^3 ha^{-1} yr^{-1}, mean medium-and long-term gully erosion rates can be almost an order of magnitude larger. The scarce data available for the Mediterranean indicate that the contribution of gully erosion to total soil loss from agricultural catchments can vary considerably in space and time. More research is needed to elucidate how various factors such as precipitation, topography, soils and land use affect this contribution. Gullying contributes not only to on-site problems (e.g. soil profile truncation and loss of fertile soil) but also to off-site problems such as siltation of reservoirs. While gullies are important sediment sources, they are also efficient sediment transporting paths, and too little attention has been given to this land degradation process. Data discussed in this chapter clearly indicate that land use changes affect gully erosion rates. For instance, it is clear that the expansion of almond groves in southeast Spain have a clear impact on the reactivation or stabilisation of bank gullies. This involves subprocesses such as tension crack development, piping, plunge-pool erosion, fluting, bifurcation, mass wasting on gully sidewalls, etc. Existing gully erosion models do not adequately forecast gully erosion rates at different temporal scales, and this may in part be due to a poor understanding of the subprocesses involved, such as the interaction between water erosion and mass movement processes. Hence, there is still a need for more detailed monitoring, experimenting and modelling of the development and infilling of both ephemeral gullies, permanent gullies and bank gullies in a variety of environments, in order to increase our capacity to predict impacts of land use changes (e.g. agricultural intensification, land levelling or land abandonment) on the location, the total length and the cross-section (size and shape) of various gully types. Such models are also needed for predicting sediment sources and sediment volumes in upland areas, and for evaluating the efficiency of measures for gully erosion control. Appropriate and standardised monitoring techniques need to be developed in order to study gully channel development with a higher precision than that obtained by current techniques. The topographic threshold concept could be a useful tool to help to locate gullies in the landscape. However, more research is needed to establish threshold conditions for incipient gullying and gully trajectories for a range of climatological, topographic, pedologic and land use conditions.

ACKNOWLEDGEMENTS

The research for this study was carried out as part of the MEDALUS III (Mediterranean Desertification and Land Use Project) collaborative research project. MEDALUS III was funded by the European Commission and Climate Research Programme (contract ENV4-CT95-0118, Climatology and Natural Hazards) and this support is gratefully acknowledged.

REFERENCES

Alonso, C.V. and Combs, S.T. 1990. Streambank erosion due to bed degradation – a model concept. *Transactions of the ASAE*, 33(4), 1239–1248.

Andreu, V., Rubio, J.L. and Cerni, R. 1998. Effects of Mediterranean shrub cover on water erosion (Valencia, Spain). *Journal of Soil and Water Conservation*, 53(2), 112–120.

Avendaño Salas, C., Sanz Montero, E., Cobo Rayán, R. and Gómez Montaña, J.L. 1997. Sediment yield at Spanish reservoirs and its relationship with the drainage basin area. *Commission Internationale des Grands Barrages, Dix-neuvième Congrès des Grands Barrages, Florence, Italy*, 863–873.

Beer, C.E. and Johnson, H.P. 1963. Factors in gully growth in the deep loess area of western Iowa. *Transactions of the ASAE*, 6, 237–240.

Bennett, S.J., Alonso, C.V., Prasad, S. and Römkens, M.J.M. 2000. Experiments on headcut growth and migration in concentrated flows typical of upland areas. *Water Resources Research*, 36, 1911–1922.

Berael, C. 1994. *Ruimtelijke spreiding van gesteentefragmenten en hun invloed op geul- en ravijnerosie in Zuidoost Spanje*. Unpublished MSc. thesis, Department of Geography, K.U. Leuven, 176pp.

Beuselinck, L., Steegen, A., Govers, G., Nachtergaele, J., Takken, I. and Poesen, J. 2000. Characteristics of sediment deposits formed by intense rainfall events in small catchments in the Belgian Loam Belt. *Geomorphology*, 32(1-2), 69–82.

Brice, J.B. 1966. Erosion and deposition in the loess-mantled Great Plains, Medicine Creek drainage basin, Nebraska. *US Geological Survey Professional Paper*, 352H, 235–339.

Bufalo, M. and Nahon, D. 1992. Erosional processes of Mediterranean badlands; a new erosivity index for predicting sediment yield from gully erosion. *Geoderma*, 52, 133–147.

Bull, L.J. and Kirkby, M.J. 1997. Gully processes and modelling. *Progress in Physical Geography*, 21(3), 354–374.

Burkard, M.B. and Kostaschuk, R.A. 1995. Initiation and evolution of gullies along the shoreline of lake Huron. *Geomorphology*, 14, 211–219.

Burkard, M.B. and Kostaschuk, R.A. 1997. Patterns and controls of gully growth along the shoreline of lake Huron. *Earth Surface Processes and Landforms*, 22, 901–911.

Casali, J., López, J.J. and Giraldez, J.V. 1999. Ephemeral gully erosion in southern Navarra (Spain). *Catena*, 36, 65–84.

Castillo, V.M., Martinez-Mena, M. and Albaladejo, J. 1997. Runoff and soil loss response to vegetation removal in a semiarid environment. *Soil Science Society of America Journal*, 61(4), 1116–1121.

Clarke, M.L. and Rendell, H.M. 2000. The impact of the framing practice of remodelling hillslope topography on badland morphology and soil erosion processes. *Catena*, 40, 229–250.

Collison, A. 1996. Unsaturated strength and preferential flow as controls on gully head development. In M.G. Anderson and S.M. Brooks (Eds) *Advances in Hillslope Processes*, 2, 753–769.

Coronato, F.R. and del Valle, H.F. 1993. Methodological comparison in the estimate of fluvial erosion in arid closed basin of north-eastern Patagonia. *Journal of Arid Environments*, 24, 231–239.

Crouch, R.J. 1983. The role of tunnel erosion in gully head progression. *Journal of Soil Conservation of New South Wales*, 39, 148–155.

De Ploey, J. 1974. Mechanical properties of hillslopes and their relation to gullying in central and semi-arid Tunisia. *Zeitschrift für Geomorphologie Supplement Band*, 21, 177–190.

De Ploey, J. 1984. Hydraulics of runoff and loess loam deposition. *Earth Surface Processes and Landforms* 9, 533–539.

De Ploey, J. 1989. A model for headcut retreat in rills and gullies. *Catena Supplement*, 14, 81–86.

Desmet, P.J.J., Poesen, J., Govers, G. and Vandaele, K. 1999. Importance of slope gradient and contributing area for optimal prediction of the initiation and trajectory of ephemeral gullies. *Catena*, 37, 377–392.

Dietrich, W.E. and Dunne, T. 1993. The channel head. In K. Beven and M.J. Kirkby (Eds) *Channel Network Hydrology*. Wiley, Chichester, 175–219.

Donker, N.H.W. and Damen, M.C.J. 1984. Gully system development and an assessment of gully initiation risk in Miocene deposits near Daroca, Spain. *Zeitschrift für Geomorphologie Supplement Band*, 49, 37–50.

Faulkner, H. 1995. Gully erosion associated with the expansion of unterraced almond cultivation in the coastal Sierra de Lujar, S. Spain. *Land Degradation and Rehabilitation*, 9, 179–200.

Flanagan, D.C. and Nearing, M. 1995. *USDA – Water Erosion Prediction Project Hillslope Profile and Watershed Model Documentation*. National Soil Erosion Research Laboratory, West Lafayette, Indiana Report No. 10.

Foster, G.R. 1986. Understanding ephemeral gully erosion. In *Soil Conservation*, National Academy of Science Press, Washington D.C., 2, 90–125.

Gabbard, D.S., Huang, C., Norton, L.D. and Steinhardt, G.C. 1998. Landscape position, surface hydraulic gradients and erosion processes. *Earth Surface Processes and Landforms*, 23, 83–93.

García-Ruíz, J.M., Lasanta, T. and Alberto, F. 1997. Soil erosion by piping in irrigated fields. *Geomorphology*, 20, 269–278.

Govers, G. 1985. Selectivity and transport capacity of thin flows in relation to rill erosion. *Catena*, 12, 35–49.

Govers, G. and Poesen, J. 1988. Assessment of the interrill and rill contributions to total soil loss from an upland field plot. *Geomorphology*, 1, 343–354.

Graf, W.L. 1979. The development of montane arroyos and gullies. *Earth Surface Processes and Landforms*, 4, 1–14.

Grissinger, E. 1996a. Rill and gullies erosion. In M. Agassi (Ed.) *Soil Erosion, Conservation, and Rehabilitation*. Marcel Dekker, New York, 153–167.

Grissinger, E. 1996b. Reclamation of gullies and channel erosion. In M. Agassi (Ed) *Soil Erosion, Conservation, and Rehabilitation*. Marcel Dekker, New York, 301–313.

GTZ. 1996. *L'Aménagement des zones marneuses dans les bassins-versants des montagnes de l'Atlas Tellien semi-aride*. Deutsche Gesellschaft für Technische Zusammenarbeit (GTZ), Eschborn, Germany, 142pp.

Gutiérrez, M., Sanco, C., Benito, G., Sirvent, J. and Desir, G. 1997. Quantitative study of piping processes in badland areas of the Ebro Basin, NE Spain. *Geomorphology*, 20, 237–253.

Haan, C.T., Barfield, B.J. and Hayes, J.C. 1994. *Design Hydrology and Sedimentology for Small Catchments*. Academic Press, London, 239pp.

Harvey, A. 1982. The role of piping in the development of badlands and gully systems in south-east Spain. In R. Bryan and A. Yair (Eds) *Badland Geomorphology and Piping*. GeoBooks, Norwich, UK, 317–335.

Harvey, A.M. 1987. Patterns of Quaternary aggradational and dissectional landform development in the Almeria Region, Southeast Spain: a dry-region, tectonically active landscape. *Die Erde*, 118, 193–215.

Hauge, C. 1977. Soil erosion definitions. *California Geology*, 30, 202–203.

Heusch, B. 1970. L'érosion du Pré-Rif. Une étude quantitative de l'érosion hydraulique dans les collines marneuses du Pré-Rif Occidental. *Annales de la Recherche Forestière au Maroc*, 12, 9–176.

Heusch, B. 1980. Erosion in the Ader Dutchi Massif (Niger). In M. De Boodt and D. Gabriels (Eds) *Assessment of Erosion*. Wiley, Chichester, 521–529.

Imeson, A.C. and Kwaad, F.J.P.M. 1980. Gully types and gully prediction. K.N.A.G. *Geografisch Tijdschrift*, XIV 5, 430–441.

Ireland, H.A., Sharpe, C.F.S. and Eargle, D.H. 1939. Principles of gully erosion in the piedmont of South Carolina. *USDA Technical Bulletin*, 63, 1–143.

Kirkby, K.J. 1988. Hill slopes and hollows. *Nature*, 336, 201.

Kirkby, M. 1994. Thresholds and instability in stream head hollows: a model of magnitude and frequency for wash processes. In M.J. Kirkby (Ed.) *Process Models and Theoretical Geomorphology*. Wiley, Chichester, 293–314.

Kirkby, M.J. and Bull, L.J. 2000. Some factors controlling gully growth in fine-grained sediments: a model applied to Southeast Spain. *Catena*, 40, 127–146.

Koninckx, X. 2000. *Studie van de ruimtelijke spreiding van sedimentexport in Spanje op basis van stuwmeeropvullingen*. Unpublished MSc. thesis, Department of Geography, K.U. Leuven.

Kosmas, C., Danalatos, N., Cammeraat, L.H., Chabart, M., Diamantopoulos, J., Farand, R., Gutiérrez, M., Jacob, A., Marques, H., Martinez-Fernandez, J., Mizara, A., Moustakas, N., Nicolau, J.M., Oliveros, C., Pinna, G., Puddu, R., Puigdefàbregas, J., Roxo, M., Simao, A., Stamou, G., Tomasi, N., Usai, D. and Vacca, A. 1997. The effect of landuse on runoff and soil erosion rates under Mediterranean conditions. *Catena*, 29, 45–59.

Knisel, W.G. (Ed.) 1980. *CREAMS: A Field-Scale Model for Chemicals, Runoff and Erosion from Agricultural Management Systems*. U.S. Department of Agriculture, Conservation Report No. 26, 640pp.

Knisel, W.G. 1993. *GLEAMS: Groundwater Loading Effects of Agricultural Management Systems*. University of Georgia, Coastal Plains Experiment Station, Biological and agricultural Engineering Department Publication No. 5, 260pp.

Lane, L. and Foster, G. 1980. Concentrated flow relationships. In W.G. Knisel (Ed.), *CREAMS: A Field-Scale Model for Chemicals, Runoff and Erosion from Agricultural Management Systems*. U.S. Department of Agriculture, Conservation Report No. 26, 474–485.

Laouina, A., Chaler, M., Naciri, R. and Nafaa, R. 1993. L'Erosion anthropique en pays Méditerranéen: le cas du Maroc septentrional. *Bull. Assoc. Géogr. Franç., Paris*, 1993–5, 384–398.

Lecompte, M., Chodko, J., Lhenaff, R. and Marre, A. 1997. Marls gullying: canonical analysis of erosion and sedimentation sequences (Baronnies, France). *Geodinamica Acta*, 10(3), 115–124.

Lhénaff, R., Coulmeau, P., Lecompte, M. and Marre, A. 1993. Erosion and transport processes on badland slopes in Baronnies mountains (French Southern Alps). *Geografia Fisica e Dinamica Quaternaria*, 16(1), 65–73.

López-Bermúdez, F. 1990. Soil erosion by water and the desertification of a semi-arid Mediterranean fluvial basin: the Segura basin, Spain. *Agriculture, Ecosystems and Environment*, 33, 129–145.

López-Bermúdez, F. and Romero-Díaz, M.A. 1989. Piping erosion and badland development in South-East Spain. *Catena Supplement*, 14, 59–73.

Martín-Penela, A.J. 1994. Pipe and gully systems development in the Almanzora basin (southeast Spain). *Zeitschrift für Geomorphologie*, 38, 207–222.

Martínez-Casasnovas, J.A. 1998. A methodology to compute the rate of gully erosion from multitemporal analysis of aerial photographs and digital elevation models. *Proceedings of the First National Congress on Geographic Information, Valladolid, Spain*, 6–8 Oct. 1998, CD-ROM, 12pp.

Merkel, W.H., Woodward, D.E. and Clarke, C.D. 1988. Ephemeral gully erosion model (EGEM). In *Modelling Agricultural, Forest, and Rangeland Hydrology*. American Society of Agricultural Engineers Publication 07-88, 315–323.

Meyer, A. and Martínez-Casasnovas, J.A. 1999. Prediction of existing gully erosion in vineyard parcels of NE Spain: a logistic modelling approach. *Soil and Tillage Research*, 50, 319–331.

Montgomery, D.R. and Dietrich, W.E. 1994. Landscape dissection and drainage area-slope thresholds. In M.J. Kirkby (Ed.) *Process Models and Theoretical Geomorphology*. Wiley, Chichester, 221–246.

Morgan, R.P.C., Quinton, J.N., Smith, R.E., Govers, G., Poesen, J.W.A., Auerswald, K., Chisci, G., Torri, D. and Styczen, M.E. 1998. The European Soil Erosion Model (EUROSEM): a dynamic approach for predicting sediment transport from fields and small catchments. *Earth Surface Processes and Landforms*, 23, 527–544.

Moore, I.D., Burch, G.J. and Mackenzie, D.H. 1988. Topographic effects on the distribution of surface soil water and the location of ephemeral gullies. *Transactions of the ASAE*, 31(4), 1098–1107.

Muxart, T., Cosandey, C. and Billard, A. 1990. L'Erosion sur les Hautes Terres du Lingas: un processus naturel, une production sociale. *Mémoires et documents de Géographie*, CNRS, Paris, 140pp.

Nachtergaele, J., Poesen, J., Vandekerckhove, L., Oostwoud Wijdenes, D. and Roxo, M. 2001a. Testing the ephemeral gully erosion model (EGEM) for two Mediterranean environments. *Earth Surface Processes and Landforms*, 26, 17–30.

Nachtergaele, J., Poesen, J., Steegen, A., Takken, I., Beuselinck, L., Vandekerckhove, L. and Govers, G. 2001b. The value of a physically based model versus an empirical approach in the prediction of ephemeral gully erosion for loess-derived soils. *Geomorphology*, 40, 237–252.

Nir, D. and Klein, M. 1974. Gully erosion induced in land use in a semi-arid terrain (Nahal Shiqma, Israel). *Zeitschrift für Geomorphologie Supplement Band*, 21, 191–201.

Nogueras, P., Burjachs, F., Gallart, F. and Puigdefàbregas, J. 2000. Recent gully erosion in El Cautivo badlands (Tabernas, SE Spain). *Catena*, 40(2), 203–215.

Olivry, J.C. and Hoorelbeck, J. 1989–90. Erodabilité des terres noires de la vallée du Buëch (France, Alpes du Sud). *Cahiers ORSTOM, Sér. Pédol.*, 25(1-2), 95–110.

Oostwoud Wijdenes, D.J. and Bryan, R.B. 1994. Gully headcuts as sediment sources on the Njemps Flats and initial low-cost gully control measures. *Advances in Geoecology*, 27, 205–229.

Oostwoud Wijdenes, D., Poesen, J., Vandekerckhove, L., Nachtergaele, J. and De Baerdemaeker, J. 1999. Gully-head morphology and implications for gully development on abandoned fields in a semi-arid environment, Sierra de Gata, Southeast Spain. *Earth Surface Processes and Landforms*, 24, 585–603.

Oostwoud Wijdenes, D., Poesen, J., Vandekerckhove, L. and Ghesquiere, M. 2000. Spatial distribution of gully head activity and sediment supply along an ephemeral channel in a Mediterranean environment. *Catena*, 39, 147–167.

Osborn, H.B. and Simanton, J.R. 1989. Gullies and sediment yield. *Rangelands*, 11(2), 51–56.

Patton, C. and Schumm, S.A. 1975. Gully erosion, Northwestern Colorado: a threshold phenomenon. *Geology*, 3, 88–90.

Perath, I. and Almagor, G. 2000. The Sharon escarpment (Mediterranean coast, Israel): stability, dynamics, risks and environmental management. *Journal of Coastal Research*, 16(1), 207–224.

Piest, R.F. and Spomer, R.G. 1968. Sheet and gully erosion in the Missouri Valley Loessial Region. *Transactions of the ASAE*, 11, 850–853.

Plata Bedmar, A., Cobo Rayan, R., Sanz Montero, E., Gómez Montaña, J.L. and Avendaño Salas, C. 1997. Influence of the Puentes reservoir operation procedure on the sediment accumulation rate between 1954–1994. *Commission Internationale des Grands Barrages, Proc. 19th Congress Grands Barrages, Florence, Italy, 1997*, Q.74, R.52, 835–847.

Poesen, J. 1993. Gully typology and gully control measures in the European loess belt. In S. Wicherek (Ed.) *Farm Land Erosion in Temperate Plains Environment and Hills*. Elsevier Science Publishers, Amsterdam, 221–239.

Poesen, J., De Luna, E., Franca, A., Nachtergaele, J. and Govers, G. 1999a. Concentrated flow erosion rates as affected by rock fragment cover and initial soil moisture content. *Catena*, 36, 315–329.

Poesen, J. and Govers, G. 1990. Gully erosion in the loam belt of Belgium: typology and control measures. In J. Boardman, D.L. Foster and J.A. Dearing (Eds) *Soil Erosion on Agricultural Land*. Wiley, Chichester, 513–530.

Poesen, J.W.A. and Hooke, J.M. 1997. Erosion, flooding and channel management in Mediterranean environments of southern Europe. *Progress in Physical Geography*, 21(2), 157–199.

Poesen, J., Oostwoud Wijdenes, D. and Vandekerckhove, L. 1999b. Empirical evidence for the impact of human activities and climate on headwater channel changes. In *MEDALUS III, Project IV Final Report*. Contract ENV-CT95-0118.

Poesen, J., Vandaele, K. and van Wesemael, B. 1996. Contribution of gully erosion to sediment production in cultivated lands and rangelands. *IAHS Publication*, 236, 251–266.

Poesen, J., Vandaele, K. and van Wesemael, B. 1998. Gully erosion: importance and model implications. In J. Boardman and D. Favis-Mortlock (Eds) *Modelling Soil Erosion by Water*. Springer-Verlag, Berlin Heidelberg, NATO ASI Series I-55, 285–311.

Prosser, I.P. (in press). Gully erosion, land-use and climate change. In J. Boardman and D. Favis-Mortlock (Eds) *Climate Change and Soil Erosion*. Imperial College Press, London.

Prosser, I.P. and Abernethy, B. 1996. Predicting the topographic limits to a gully network using a digital terrain model and process thresholds. *Water Resources Research*, 32(7), 2289–2298.

Puigdefàbregas, J., Solé, A., Gutíerrez, L., del Barrio, G. and Boer, M. 1999. Scales and processes of water and sediment redistribution in drylands: results from the Rambla Honda field site in Southeast Spain. *Earth Science Reviews*, 48, 39–70.

Radoane, M., Ichim, I. and Radoane, N. 1995. Gully distribution and development in Moldavia, Romania. *Catena*, 24, 127–146.

Renard, K.G., Foster, G.R., Weesies, D.K., McCool, D.K. and Yoder, D.C. 1997. *Predicting Soil Erosion by Water: A Guide to Conservation Planning with the Revised Universal Soil Loss Equation (RUSLE)*. US Department of Agriculture, Agriculture Handbook No. 703, 404pp.

Robinson, K.M. and Hanson, G.J. 1996. Gully headcut advance. *Transactions of the ASAE*, 39(1), 33–38.

Romero-Díaz, M.A., Cabezas, F. and López-Bermúdez, F. 1992. Erosion and fluvial sedimentation in the River Segura basin (Spain). *Catena*, 19, 379–399.

Romero-Díaz, A., Cammeraat, L.H., Vacca, A. and Kosmas, C. 1999. Soil erosion at three experimental sites in the Mediterranean. *Earth Surface Processes and Landforms*, 24, 1243–1256.

Sanz Montero, M.E., Cobo Rayán, R., Avendaño Salas, C. and Gómez Montaño, J.L. 1996. Influence of the drainage basin area on the sediment yield to Spanish reservoirs. First European Conference and Trade Exposition on Erosion Control, IECA, Barcelona, May 1996 (in press).

Schumm, S.A., Harvey, M.D. and Watson, C.C. 1984. *Incised channels. Morphology, dynamics and control*. Water Resources Publications, Littleton, Colorado, USA, 200pp.

Seginer, I. 1966. Gully development and sediment yield. *Journal of Hydrology*, 4, 236–253.

Selby, J.M. 1993. *Hillsope Materials and Processes*. Oxford University Press, Oxford, 451pp.

Sneddon, J., Williams, B.G., Savage, J.V. and Newman, C.T. 1988. Erosion of a gully in Duplex Soils. Results of a long term photogrammetric monitoring program. *Australian Journal of Research*, 26, 401–408.

Soil Conservation Service 1966. Procedure for determining rates of land damage, land depreciation and volume of sediment produced by gully erosion. Technical Release 32. In FAO (Ed.) *Guidelines for Watershed Management*. FAO Conservation Guide 1, Rome, 125–141.

Soil Conservation Service 1989. *Urban Hydrology for Small Watersheds*. Technical Release 55, US Department of Agriculture, Washington, DC.

Soil Science Society of America, 1996. *Glossary of Soil Science Terms*. Soil Science Society of America, Madison, 134pp.

Stein, O.R. and Julien, P.Y. 1993. Criterion delineating the mode of headcut migration. *Journal of Hydraulic Engineering*, 119(1), 37–50.

Stocking, M.A. 1980. Examination of factors controlling gully growth. In M. De Boodt and D. Gabriels (Eds) *Assessment of Erosion*. Wiley, Chichester, 505–520.

Ternan, J.L., Williams, A.G. and Gonzalez del Tanago, M. 1994. Soil properties and gully erosion in the Guadalajara Province, central Spain. In R.J. Rickson (Ed.) *Conserving Soil Resources*. Wallingford, CAB International, 56–69.

Thompson, J.R. 1964. Quantitative effect of watershed variables on rate of gully-head advancement. *Transactions of the ASAE*, 7, 54–55.

Thornes, J. 1984. Gully growth and bifurcation. In Erosional control – Man and Nature. *Proceedings of the XVth Conference of the International Erosion Control Association, Denver, USA*, 131–140.

Tomas, P. 1992. *Estudo da erosao hidrica em solos agricolas. Aplicaçao a regiao Sul de Portugal*. MSc. thesis, Instituto Superior Technico, Lisbon.

Torri, D. and Bryan, R. 1997. Micropiping processes and biancana evolution in southeast Tuscany, Italy. *Geomorphology*, 20, 219–235.

Tucker, G.E. and Bras, R.L. 1998. Hillslope processes, drainage density, and landscape morphology. *Water Resources Research*, 34(10), 2751–2764.

Vandaele, K., Poesen, J., Marques da Silva, J.R., Govers, G. and Desmet, P. 1997. Assessment of factors controlling ephemeral gully erosion in southern Portugal and central Belgium using aerial photographs. *Zeitschrift für Geomorphologie*, 41(3), 273–287.

Vandekerckhove, L., Poesen, J., Oostwoud Wijdenes, D. and de Figueiredo, T. 1998. Topographical thresholds for ephemeral gully initiation in intensively cultivated areas of the Mediterranean. *Catena*, 33, 271–292.

Vandekerckhove, L., Poesen, J., Oostwoud Wijdenes, D., Gyssels, G., Beuselinck, L. and de Luna, E. 2000a. Characteristics and controlling factors of bank gullies in two semi-arid Mediterranean environments. *Geomorphology*, 33, 37–58.

Vandekerckhove, L., Poesen, J., Oostwoud Wijdenes, D., Nachtergaele, J., Kosmas, C., Roxo, M.J. and de Figueiredo, T. 2000b. Thresholds for gully initiation and sedimentation in Mediterranean Europe. *Earth Surface Processes and Landforms*, 25, 1201–1220.

Vandekerckhove, L. 2000. Gully initiation and development in Mediterranean environments. Unpublished PhD thesis. Department of Geography-Geology, K.U. Leuven, Belgium, 319pp.

Vandekerckhove, L., Poesen, J., Oostwoud Wijdenes, D. and Gyssels, G. 2001. Short-term bank gully retreat rates in Mediterranean environments. *Catena*, 44, 133–161.

van der Poel, P. and Schwab, G.O. 1988. Plunge pool erosion in cohesive channels below a free overfall. *Transactions of the ASAE*, 31(4), 1148–1153.

Veness, J.A. 1980. The role of fluting in gully extension. *Journal of Soil Conservation of New South Wales*, 36, 100–108.

Wainwright, J. 1996. Hillslope response to extreme storm events: the example of the Vaison-la-Romaine event. In M.G. Anderson and S.M. Brooks (Eds) *Advances in Hillslope Processes*. Wiley, Chichester, 997–1026.

Woodward, D.E. 1999. Method to predict cropland ephemeral gully erosion. *Catena*, 37, 393–399.

9 Channel Heads and Channel Extension

LOUISE J. BULL[1] AND MIKE J. KIRKBY[2]
[1]*Department of Geography, University of Durham, UK*
[2]*School of Geography, University of Leeds, UK*

9.1 INTRODUCTION

Gully erosion, caused by the instability of channel heads, is a serious problem in dryland environments and is responsible for the destruction of agricultural land and structures such as roads, bridges and pipelines. Gully erosion produces large volumes of sediment that are transported downstream with detrimental effects on water quality, reservoir capacity and flood-plains. In any channel network, approximately half of the total length of channels is in unbranched (first-order) fingertip tributaries. Environmental changes that promote channel extension have therefore a very large potential impact on the landscape. During discharge events channel heads may advance great distances upslope, or retreat downslope if the hollow refills. In extreme cases, gullies can grow in length by tens of metres per year, and may also incise their channels, creating steep ravine banks. One possible end result of these processes is the creation of badland areas, where there is little or no remaining land that is suitable for agriculture.

In terms of landscape dynamics, the channel head is one of the most important elements of the coupled hillslope–channel system. The location of the channel head controls the distance to the catchment divide and therefore influences the drainage density and average hillslope length of a catchment, although bifurcation frequency, confluence angles and tributary spacing are also important. The position of the channel head is controlled by the balance of sediment supply and sediment removal (Kirkby, 1980, 1994; Dietrich and Dunne, 1993). A change in any factor which influences this balance of forces, such as fluctuations in climate or land use, alter the surface erodibility, sediment supply and runoff rates and may therefore result in movement of the channel head. The channel head is thus the element of the landscape that is most sensitive to change in external factors. Understanding the dynamics of channel heads is becoming increasingly important as the incidence of large storms in the dryland areas increases (Llasat and Puigcerver, 1994; Poesen and Hooke, 1997) and as the importance of gullies as a dominant sediment source has been confirmed (Poesen and Hooke, 1997; Bull and Kirkby, 1997). The study of the channel head and channel extension is thus essential to our under-standing of natural landscapes and provides a possible practical opportunity to control erosion and sedimentation.

This chapter considers channel heads and channel extension for all dryland environments around the world. By this we mean the Mediterranean basin and regions that exhibit similar characteristics for plant growth, i.e. temperature, rainfall and soils. The Mediterranean basin can be demarcated by the watersheds of the rivers that border and drain to the Mediterranean

Dryland Rivers: Hydrology and Geomorphology of Semi-arid Channels. Edited by L.J. Bull and M.J. Kirkby.
© 2002 John Wiley & Sons, Ltd.

Sea (Macklin et al., 1992), with the present-day configuration resulting from the interplay between crustal mobility, climate and sea-level change and human interference (Macklin et al., 1992). Notable regions with similar characteristics to the Mediterranean basin include North Africa, Greater California, Chile, the southwestern Cape of South America and Southern Australia.

Dryland environments have been central to investigations of channel heads and channel extension because the circumstances that favour gully development are frequently found in these regions. Drylands encourage gully development due to the combination of climate, relief and vegetation special to this environment. The combination of climate and steep relief results in high-energy environments and leads to high erosivity on steep slopes with shallow soils, especially in areas with poor surface vegetation. Water infiltration rates are also often low due to stony or compacted soil surfaces, and water-repellent soils (often due to the effect of fire). These conditions result in high rates of surface runoff and erosion.

This chapter aims to summarise the research of channel heads and channel extension undertaken in different dryland environments. This will be carried out by discussing the different types of channel heads, evaluating the processes operating at channel heads and promoting channel advance, and summarising the theories for channel head initiation. During these discussions both theoretical and computational models of channel initiation and channel extension will be evaluated. Finally the chapter will consider the basin wide factors promoting channel initiation and extension and examine the future of research necessary to improve our understanding of the channel head.

9.2 RESEARCH INTO CHANNEL HEADS AND CHANNEL EXTENSION

Research into channel heads and channel extension has been carried out in many dryland regions; however, much key work was carried out in California, Colorado and New Jersey. Initial investigations into badland topography related channel initiation to relief, lithology and climate and cycles of incision and infilling (Schumm, 1956a, 1956b, 1964). Working in New Jersey, Schumm (1956a) proposed that the rate of erosion was a function of slope angle, and that slope retreat may not conform to accepted theories of runoff action as a function of depth and distance downslope. The importance of process dominance by creep and splash was also noted (Schumm, 1956a; Schumm and Hadley, 1957). Work on the Mancos shales in Colorado highlighted the importance of seasonal changes in the soil, which in turn caused a seasonal change in infiltration capacity. These changes controlled not only soil erosion, but also the hydrologic characteristics of small basins (Schumm and Lusby, 1963). Patton and Schumm (1975) proposed that the critical slope for entrenchment was probably related to the magnitude of runoff, but since discharge measurements were not available they used the drainage basin area as the most representative measure of discharge. An inverse relationship was found between drainage basin area and critical slope of entrenchment (Figure 9.1) (Patton and Schumm, 1975). The lower limit of the scatter of the data was used to establish the slope–area relationship, which could then be used to identify potentially unstable valley floors. However, this research was only valid for the study basins in northwestern Colorado where the investigations were undertaken.

This work was continued by Dietrich and co-workers who investigated types of channel heads and processes controlling channel initiation (Dietrich et al., 1992; Montgomery and Dietrich, 1989, 1994; Dietrich and Dunne, 1993), although not all of this work was undertaken

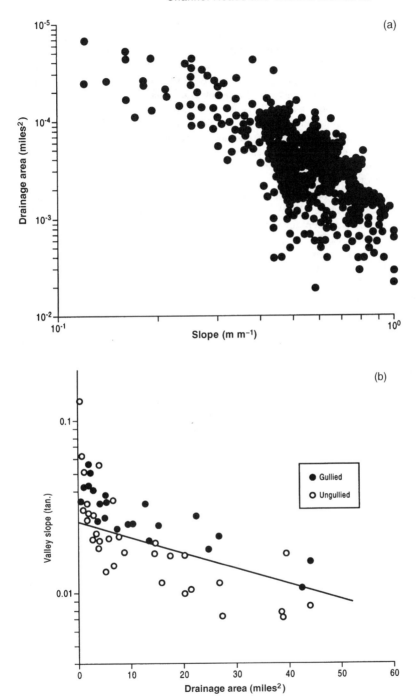

Figure 9.1 Slope–area relationships for gullied and ungullied valleys in (a) Nevada reproduced from Montgomery and Dietrich (1994) and (b) the Piceance basin taken from Patton and Schumm (1975)

in drylands. It was proposed that the local slope at the channel head is inversely related to source area, source basin length, and contributing area per unit contour length (Abrahams, 1984; Dietrich et al., 1986; Montgomery and Dietrich, 1989). This work also showed that above the channel head the valley slope tended to be constant. For study sites in California, the location of heads on steep slopes is controlled by subsurface flow inducing instability of colluvial fill, whereas on gentle slopes head location is governed by overland flow (Montgomery and Dietrich, 1989). Other research by Bradford and Piest (1977, 1978, 1985) investigated the timing and nature of gully sidewall failure. Erosion was related to thresholds for mass wasting and debris transport and a model of cyclic growth proposed. There have also been studies of historical channel initiation and extension, such as Cooke (1974) and Heede (1974), and modelling has also been developed with western American drylands in mind.

The models of Willgoose et al. (1991a) and Howard (1994) illustrate recent attempts at high-resolution process-based simulation of the slope and channel development at the basin scale. The model developed by Willgoose et al. (1991a, 1991b, 1991c, 1994) is a large-scale representation of catchment evolution involving channel network growth and elevation evolution. It assumes that overland flow, as well as flow in channels, is transport limited. It brings together a model of erosion processes that has been theoretically and experimentally verified at small scales and a physically based conceptualisation of channel growth processes (Willgoose et al., 1991a). The drainage basin model of Howard (1994) combines diffusive plus advective processes and assumes that headwater channels are detachment limited. Potential erosion or deposition due to diffusive processes is given by the spatial divergence of the vector flux of regolith movement. The rate of movement is expressed as two terms, one for creep and/or splash and one for mass movement. Fluvial erosion can be classified using two methods. Firstly, erosion is considered to be detachment limited in steep channels flowing on bedrock or regolith in which the bedload sediment flux is less than a capacity load. Secondly, in low-gradient alluvial channels erosion is transport limited and the erosion rate due to detachment is assumed to be proportional to the shear stress exerted by the dominant discharge. One advantage of this model is that both fluvial and slope processes can occur in each cell and that the location of channel heads is defined by a morphometric criterion.

Research in the Mediterranean basin has focused on investigating different types of channel heads (De Ploey, 1974; Imeson et al., 1982), determining the influence of piping (Gutiérrez et al., 1988, 1997; Torri and Bryan, 1997), and examining the effect of structural controls (La Roca Cervignon and Calvo Cases, 1988; Martin-Penela, 1994). Meanwhile early investigations in Southern Australia focused on tunnel erosion (Crouch, 1976) and then turned to the impact of sidewall erosion on channel head advance (Veness, 1980; Blong et al., 1982; Crouch and Blong, 1989). The historical development of gully heads has also been investigated (Prosser and Winchester, 1996).

Research into channel heads and extension has also been undertaken in South Africa and Israel. Work in South Africa has investigated the long-term development and movement of channel heads (Dardis, 1989; Botha et al., 1994), types of channel heads (Bryan and Oostwoud Wijdenes, 1992) and processes acting (Bryan et al., 1988; Liggitt and Fincham, 1989; Oostwoud Wijdenes and Bryan, 1991, 1994; Oostwoud Wijdenes and Gerits, 1994). In South Africa the badlands are dominated by surface piping, subsurface cavitation and tunnel erosion. Gullies have been estimated to have formed in the last 2000 years, but similar forms were observed up to 25 000 years BP. Sequences older than this were thought to have been formed by sheetwash and sediment gravity flow rather than piping (Dardis, 1989). Investigations in Israel have studied the long-term development of channels (Nir and Klein, 1974), processes acting (Seginer, 1966), and the characteristics of channel heads (Yair et al., 1980).

9.3 TYPES OF CHANNEL HEAD

The channel head is a distinct morphological feature and detailed descriptions are necessary to prompt more exact physical hypotheses (Dietrich and Dunne, 1993). The transition from the upstream valley floor to a channel may occur without a significant topological break, but it is more common to observe a distinct headcut or step. Dietrich and Dunne (1993) presented a detailed classification of channel heads based on fieldwork experience in Oregon, Washington and California (Figure 9.2). This is a simple classification of channel head types based on the depth of incision at the channel head because the processes involved in channel initiation and the dynamics of the channel head vary with the height of the headcut. The classification does not include the planform of the head, but by definition the channels have definable banks and Dietrich and Dunne (1993) proposed a direct relationship between the height of the headcut and channel width. However, the relationship between height of headcut and channel width will also be influenced by soil erodibility, soil profile characteristics and the shape of the valley upslope of the headcut, and hence, the relationship may be more complex and some workers have found inverse relationships (Bryan, pers. comm.). By Dietrich and Dunne's (1993) classification, channel heads are gradual, stepped or headcut. Types of channel head are also distinguished by dominant flow processes which helps to emphasise the relationship between runoff, mechanics of erosion, and channel head form. The two extremes in this classification are erosion driven by Hortonian overland flow and exfiltrating subsurface flow (Dietrich and Dunne, 1993).

According to Dietrich and Dunne (1993) the gradual and small-step channel heads are sufficiently subtle that vegetation may grow across the channel, making it difficult to define the channel head on the ground, and also blocking sediment transport which may make the transition from hillslope to channel discontinuous. On steeper slopes the channel heads tend to be large steps with the height roughly equal to the soil depth, and with the channel bed frequently on rock. However if incision at the channel head is triggered by base-level drop

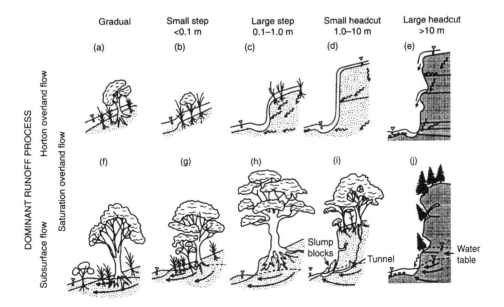

Figure 9.2   Typology of channel heads following Dietrich and Dunne (1993)

and rejuvenation, steep slopes are not necessary for steep headcuts to develop. Headcuts can incise up to several metres and tend to be vertical in profile and concave in plan view.

Alternative channel head classifications include work by De Ploey (1973, 1989), Imeson et al. (1982) and Bryan (1990). De Ploey (1973) noted three types of gullying in Tunisia based on morphology. The first were V- or U-shaped axial gullies, marked by a single headcut and with constant width; the second were digitate gullies, which showed complex patterns of retreat and involved the development of several headcuts; and the final type were frontal gullies with pedimentation that started from riverbanks and vertical stabilised gullies. De Ploey (1989) investigated the coalescence of many small headcuts into larger headcuts by developing a simple model based on width, depth and discharge. Imeson et al. (1982) found three types of gullying in Morocco – U-shaped, V-shaped and U-shaped caused by pipe collapse – which are likely to be similar to the first category defined by De Ploey (1973). During laboratory flume investigations Bryan (1990) found that digitate headcuts were developed as flow concentration and microrill incision occurred. Oostwoud Wijdenes and Bryan (1991) also observed three different systems in Kenya. The first were deep badlands systems due to basal level decline and enhanced by dispersive soils and localised piping; the second were shallow systems of small gullies, rapidly dissecting a level grassed surface due to soil cracking, tunnel erosion and scour; and the third type were shallow, highly active systems, reflecting high runoff from devegetated slopes. Finally, Montgomery and Dietrich (1989) classified channel heads as either gradual or abrupt. For either type of channel head the channel reach immediately downstream may be contiguous with the channel network or may consist of a series of short, discontinuous channel segments.

In the Mediterranean the classification of channel heads can be simplified to just two types: abrupt and gradual. Abrupt channel heads tend to be associated with cohesive soils and lithologies such as marls. These channel heads tend to be steep and incised up to about 3 m and frequently mark the upstream point of channel flow in ephemeral river systems. However, smaller abrupt heads also occur within the ephemeral channel system where they indicate a point in the coupled channel–hillslope system where active incision is taking place. This is related to the topography, land use and lithology that combine to produce flood runoff above the erosion threshold. Gradual channel heads tend to be developed in coarse, uncohesive sediments such as schists.

9.4 PROCESSES ACTING AT CHANNEL HEADS

Channel initiation and channel extension results from a complex array of processes. Within any given area several channel initiation processes may be active, reflecting variations in slope, soil type, soil thickness, vegetation type and vegetation density. Processes controlling each channel head initiation mechanism can be described as threshold phenomena (Kirkby, 1980; Montgomery and Dietrich, 1994). Based on this concept Montgomery and Dietrich (1994) divided the landscape into process regimes by plotting erosional thresholds for different processes of channel incision (Figure 9.3). The processes discussed below are responsible for channel head initiation and also channel extension. It must be remembered that channel heads are dynamic and the range of channel head locations defines the zone over which sediment transport occurs.

9.4.1 Overland Flow

Overland flow occurs in small rills or as sheets of moderate depth over large surfaces. For erosion to occur the rate of rainfall must be sufficient to produce runoff, and the shear stress

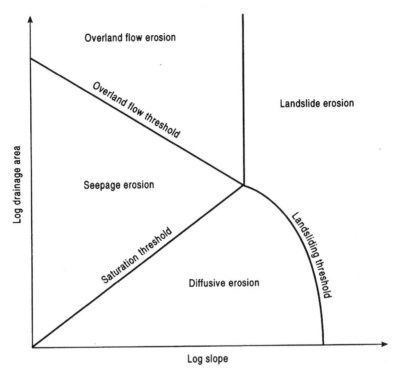

Figure 9.3   Thresholds for channel initiation after Montgomery and Dietrich (1994). Below the thresholds of transport by soil saturation, overland flow and landsliding only diffusive sediment transport processes operate and this region should correspond to slope stability

produced by the moving water must exceed the resistance of the soil surface. Erodibility is a function of the permeability of the surface, the physical and chemical properties that determine the cohesiveness of the soil, and the vegetation. The shear stress produced by overland flow depends on raindrop impact, slope and depth. Depth is controlled by the relative rates of rainfall and infiltration, the velocity of flow and the length of slope. Raindrop impact can either dislodge particles or seal the surface to reduce infiltration and increase runoff. Yair and Lavee (1974) showed that natural rainfall events are capable of generating runoff on slopes of the Sinai screes. Partial area runoff was observed on debris slopes, and the contributing area varied vertically and laterally with variations in properties of the scree mantle. Savat and Poesen (1981) proposed that rills only form if hydraulic conditions allow coarse particles to be transported as easily as fines, but this may be experiment and site dependent.

There are two styles of overland flow generation: Hortonian overland flow and saturation overland flow. Hortonian overland flow extends to the catchment divide, whereas saturation overland flow is usually confined to slope base concavities and hollows. The relationship between discharge and the distance from the divide is strongly affected by the soil profile at a particular elevation, soil depth, hydraulic conductivity and plan-form conductivity. Hortonian overland flow only occurs during periods of rainfall at an intensity exceeding the infiltration capacity. Sediment transport in thin flows may be stimulated by raindrop impact (Moss et al., 1979), but part of the transporting capacity may be satisfied by material entering the flow as rainsplash.

Saturation overland flow includes a contribution from rain falling onto saturated areas, but also includes exfiltration of subsurface flow, which can continue after rainfall has ceased, and is therefore unaffected by drop impact and rainsplash. Where exfiltration forms an important fraction of the overland flow, the shear force is augmented by a small lift force associated with the emergence of water from the soil (Kochel et al., 1982). Saturation overland flow usually occurs over surfaces stabilised by a dense groundcover and root mat. If soil transmissivity or gradient decreases downslope, or if topography converges in plan-form, there will be some exfiltration. Kilinc and Richardson (1973) found that in thin flows entrainment efficiency increased with rainfall rate and that rainsplash was an important factor effecting particle entrainment in sheetflow. Savat (1979) then suggested that rainsplash has little effect on flow velocities and net drag force, but it increased transport capacity and wash sediment concentrations, which in turn increased kinematic viscosity by up to 30%. This was confirmed by Torri et al. (1987).

Incision occurs where the total shear stress of the underloaded flow is high enough to disrupt the root mat or high enough to incise bare patches of soil between vegetation. It also occurs where sheetwash transport through vegetation is available to evacuate sediment loosened by biogenic disruption, or where seepage erosion triggers the initial incision through the vegetated surface, allowing the bare soil to be exploited by saturation overland flow. The ability of overland flow to transport material is usually expressed in terms of flow power (Bagnold, 1966) or tractive shear stress (Horton, 1945; Yalin, 1971). Incision occurs at a critical shear stress (τ_{cr}) estimated by:

$$\tau_{cr} = \rho U_{*cr}^2 = (pgRS)_{cr} \qquad (9.1)$$

where ρ = density of water
          U_* = shear velocity
          g = gravitational constant
          R = hydraulic radius
          S = slope.

This expression is used to link critical unit discharge to critical shear stress and slope at the point of incision (De Ploey, 1990). The initiation of motion is related to a critical threshold, determined by surface properties. On vegetation or fine sediment, the threshold is set by soil cohesion or vegetation mat strength. For unconsolidated sand or gravels, the threshold is related to the forces required to move the grain layer, perhaps assuming equal mobility. Channel form is strongly influenced by vertical differences in soil strength, caused by, for example, a tougher surface or a plough pan or other resistant layer at depth. Shield's criterion is not well suited to predicting the inception of sediment transport in overland flow since it over-predicts the requirements for transport inception in rain-impacted flow (Guy and Dickinson, 1990).

Channel initiation by overland flow is affected by characteristics of the lithology. Texture, aggregate stability and soil shear strength are the main soil characteristics that are relevant for soil detachability (Arulanandan et al., 1975; Torri, 1987). De Ploey (1971) found that liquefaction in sandy sediments facilitates and even starts erosion, and was especially important during sheet erosion. However, the main property of clay governing its susceptibility to erosion is the ratio of dissolved sodium ions to other basic cations in pore water that helps to disperse material (Sherard et al., 1972). Important changes in aggregation and crust characteristics can result from changes in the balance of the soil and pore fluid chemistry (Sargunam et al., 1973; Gerits, 1986; Gerits et al., 1986). Yair et al. (1980) found that in the Zin Valley, Israel, due to the high stability and strong flocculation of clay-rich aggregates, rainsplash is ineffective in surface sealing. Therefore infiltration capacities remain high despite prolonged periods of rain. Rill system development may also be closely related to regolith hydrologic properties, and influenced by consistency, shrink–swell capacity and dispersibility (Gerits et al., 1987; Imeson and Verstraeten, 1988, 1989). Organic matter content also exerts a strong influence on aggregate stability (Imeson et al., 1982).

Overland flow may play a significant role in extending steep headcuts, by creating a waterfall and plunge pool that undermines the back wall of the headcut. Part of the water falling from the lip forms a backward eddy that directs flow strongly against the foot of the headwall. Sediment carried forward out of the plunge pool creates a ridge at the downstream end of the plunge pool, defining its form. De Ploey (1989) suggests that the rate of headcut recession may be related to gross hydraulic properties of the flow across a wide range of scales.

In dryland environments Hortonian overland flow is much more important than saturated overland flow because of storm and hillslope characteristics. Storms in drylands are infrequent, of short duration, spatially localised and have high rainfall intensities. Thus antecedent conditions are usually dry and catchments rarely fully wet up because of the short duration of rainfall, thus limiting overland flow produced by saturation excess. However, in some dryland environments, such as south California, the cumulative effect of prolonged winter rainfall can be important. In the Mediterranean basin the same area of a catchment is unlikely to be wetted during successive events because of the limited spatial extent of storms. However, in some drylands there is evidence of systematically higher rainfall (Yair et al., 1978). Bare patches that are subject to crusting also dominate runoff characteristics of dryland catchments. This lowers the infiltration capacity of the soil and promotes Hortonian runoff during intense storms. Key areas for runoff production in this way are areas where the soil or regolith has a low infiltration capacity, combined with steep slopes and sparse vegetation.

9.4.2 Channel Infilling by Hillslope Processes

For an incision to develop, the threshold of resistance needs to be exceeded, but the forces promoting incision must also exceed those opposing erosion. Runoff and overland flow tend to promote erosion, but other processes such as rainsplash, wetting and drying cycles and frost action tend to fill these incisions. For incisions to grow the rate of sediment transport out of an incision must also exceed the rate of sediment input at the same point, otherwise filling will occur. The interrill hillslope processes involved have been described as rainsplash and rainflow. Both processes depend on raindrop impact to detach soil material, but they differ in the way in which detached material is transported. In rainsplash, transport is by saltation through the air, while in rainflow transport is by the flow (which is still too weak to entrain sediment). As the flowing films become deeper, the effect of raindrop impact on the soil surface is attenuated, and declines appreciably for depths greater than the raindrop diameter. On the basis of field observations of hillslopes in southern Kenya, Dunne (1980) suggested that whether or not the hillslope has attained constant form, the diffusive action of rainsplash might protect a surface against incision by rillwash, and the relative intensities of diffusive rainsplash and incisive rillwash would determine the position of the channel head. The effect of rainflow, on the other hand, appears to be more or less neutral, neither infilling nor enlarging small incisions.

Gradual channel heads can also be destroyed by frost action and creep (Schumm and Lusby, 1963). Granular ice crystals were observed in the upper soil zone during the winter. The crystals serve as growth centres that attract water from the surrounding particles. Several periods of freezing and thawing changed the less permeable rilled surface into a highly permeable surface composed of aggregates without rills. Frost action and compaction cause creep in the upper two inches of the lithosols. Desiccation (Engelen, 1973) and shallow mass movements may also destroy channel heads (Bryan, 1987; Bryan and Price, 1980).

9.4.3 Pipe Initiation

Key research has been carried out in dryland environments because pipes are more obvious, and also because the largest pipes tend to be located in semi-arid environments (Figure 9.4) (Bryan and Jones, 1997). However, some of the earlier theories were also developed for temperate areas. In some dryland areas there is a very close association between piping and gully heads, and the erosive effects of subsurface flow through tunnel collapse and gully formation are well known (Buckham and Cockfield, 1950; Berry and Ruxton, 1960; Berry, 1970; Heede, 1971; Morgan, 1976). Leopold et al. (1964) proposed that pipes were found mostly in badlands areas, and often at the head of drainage networks (although this is now known to be incorrect). De Ploey (1974) also highlighted the importance of piping in producing gully systems during work in Tunisia, especially in loamy sands. Bocco (1991, 1993) found that subsurface erosion, especially piping, was an important factor in nearly 60% of gully initiation reviewed, and that subsurface erosion on its own was the most rapid cause of gully extension. Researchers in Africa and California have also noted piping as the major cause of channel initiation (Higgins, 1990; Dardis and Beckedahl, 1988).

Piping intensity reflects a critical interaction between climate conditions, soil/regolith characteristics and local hydraulic gradients. Formation and development of pipes is controlled by the presence of silt–clay material containing cracks, fissures and other discontinuities. Other factors encourage the action of subsurface flows and the growth of the pipe network such as

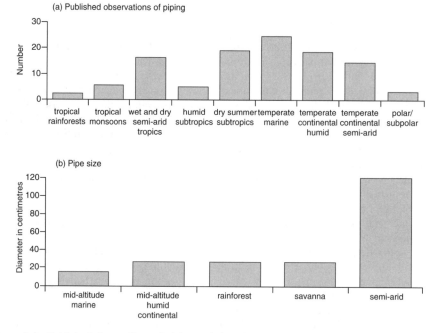

Figure 9.4   Published observations of piping and pipe size taken from Bryan and Jones (1997)

high hydraulic gradients, the existence of soluble ions (especially sodium), prolonged periods of drought to encourage desiccation cracks, and irregular heavy storms to activate the pipes (Martin-Penela, 1994). Sparse or non-existent vegetation also increases the erosional power of surface and subsurface runoff. Piping may be initiated by concentration in hollows creating percolines, from increased infiltration through desiccation cracks or tension cracks, by sub-surface seepage, and by slaking.

For piping initiated in hollows or percolines, accelerated weathering and destruction of the soil structure by persistent saturation increases surface and subsurface height differentials between seepage zones and inter-cols, this enlarges the soil–water catchment and improves the capture of surface runoff. This results in headwater sapping and downslope erosion by inertia as and when more efficient drainage develops. Where cracks or biotic voids provide 'efficient drainage' in the form of concentrated flow or 'combining flow', similar extension and enlargement may occur. Bryan et al. (1978) produced a more detailed method of pipe formation following research in Dinosaur Park: pipe flow originated from rill flow infiltrating to a shard layer, which instantly slakes on contact with the water. This creates an impermeable surface over which subsurface flow occurs. For the Apennines in Italy, Torri and Bryan (1997) suggested that the development of the Biancane badlands was caused by micropipes that formed main routes for the evacuation of material during pseudokarstic enlargement of pores and cracks. Other work showed that rilling and gullying might develop from subsurface seepage without the presence of open pipes (Fenneman, 1922; Bunting, 1961; Howard and MacLane, 1988). Early work by Haworth (1897) proposed that pipes were produced by the creeping of wet soil below the sod, caused by irregularities in the compactability and solubility of the underlying bedrock, while Johnson (1901) proposed that pipes were caused by sinking as a result of compacting and settling of open textured soil by water percolating downward to the level of permanent saturation.

The development of pipes depends on the material being excavated to leave pipes open to flow. Two alternative theories of pipe development are based on this idea. Firstly, Rubey (1928) and Buckham and Cockfield (1950) suggested that sinkholes formed by saturated silt sliding out to create tunnels. For this to occur a free face is needed, which is not necessarily present in many areas of badlands. Secondly, Howard and MacLane (1988) and Howard (1988) proposed that pipes and tunnels were initiated by seepage erosion at a free face. The excavation of sediment was produced by seepage erosion, leaving tunnels for preferential flow. Dietrich and Dunne (1993) consider that pipe formation requires high hydraulic gradients in the neighbourhood of a potential pipe outlet; others have disagreed with this, suggesting that high hydraulic gradients are not necessary, especially if the material is saturated and of low cohesion (Howard and MacLane, 1988; Bryan et al., 1998). Outflow of groundwater must have sufficient velocity to entrain the material at the outlet point and create a pocket that further concentrates flow, so that the pipe grows by evacuation of sediment until a macroscopic through connection is made. In fine-grained materials, flow velocities are generally low, and similar to those in rills (Bryan et al., 1978; Bryan, 1996), so that the dispersive properties of the soil are likely to be critical to pipe initiation. The necessary high hydraulic gradients are frequently achieved by the opening of deep cracks upslope of the eventual pipe outlet. Such cracks may be due to desiccation along root pathways or tension cracks, and are particularly widespread in horizontally bedded alluvial silt–clay fills. Once the through connection exists, a pipe allows transport of large volumes of water and sediment, and their collapse can lead directly to channel extension.

Problems exist as to the stage at which rudimentary networks can be called 'pipe networks'. The established definition of soil pipes is 'subterranean channels developed as a consequence of the movement of water in currents' (Parker, 1963; Jones, 1981). In other words they are sculpted by water and it would be expected that their geometry would reflect hydraulic conditions. Jones (1971) introduced the term 'pseudo pipe' to refer to features which act like pipes inasmuch as they carry elements of combining flow, but do not owe their form to flowing water. Another criterion is whether the 'pipes' are initially full of water and therefore modified along the wetted perimeter (true pipes), or only partially filled and modified along the base (tunnel flow) (Bryan, pers. comm.). However, in practice it is difficult to decide how much modification by erosion is needed before calling a feature a pipe.

Harvey (1982) produced a more comprehensive approach to the development of gullies by pipe collapse. Pipes were categorised by size, depth, and their influence on gully development. A variety of factors producing pipes was also proposed, including (a) removal of weak material to form small pipes which then merged to form gullies, (b) pipes which developed with respect to differential porosity, cementation or solubility, and (c) pipes which form with respect to tension cracks. All forms of pipes would then collapse to form gullies. Surface crusting was highlighted as being conducive to pipe formation.

9.4.4   Pipe Enlargement by Flow

The extension and integration of networks depends on the balance between opposing components of the hillslope system, especially soil creep and other forms of mass movement (Dalrymple et al., 1968; Carson and Kirkby, 1972; Kirkby and Chorley, 1967). Once a recognisable pipe exists then the problems of soil displacement within that pipe are similar to corrosion and transportation in open channels. One possible difference is that the gradient term in the Darcy–Weisbach equation may be the hydraulic gradient for full pipes, rather than the water surface slope for open channels of partially filled pipes (Jones, 1981), although it is difficult to determine the frequency with which pipes are filled. Erosion of pipes results from the application of shear stress to the margins of the conduit, and Dunne (1990) proposed the term 'tunnel scour' to emphasise the difference between this process and other forms of subterranean erosion. When the flow becomes non-Darcian as it enters macropores, it accelerates, especially where hydraulic gradients are steep in mountainous terrain or perpendicular to high channel heads or banks. The important characteristics governing erosion are then the flow velocity and discharge rather than the hydraulic gradient. No measurements or calculations have been made of the fluid shear on the margins, but sediment concentrations in the effluent indicate that some combination of shear and collapse mobilises considerable volumes of soil particles and enlarges the conduits (Jones, 1982; Tanaka et al., 1982; Tsukamoto et al., 1982; Dietrich and Dunne, 1993). The critical shear stress of the boundary material may be substantially reduced in the presence of soluble cements or dispersible clay minerals (Holmgren and Flanagan, 1977), and many authors have noted the association between tunnel scour and high concentrations of salts (Heede, 1971; Bryan et al., 1978; Imeson et al., 1982).

Enlargement of tunnels and thinning of the roof eventually causes lines of pits to coalesce into stream channels (Dietrich and Dunne, 1993). The channel heads are usually vertical with steep banks and sharp edges, and the channels have low gradients and tend to be straight for long reaches between sharp bends. Steps occur in some channels where bedding planes localise changes in resistance. The scale of headcuts, sidewalls and related features depend on the depth of incision, and the cohesive strength of the undispersed, unweathered sedimentary rock (Dietrich and Dunne, 1993).

9.4.5   Mass Failures

Mass failures occur at channel heads but also along steep sidewalls of gullies. Failure of near vertical walls at the stream head results in channel extension, while widening of gullies due to bank slumping contributes to total gully sediment yield (Bradford and Piest, 1977). Sidewall erosion has been shown to be responsible for more than half of the gully volume in New South Wales (Blong et al., 1982). Failure of steep channel heads and sidewalls occurs when the driving forces exceed resisting forces. Gully heads and walls are loaded by three different forces: (1) the weight of the soil, (2) the weight of water added by infiltration or a rise in the water table and (3) seepage forces of percolating water (Bradford and Piest, 1977). The change in water content is important because it has a strong influence on the shearing resistance of the soil, and the shear strength is also influenced by freeze–thaw and wetting–drying cycles. Vertical tension cracks tend to decrease the overall stability by reducing cohesion, and when these cracks are filled with water the pore water pressure increases dramatically, often resulting in failure. The stability of sidewalls can also be influenced by undercutting during periods of flow, even if flows do not reach bankfull.

The stability or instability of a sharp headcut or sidewall depends on whether the driving forces are greater than the forces resisting erosion. These depend on such factors as slope angle, soil characteristics, moisture conditions and the amount of debris at the base of the channel head or wall (Bradford and Piest, 1985). Failure occurs when:

$$F_s = \tau_t / \tau \tag{9.2}$$

where F_s = the factor of safety
  τ_t = shear strength along a surface
  τ = equilibrium shear strength along the same shear surface.

Changes in the driving forces occur due to changes in energy, either associated with climate (changes in temperature or moisture) or gravity. Sidewall failure usually occurs by energy changes associated with variations in moisture content (Bradford and Piest, 1985). Slope failure by the addition of water occurs because of increased weight, increased seepage and a decrease in apparent cohesion.

Resisting forces are due to the shear strength of the soil mass, which is expressed by the Coulomb equation:

$$\tau_t = c' + (\sigma_n - \mu) \tan \phi' \tag{9.3}$$

where τ_t = shear strength along a surface
  c' = cohesion
  ϕ' = friction angle
  σ_n = total normal stress on the failure plane at the time of failure
  μ = pore water pressure.

There are three modes of failure: deep-seated circular arc failures, slab failure or pop out failure followed by slumping of overhangs (Bradford and Piest, 1985). Rotational slip failure may be a base, toe or slope failure depending on where the failure arc intersects the ground surface. Non-circular slips are associated with heavily fissured material, the presence of a soft layer low in the bank, multilayered banks, and cases of unusual drainage. Rotational slips are also limited to clays because the shear stress increases more rapidly with depth than the strength of the clay, and at greater depths, is larger than the shear strength. The failure surface is curved because valley sides are not vertical, so the principal axis is not constant within the soil mass, and the direction of the major principal plane is also continually changing.

Slab failure is caused by the development of tension cracks, and is important on cohesive soils (Figure 9.5). Tension cracks develop downwards from the ground surface some distance behind the edge of the bank because of tensile stress in this region. The maximum depth to which they develop can be predicted from engineering properties of the soil (Terzaghi and Peck,

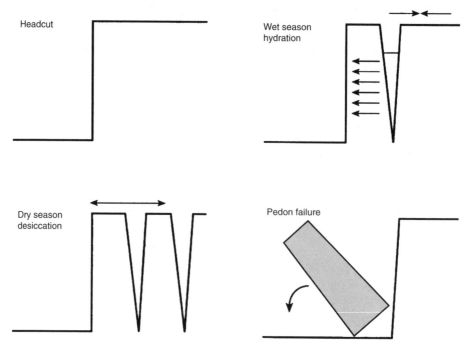

Figure 9.5 Slab failure at a gully head

1948). Tension cracks reduce the effective length of the potential failure surface and decrease bank stability. They also have a significant hydraulic effect, by allowing high interior pore pressures when the cracks are filled with standing water. In low banks the tension crack may occupy a significant proportion of the bank height, and failure then takes place by the shearing or toppling of a block forward into the channel, rather than back tilting of the upper surface. The stability of such banks depends on the tensile strength rather than the shear strength of the soil.

Thresholds encountered in processes of mass wasting of gully walls and heads are related to stratigraphy rather than to the shape of the land surface (Bradford and Piest, 1985). Thresholds encountered in the debris transport process are related to climate and vegetation. Prolonged wet periods increase the likelihood of gully wall failure, but only when runoff exceeds some critical value does the rate of gully development rapidly accelerate. Years with above-average rainfall and periods of high intensity result in increased sediment loss from gullies (Bradford and Piest, 1985). Land use and vegetation overshadow climate effects on present-day gullying. Locally land management techniques, such as tillage and terracing, tend to prevent the clean out of sediment, yet if land is abandoned changes in roughness and infiltration capacity can increase runoff and erosion.

De Ploey (1989) developed a model for headcut retreat by a combination of erosion and splash which leads to undercutting and failure of material through the operation of the plunge pool in steep headcuts. The plunge pool plays an important role in influencing mass failure due to the interaction of runoff and seepage (Bradford and Piest, 1978). The model is deterministic and encompasses material properties and fluid dynamics. The material properties are incorporated as an erodibility factor that includes cohesion, resistance to penetration, and bulk unit

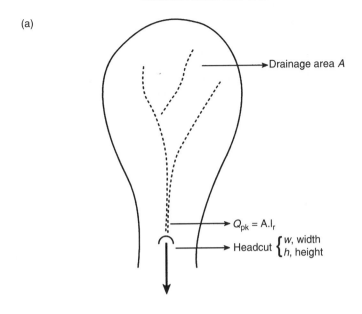

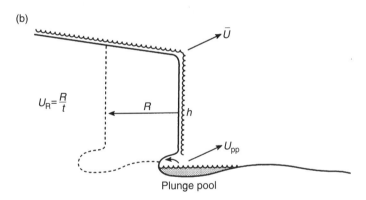

Figure 9.6 The effect of a single rainstorm event on headcut retreat (a) drainage area and (b) the headcut after De Ploey (1989)

weight. The model studies the effect of a single storm in a drainage area (A) (Figure 9.6) by examining a headcut with width, w, and height, h, which will move upstream a distance, R, with a rate U_R so that:

$$U_R = \frac{R}{t} \qquad (9.4)$$

where U_R = rate of advance upstream
R = distance moved
t = storm duration.

Rainfall intensity, I_r, occurs over the time period, t, producing a flow, Q. The total mass of water, m, is:

$$m = Qt \tag{9.5}$$

Assuming the mean velocity of water over the headcut is $\bar{U}$, the speed at which water strikes the plunge pool U_{pp} can be calculated using fluid dynamics:

$$U_{pp} = (2gh + u^2)^{0.5} \tag{9.6}$$

The corresponding kinetic energy is:

$$m\frac{U_{pp}^2}{2} = \frac{Qt(2gh + u^2)}{2} \tag{9.7}$$

To analyse the stability of the headcut wall:

$$H_c = \frac{N_s c'}{U_w} \tag{9.8}$$

where H_c = critical height
       N_s = stability factor
       c' = cohesion
       U_w = bulk density.

The volume of material eroded by the headcut advance is:

$$V = Rwh \propto \frac{1}{H_c} = \frac{U_w}{N_s c'} \tag{9.9}$$

If the resistance and internal stability of the banks is introduced:

$$V \propto \frac{Qt(2gh + u^2)U_w}{2N_s c'} \tag{9.10}$$

An erodibility factor (E_R) is introduced so:

$$V = E_R Qt\left(2gh + \frac{u^2}{2}\right) \tag{9.11}$$

The distance of gully head advance during a storm is therefore:

$$R = \frac{E_R Qt[g + (u^2/2h)]}{w} \tag{9.12}$$

This supported the theory that the rate of retreat of a headcut has a linear relationship with discharge (De Ploey, 1989) and was supported by fieldwork carried out by Piest et al. (1975) in Iowa.

The model by De Ploey (1989) was simplified by Temple (1992) to:

$$X = CqaH^b \tag{9.13}$$

where X = amount of retreat
       q = discharge
       H = height of headcut
       C = material dependent coefficient measured in the field

and a and b are constants, expected to be $\frac{1}{3}$ and $\frac{1}{2}$ respectively.

Robinson and Hanson (1995) tested these models of mass failures for steep headcuts. Discharge, height and backwater levels were kept constant and soil properties were varied. Results showed that rates of headcut advance varied more than one hundred times depending on the placement conditions. Plots of headcut position against time displayed a linear advance, with advance rate decreasing as the average density and average unconfined compressive strength increased. Soil strength increased with soil density (Robinson and Hanson, 1995).

A classification of sidewalls has also been developed which has been used to predict major gully sediment sources (Crouch and Blong, 1989). This ties in processes of mass failure with the other processes acting on sharp headcuts. Vertical sides were assumed to be produced by mass failure after pop-out failure by basal saturation or abrasion at the base, fluted sidewalls are induced by rainsplash and wash, and sloping sides are formed by rainwash, creep and rill development. Fluting is where small ridges and depressions develop, caused by differential erosion. Rates of retreat were predicted by classifying the lengths of sidewalls affected by various processes, multiplied by the erosion rate to determine the volume of soil eroded, and then summed to calculate gully sidewall loss (Crouch and Blong, 1989). Veness (1980) also highlighted the importance of processes acting on steep gully sidewalls and proposed a cycle of fluting development. Initially a free face is produced by mass failures that has no rills. Rilling is then initiated as long as the sidewalls are steep enough to allow running water to dislodge material. Rills then act as lines of preferential flow and a transition occurs from rills to flutes, assuming that there is sufficient water to continue the erosive process and that the channel can remove eroded material from the base of the sidewall. Development of the flutes continues until they are destroyed by undermining, which produces collapse of the flutes. If material is not removed from the base of the sidewall the flutes can in effect choke themselves with eroded sediment, again resulting in destruction.

Mass failures also include the collapse of underground pipe networks. This tends to occur when the pipe has been sufficiently eroded by the shear stress of the water so that the roof can no longer be supported. Early work was conducted by Haworth (1897) and Johnson (1901), and Downes (1949) published a more detailed account of piping and reported gully formation in Victoria when pipes in solonetzic soils reached 1.5 to 1.8 m wide and collapsed. Other supporters of gully formation by pipe collapse include Cumberland (1944), Colclough (1965) and Aghassy (1973), but none of these workers suggested reasons for pipe development. Ingles and Aitchinson (1969) developed the theory of Johnson (1901) by investigating the soil properties promoting the development of pipes. They concluded that failure by piping was due to the effects of the force of flowing or percolating water on cohesive soil in which the interparticle forces are reduced by decreases, or exchanges of pore fluid cations.

9.5 THEORY OF CHANNEL HEAD INITIATION

Traditionally there are two conceptual approaches to understanding channel head initiation: the stability approach (Smith and Bretherton, 1972) and the threshold approach (Horton, 1945). The stability approach is derived from the conservation of sediment mass and emphasises that the channel head represents the point where sediment transport increases faster than linearly downslope. This usually requires wash processes to dominate. The threshold approach takes the view that the channel head represents a point at which processes not acting upslope become important. The balance of sediment still determines whether the channel head becomes stable or migrates, but changing process domains drive incision. However, it is not clear whether there is always a change in process at the headcut, or whether there is a change in the intensity of the process operating, or if a variation in the spatial distribution causes incision.

More recently the two approaches were combined in a single approach published by Kirkby (1994). The different approaches tend to be better suited to different environments and determine the two extremes of a range of factors that combine to produce channel heads. Current work has produced a family of continuous models that provide an elementary theory of evolution of fluvial landscapes including the emergence of channelised flows, development of stable valleys, erosion of surfaces, relationships between surface forms and flow and landform variability (Smith et al., 1997a, 1997b).

In Mediterranean environments, where sediment transport is dominated by sheetwash and rainsplash, the development of incision may reflect the dominance of overland flow over rainsplash that fills incipient channels (Montgomery and Dietrich, 1994). This contrasts with channel heads in humid, soil-mantled areas that represent a change in sediment transport processes compared with a change in process dominance (Montgomery and Dietrich, 1994).

Smith and Bretherton (1972) treated the landscape as a continuous surface on which sediment transport is a function of drainage area (used as a surrogate for water discharge) and slope gradient. The surface is then defined as stable or unstable depending on whether a small initial perturbation on the surface will grow or not. Under stable conditions the initial perturbation does not grow, but during unstable conditions the additional sediment transport capacity generated by flow convergence exceeds the sediment brought in with the flow and results in erosion. Smith and Bretherton (1972) showed that the area of instability corresponds to the area of slope profile concavity where the whole profile is being eroded at a uniform rate. Deviations from this condition allow the stable zone to encroach into the lower slope concavity where the channel is downcutting more slowly than the divide and *vice versa*. This analysis is most appropriately applied to rills on smooth slopes of badland surfaces (Smith and Bretherton, 1972; Dietrich and Dunne, 1993).

Horton (1945) proposed that when rainfall intensity exceeds infiltration capacity, water accumulates in surface depressions, then spills over to run downslope as an irregular sheet of turbulent flow that imposes a shear stress on the surface. Close to the divide the shear stress is less than the critical value that needs to be exceeded for entrainment, but downslope the flow thickens and accelerates, exceeding the critical threshold. This flow is then able to erode the surface and incise. In effect this produces a 'belt of no erosion' which is a function of the difference between rainfall intensity and infiltration capacity, hydraulic roughness, critical tractive force of the surface and the local gradient. Beyond the 'belt of no erosion' the erosivity of sheetwash is proportional to the shear stress and, therefore, the local product of flow depth and water surface gradient.

Willgoose et al.'s model of channel initiation (Willgoose et al., 1991a, 1991b, 1991c) is based on the Hortonian threshold concept. The model is of the erosional development of catchments and channel networks, and differentiates between the dominant transport processes on hillslopes and in channels as functions of observable mechanisms. The initial surface was subject to random variations to perturb overland flow following conclusions from Bowyer-Bower and Bryan (1986). The model describes long-term changes in elevation due to tectonic uplift, fluvial erosion, creep, rainsplash and landsliding. Channels form when the channel initiation function exceeds a threshold that is dependent on slope and discharge. In this model elevation changes are determined from continuity equations for flow and sediment transport, with sediment transport related to discharge and slope. Flow and sediment transport equations are coupled for channel and hillslopes, and channel network extension results from physically based flow interactions on hillslopes. Channel heads advance when thresholds are exceeded. Highly non-linear interactions between channel head advance and contributing hillslope leads to a network growth process that is highly sensitive to initial conditions.

Kirkby (1994) combined the Smith and Bretherton instability criterion and the Horton tractive stress criterion in a comprehensive model of stream head location. Wash and rainsplash are disaggregated on a storm basis to provide explicit long-term integrals over the frequency distribution, as well as a magnitude and frequency interpretation of stream head behaviour. The theory forecasts a constant relationship between slope and contributing area at stream head locations where instability is dominant, such as in semi-arid environments with sparse vegetation, a stony regolith or without strong sub-surface piping. Where tractive thresholds dominate, there is a strong inverse relationship with the distance to the stream head, which can be masked by strong subsurface flow. These conditions are best met in temperate environments with a strong turf cover and a continuous transition exists between these two types. Figure 9.7 shows slope-area relationships for an example set of parameter values. In general, the condition is critical at up-valley sites, corresponding to large critical storms, while the rill erosion threshold is critical at downstream sites for smaller critical storms.

The drainage basin model of Howard (1994) combines diffusive (mass wasting and rain-splash) plus advective (fluvial erosion) processes and is also based on both the threshold and mass conservation concepts. This model differs from previous advection–dispersion models (Ahnert, 1976, 1987; Kirkby, 1971, 1986; Willgoose et al., 1991a, 1991b) by assuming that headwater channels are detachment limited. Potential erosion or deposition due to diffusive processes is given by the spatial divergence of the vector flux of regolith movement. The rate of movement is expressed as two terms, one for creep and/or splash and one for mass movement. Processes are considered to be continuous in time, representing an implicit integration over the frequency distribution. One advantage of this model is that both fluvial and slope processes can occur in each cell and that the location of channel heads is defined by a morphometric criterion.

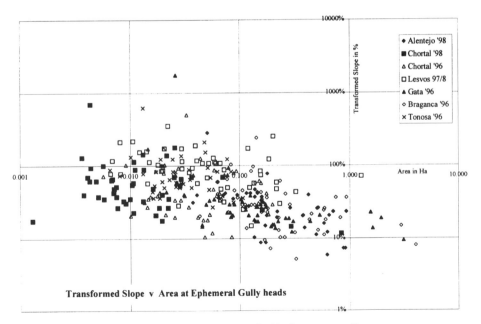

Figure 9.7   The relationship between slope and area for Mediterranean gullies

The governing equations can be written in the following forms (Kirkby and Bull, 2000). For a flow strip of variable width, w:

1: The continuity equation allows for the conservation of mass. The expression is in volume terms, so that elevation should be corrected to rock-equivalent units.

$$\frac{\partial z}{\partial t} + \frac{1}{w}\frac{\partial(Sw)}{\partial x} = 0 \qquad (9.14)$$

where z = elevation,
        w = flow strip width,
        S = actual sediment transport (generally at less than the capacity, C),
        x = horizontal distance measured along the flow strip.

2: The sedimentation balance equation gives the total rate of sediment pick-up from its rate of detachment and re-deposition. If the material is consolidated, then the detachment rate is reduced by a fraction $\alpha < 1$. The difference between total and partial differentials on the left-hand side is only important in transient flows, and is not considered to be relevant in this application.

$$\frac{1}{w}\frac{d(Sw)}{dx} = \alpha\left(D - \frac{S}{h}\right) \qquad (9.15)$$

where D = the rate of sediment detachment
        h = the sediment travel distance.

3: For wash and splash, this equation has been simplified using the concept of the effective bedload fraction (EBF). This is formally defined as:

$$\text{EBF} = \frac{\sum_d S/h}{\sum_d C/h} = \frac{\sum_d S/h}{\sum_d D} \qquad (9.16)$$

where the summations are taken over each grain-size class and C is the transporting capacity. For fines, $S \ll C$ and the contribution to the EBF is small, while for coarser material for which movement is transport limited, the contribution to the EBF is almost total.

4: The sediment transport law for rainsplash, rainflow and rill erosion is taken from the family of empirical expressions for sediment transport in a storm:

$$C = k_1 r^2 \Lambda^a + k_2 \Lambda^b (q\Lambda - \Theta)^c \qquad (9.17)$$

where C = the sediment transporting capacity ($= Dh$)
        r = storm rainfall
        q = overland flow storm discharge per unit width ($\text{m}^2\,\text{s}^{-1}$)
        Λ = the local gradient
        Θ = the flow power threshold for entrainment ($\text{m}^2\,\text{s}^{-1}$)

and k_1, k_2, a, b and c are constants.

The first term represents rainsplash, and the second channelled wash (termed rillwash below).

5: The sediment transport law for mass movements. These are treated as a continuous process (Kirkby, 1994), following equations (9.3) and (9.4) above, but with a separate sediment budget, and with different rates of detachment and travel distance, as follows:

$$\begin{aligned} D_M &= D_0\Lambda(\Lambda - \Lambda_0) \quad &\text{where } S > 0 \\ h_M &= h_0/(\Lambda_T - \Lambda) \quad &\text{where } \Lambda < \Lambda_T \end{aligned} \qquad (9.18)$$

In this formulation, the second equation is a sedimentation balance, and the transporting capacity, $C (= Dh)$, is not generally equal to the sediment transport rate, S. Over the range of grain sizes, the sediment travel distance increases as size decreases, so that transport is effectively flux-limited for coarse debris, and supply-limited for fine material. The response and evolution of the coarse- and fine-grained rambla systems studied in southeast Spain is thus very different. For coarse-grained systems sediment transport is largely flux-limited, and headwaters remain well connected with their source areas. For fine-grained systems, gully enlargement produces sediment that travels far from its source, and shows a delicate dependence on hydraulic and hydrological conditions. These cases are usefully distinguished by the EBF, without requiring a full analysis of the selective transportation. For the simplest case, with a uniform sediment transport environment (constant D and h above) downstream, the actual sediment transport downstream follows the approximate form:

$$S = Dh[1 - \exp(-x/h)] \qquad (9.19)$$

This expression shows sediment transport rising to its capacity over a distance scaled to the travel distance, h. Using this form, the EBF may be estimated in the form:

$$\text{EBF} = \frac{\sum D_i[1 - \exp(-x/h_i)]}{\sum D_i} \qquad (9.20)$$

This provides an explicit, though approximate, expression for the change in EBF downstream in relation to the travel distances of the various grain-size classes, and their proportions in the source material (the D_i), assuming equal mobility for particle entrainment. It may be seen that the EBF is increasingly dominated by the coarsest fractions in the downstream direction (Figure 9.8). The distance units in this figure are arbitrary, but are estimated to be of the order of 1 m, so that the rate of increase in EBF declines steadily downstream in the zone near to the channel head. It is for this reason that the use of an approximately constant value of

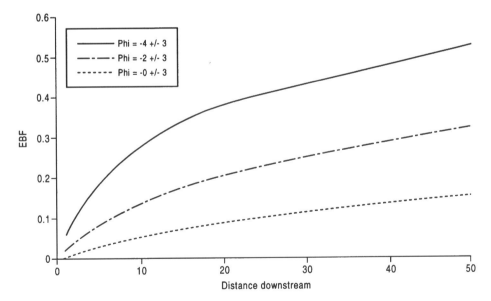

Figure 9.8 EBF weightings downstream for each grain size

the EBF is considered to provide an effective basis for a simplified operational model, and contains most of the essential elements of a model with full grain-size disaggregation. In field use, the definitions of EBF in equations (9.16) or (9.20) are not easy to apply, but a good estimate of the EBF may be obtained by comparing the grain-size distributions in the channel bed with that of the source deposits from which the bed material is derived. An enrichment factor (EF) may be calculated for each grain-size class as:

$$EF_i = \frac{\text{Frequency of } i\text{th size class in channel-bed material}}{\text{Frequency of } i\text{th size class in source material}} \quad (9.21)$$

The EBF may then be estimated as the reciprocal of the maximum enrichment factor.

This type of simulation model has much in common with other two-dimensional simulation models (Willgoose et al., 1994; Howard, 1997; Kirkby 1994, etc.), and is able to generate simulated landscapes which share many qualitative features of real landscapes. Such landscapes develop valley networks with plausible drainage densities, long profiles and hillslope profiles, even though the present model formulation does not explicitly make a demarcation between 'hillslopes' and 'channels'. The qualitative factors that are explored here are the cross-profiles of mature gullies and their evolution; and the morphology and extension of valley heads.

Preliminary tests show that steep-walled gullies are generated as the EBF falls below about 0.5, and that the introduction of a non-zero tractive power threshold appears to create less steep headcuts. Figure 9.9 gives two examples, both with a zero tractive power threshold, but with values for EBF of 1.0 and 0.1, showing the effect of this term on the morphology of gully growth. It may be seen that the reduction of EBF allows much more sediment to be evacuated, but also dramatically changes the valley form. With a lower EBF, the valley floor is flatter,

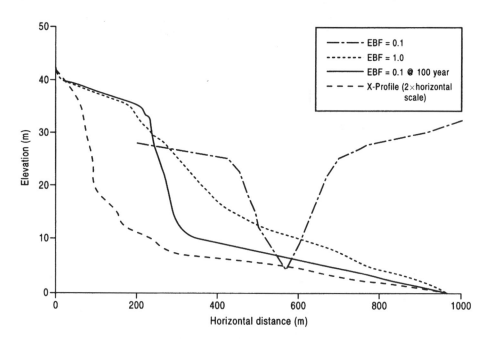

Figure 9.9 Profiles of simulated gullies. Gullies with low EBF have lower valley floor slopes and sharp valley heads. The cross profile is shown with an expanded scale

whereas the valley head remains steep, which helps to maintain its continued back-cutting. The profile form for EBF = 0.1 is also shown after 100 years, when the valley head is close to that for EBF = 1.0 after 1000 years. It may be seen that flat gully floors and steep heads are developed throughout the length of the gully. In transverse profile, the valley sides are held close to the limiting angle for mass movements (20% in this example), and there is a shallow (0.75 m) fill along the valley axis. Gullies with a low EBF have markedly lower valley floor slopes and sharp valley heads.

Smith et al. (1997a, 1997b) published two papers developing the original concept proposed by Smith and Bretherton (1972). The first paper presents a theoretical basis for modelling the evolution of drainage basins using partial differential equations, and shows that, under boundary conditions involving a fixed base level, solutions to the model exist that converge, whereby mature surface forms evolve to minimise sediment transport over the landscape surface. In contrast to the severe instability that the original equations showed when modelling the evolution of channelised flows from planar surfaces, certain classes of non-planar surfaces provide stable solutions to the equations (Smith et al., 1997a). The second paper proposed numerical solutions that suggested that the evolution of badlands is a process involving (1) a transient stage in which branching valleys emerge from unchannelled surfaces, (2) an equilibrium phase where this surface declines in a stable, self-similar mode, and (3) a final dissipative stage in which regularities in the landscape surface break down (Smith et al., 1997b). It was proposed that the family of models presented could be interpreted in terms of existing geomorphological concepts and observations, can explain variation in form within different environments, and unify aspects of continuous, discrete and variational approaches to landscape modelling presented above (Smith et al., 1997b).

9.6 THE INFLUENCE OF CATCHMENT SCALE PROCESSES

Much research has also concentrated on historic channel initiation and channel extension. In the late nineteenth century, rapid and catastrophic changes in many stream channels were documented in the American southwest when ephemeral channels began to incise into alluvial valleys creating arroyos (Figure 9.10). This inspired research into the cause of incision and processes modifying the channel heads. Similar channels have been reported in other Mediterranean environments, including Africa (Champion, 1933; Belfast, 1982; Ebisemiju, 1989; Wells et al., 1991), Australia (Crouch, 1990; Prosser, 1991) and India (Sharma, 1982).

Arroyos form during periods of incision and fill during prolonged periods of aggradation (Figure 9.11). Arroyo cutting begins with rapid initial channel downcutting that is followed by exponentially decreasing rates of downcutting as the longitudinal profile approaches a new base level of erosion (Bull, 1997). Mass movement triggered by undercutting of near vertical sidewalls promotes further stream widening once the longitudinal profile approximates a stable configuration. These local sediment perturbations influence stream responses and may initiate aggradation (Bull, 1997). Incision dates have been inferred by the minimum age of the materials the channel eroded into, and by the maximum age of sediment that subsequently filled the channel (Elliott et al., 1999). Bryan (1941) proposed that Holocene alluvium in valleys in the southwest of America could be classified into three distinct stratigraphic units dated by Miller (1958) to be: deposition 1, 7800–7200 BP; deposition 2, 850–750 BP; and deposition 3, 750–100 BP. Aggradational episodes were separated by periods of channel incision and arroyo formation. Erosion between depositions 1 and 2 is correlated with the climatic and hydrologic conditions of the Alithermal (a warmer, drier period between 7500 and 4000 BP) (Antevs,

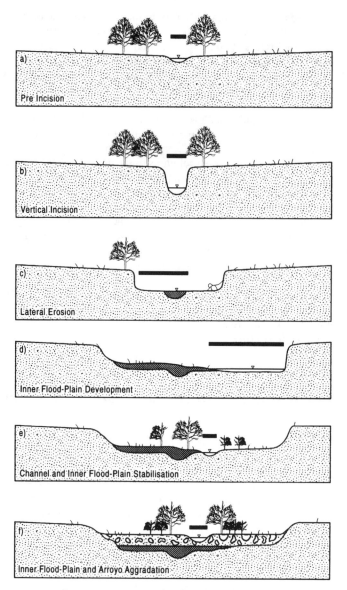

Figure 9.10   Hypothetical sequence of geomorphic evolution of an arroyo

1948). Alexander et al. (1994) determined six phases of episodic badlands development and stabilisation in southeast Spain, with ages ranging from late Pleistocene to the present day. Botha et al. (1994) found four geomorphic cycles comprising gully cut and fill in the last 135 Ka in South Africa that are similar to these.

Many arguments have been presented to explain causes of channel incision. Any perturbation that either increases stream power or decreases resisting power has the potential for initiating arroyo cutting by altering the balance of forces promoting or resisting erosion.

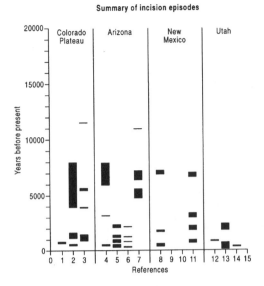

Figure 9.11 Summary of (a) incision and (b) aggradation episodes in channels of western US (after Elliott et al., 1999)

However, the main causes of channel entrenchment are a change in land use practices, climate change, large floods, and intrinsic geomorphic factors (Figure 9.12). Land use practices such as overgrazing, farming and deforestation decrease vegetative cover and infiltration rates (Cooke and Reeves, 1976). Nir and Klein (1974) suggested that high rates of channel initiation in Israel were caused by land use change from 1949 onwards. Contour ploughing decreased surface

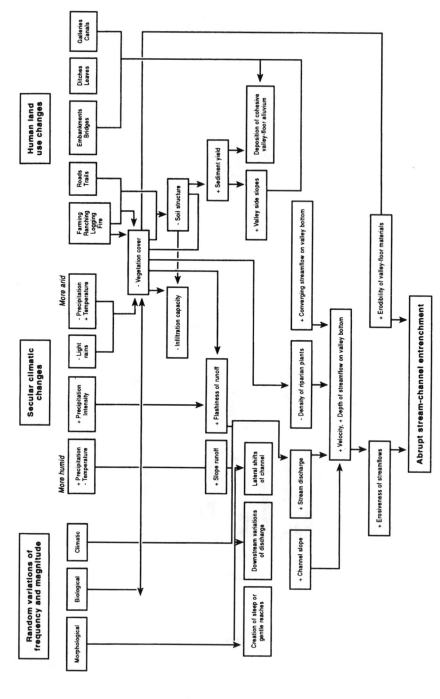

Figure 9.12   Possible causes for arroyo initiation in southern Arizona (after Bull, 1997)

runoff and stimulated interflow and the incision of gullies. Prosser and Winchester (1996) proposed that present erosion in Australia developed since European settlement, which encouraged channel initiation by the degradation of valley floor vegetation. This work also suggested that many of these deep gully systems are now approaching stability, resulting in a decrease in sediment yield (Prosser and Winchester, 1996).

Climate change has been proposed as the cause of the late Quaternary arroyo formation as well as the most recent incision and aggradation episodes in the western United States (Leopold and Snyder, 1951; Cooke and Reeves, 1976; Waters, 1985; Elliot et al., 1999). Climate change is one possible explanation because human-induced changes in land use practices could not be responsible pre-human habitation (although natural occurrences such as bison grazing, grass fires and forest fires may produce similar results). Changes in precipitation frequency and storm intensity result in changes in vegetation cover and resistance to erosion, and thus changes in production and delivery of sediment to stream channels or valley floors. Hack (1942) interpreted incision and aggradation episodes as manifestations of climatic conditions with incision occurring during dry periods and aggradation occurring during wet periods. During dry periods, vegetation dies back, and runoff volumes increase causing incision. During wet periods vegetation increases, decreasing runoff and increasing deposition. Hack (1942) and Leopold et al. (1966) suggested that arroyo incision is associated with increasing aridity and that aggradation is associated with increasing humidity, following the reasoning outlined above. Several researchers have suggested that arroyo development in the 1800s was due to an increase in large floods (Schumm and Lichty, 1963; Burkham, 1972), and large floods resulting from intense summer thunderstorms increase the amount of erosion. Periods of low annual rainfall with infrequent intense storms may also provide the trigger for arroyos (Leopold, 1951; Cooke, 1974; Elliott et al., 1999), while periods with low-intensity rainfall tend to be associated with increased vegetation density and a decrease in erosion.

Schumm and Hadley (1957) proposed a cycle of arroyo incision for semi-arid areas in eastern Wyoming and New Mexico. High transmission losses result in increasing sediment loads downstream, which lead to aggradation of the main channel and eventually the disconnection of tributaries with the trunk stream. Continued aggradation increases the gradient still further and leads to the formation of discontinuous gullies and the reintegration of the hillslope–channel system by arroyo cutting in alluvial fills. Schumm and Hadley (1957) suggested that arroyos were caused by overgrazing or climate change. Overgrazing – which was due to the introduction of cattle in 1870, with cutting being initiated in 1880 – resulted in a decline in vegetation leading to increased runoff and channel incision. Overgrazing may also result from cattle being constrained with fencing rather than an increase in cattle numbers. Climate change could be caused by a move to a drier climate, a more humid environment, or a change in rainfall intensities, but an analysis of climate change is hampered by a lack of data. The theory that cattle initiated incision has problems not only because incision began in some locations before the introduction of cattle, but also because no arroyos developed in some heavily grazed areas. Schumm and Hadley (1957) therefore concluded that the location of channel incision depended on the channel gradient and that any hypothesis of arroyo initiation needed to take into account the drainage basin character. The critical slope of entrenchment is probably related to the magnitude of runoff (Patton and Schumm, 1975).

De Ploey (1992) investigated the age of the Mediterranean badlands using an erosion susceptibility model. This model considers catchments as functional units in which geomorphic work results from process combinations such as hydraulic erosion and mass movement in terms of channel head movement. The model relates erosion to the maximum available energy, including gravitational potential as well as the kinetic energy of eroding agents. For sheet erosion the expression of erosional susceptibility (E_s) is:

$$E_s = \frac{2V_E}{APgRS} \tag{9.22}$$

where V_E = total volume (m^3) eroded within an area A (m^2)
      P = total volume of precipitation during a storm event
      g = gravitational coefficient
      R = hydraulic radius of overland flow
      S = slope gradient.

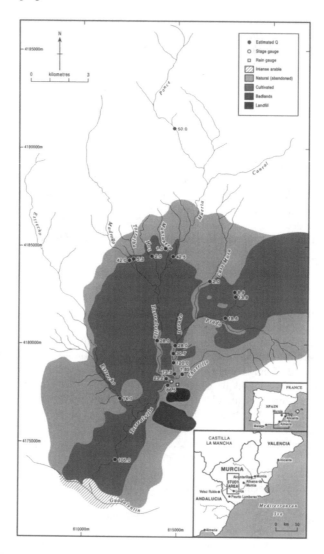

Figure 9.13 Contemporary channel incision in response to land use change in southeast Spain

For other processes:

$$E_s = \frac{2V_E}{APgh} \tag{9.23}$$

where h is the average head (m).

The model thus handles a limited set of basic and available parameters for which numerical values derive from empirical data collected in the field. The application of the erosional susceptibility model to determining the age of the Mediterranean basin badlands suggested that the badlands were formed between 2700 and 40 000 BP, which dates badlands in this region to the Upper Pleistocene era.

The influence of catchment-wide processes is also important in terms of discontinuous channels. Discontinuous channels have a repetitive channel pattern where streamflow passes through reaches characterised by headcuts, a single trunk channel that changes from moderately sinuous to straight, and braided distributary channels (Bull, 1997). This pattern reflects efficient conveyance and temporary storage of sediment by streams with limited annual stream power that is unable to flush the large amounts of supplied sediment through the channel system. Discontinuous ephemeral streams are found in semi-arid areas and are inherently unstable, changing constantly in response to self-enhancing feedback mechanisms initiated by short-term changes in climate, vegetation and land use (Bull, 1997). The most important change in these systems is the internal adjustment to variations in base level resulting from the deposition of fans (Bull, 1997). Steeper sections are the most likely sites for initiation of channel entrenchment (McGee, 1897; Schumm and Hadley, 1957; Schumm, 1973; Patton and Schumm, 1975).

Catchment-scale processes continue to have important implications for contemporary channel head initiation and extension. Recent research in the Mediterranean basin has shown that basin-wide characteristics such as topography, lithology and land use influence the resulting in-channel discharges during floods (Bull et al., 2000). Thus specific areas of Mediterranean catchments may be prone to floods and channel extension. Key hydrological response units tend to be found on steep slopes, either abandoned after agriculture or with sparse vegetation, and are composed of runoff encouraging soils such as marls (Bull et al., 2000). Areas of high discharges were also associated with areas of rilling and gullying that direct overland flow into the main channel quickly and effectively during high-intensity rainfall (Figure 9.13). These areas tend to be associated with re-incising channel heads.

9.7 RATES OF CHANNEL HEAD MOVEMENT

The development of badlands is likely to be affected by the magnitude and frequency of storm events and climatic sequences. Butcher and Thornes (1974) found that the frequency of discharge events in ephemeral channels had strong spatial variations that were useful in explaining the complex morphology. Spatial controls relate to network characteristics, width, permeability, bed material and history. It was suggested that controls varied seasonally due to subsurface storage. A particular area may tend to incise over a long period of time, but a certain climatic sequence may produce a sudden infilling. However, after this infilling, the dominant processes will again tend to incise, and an example of this can be found in some Piedmont areas (South Carolina, USA) where conditions have changed and arroyos tend to fill, and only episodically cut. A cycle of gully cutting and filling is supported by Schumm and Lusby (1963), Engelen (1973) and Bryan and Price (1980).

More recent research has suggested that channel head movement and gullying may not be as active as initially proposed, and has highlighted problems in identifying ages and rates of retreat of badlands. Blong (1970) proposed that the exact number of cut and fill cycles at any site could not be determined from the sediment record because any period of instability resulting in excessive gully widening would be able to remove evidence of earlier cut and fill cycles. For small sections of hillslopes, Campbell (1974) found that although erosion was assumed to be rapid, the surface form remained constant over a five-year period for areas in Alberta, Canada. After 10 years of monitoring there was no change in slope angle, and different slope segments were found to have different erosion rates (Campbell, 1982). This may also apply to gully heads. Wise et al. (1982) studied 40 archaeological sites in southeast Spain. Structures had survived 4000 years of denudation, implying that sheetwash and headward gully extension have been less active than the terrain would suggest, and that erosion was restricted to a few gullies and parts of major channels. These periods of development and stability may be linked to the magnitude and frequency of storm events and/or climatic sequences.

Even under active conditions of channel change, activity responds primarily to the major storms, so that the properties of the magnitude–frequency distribution are of central interest. There is at least the possibility that this distribution is not well behaved in the sense of having a well-defined mean and variance. In this case the behaviour of the channel network is controlled by the idiographic history of major events rather than by the moments of the distribution. It is presumed that one relevant way in which such 'bad' behaviour can arise is through the non-stationarity associated with externally driven climatic change.

Corresponding to this distribution over time, we also expect a distribution of forms over the channel head area. Thus the channel head area should be a zone of alternating cut and fill, and there may be single or multiple rills or ephemeral channels above the permanent channel head, particularly soon after major events. These ephemeral channels occur in a zone where the individual storm has exceeded the channel initiation threshold, even though the longer-term average threshold (represented by the permanent channel) has not been exceeded.

9.8 CONCLUSIONS

The problem of channel head and gully development is important in understanding and predicting landscape evolution, morphology and response to climate change. Much research has already been carried out but many uncertainties remain. More recent fieldwork has tended to concentrate on smaller-scale issues of gully initiation such as runoff experiments and effects of lithology, and this work has coincided with a move towards theoretical modelling. However, questions still remain concerning the location and size of gully heads and there is also a problem in determining the rate of development of gullies. Much work has published rates from short-term studies, but this does not shed light on the long-term development, i.e. whether gullies form rapidly and are then stable, or continue to retreat at the same rate.

A holistic approach to investigating gully development may help in summarising existing knowledge and directing future research. In the meantime, theoretical modelling may provide the way forward in furthering our understanding of when, where and how channel initiation occurs. Taking a longer time perspective of thousands rather than tens of years, the next direction for modelling may be based around the debate concerning the age of badlands and rates of gully development. Although badlands are generally assumed to be dynamic with high rates of erosion, this may not be the norm, and models should also be able to incorporate periods of relative stability. Modelling may also provide an intermediate approach to understanding channel head location by using simple relationships that can be predicted theoretically. However, in the long run any transport laws that are used in more complicated models need testing with field data to validate results. This also highlights the need for a comprehensive review of research to enable fieldwork and models to be developed in conjunction with each other to improve our understanding.

REFERENCES

Abrahams, A.D. 1984. Channel networks: a geomorphological perspective. *Water Resources Research*, 20, 161–188.

Aghassy, J. 1973. Man-induced badlands topography. In D.R. Coates (Ed.) *Environmental Geomorphology and Landscape Conservation (Vol. III)*. Pennsylvania, Hutchinson & Ross Inc., 124–136.

Ahnert, F. 1976. Brief description of a comprehensive three-dimensional process-response model of land-form development. *Zeitschrift für Geomorphologie Supplement Band*, 25, 29–49.

Ahnert, F. 1987. Process response models of denudation at different spatial scales. *Catena Supplement*, 10, 31–50.

Alexander, R.W., Harvey, A.M., Calvo, A., James, P.A. and Cerda, A. 1994. Natural stabilisation mechanisms on badland slopes: Tabernas, Almeria, Spain. In A.C. Millington and K. Pye (Eds) *Environmental Change in Drylands: Biogeographical and Geomorphological Perspectives*. John Wiley & Sons, Chichester, 85–111.

Antevs, E. 1948. The Great Basin, with emphasis on glacial and post-glacial times. *Bulletin*, Salt Lake City, University of Utah, 38, 168–191.

Arulanandan, K., Loganathan, P. and Krone, R.B. 1975. Pore and eroding fluid influences on surface erosion of soil. *Journal of Geotechnical Engineering Division, American Society of Civil Engineering*, 101, 51–66.

Bagnold, R.A. 1966. An approach to the sediment transport problem from general physics. *USGS Professional Paper*, 422-J.

Belfast, B.J. 1982. Effects of climate and land use on gully development: an example from northern Nigeria. *Zeitschrift für Geomorphologie*, 44, 33–51.

Berry, L. 1970. Some erosional features due to piping and sub-surface wash with special reference to the Sudan. *Geografiska Annaler*, 52A, 113–119.

Berry, L. and Ruxton, B.P. 1960. The evolution of Hong Kong Harbour basin. *Zeitschrift für Geomorphologie*, 4, 97–115.

Blong, R.J. 1970. The development of discontinuous gullies in a pumice catchment. *American Journal of Science*, 268, 369–383.

Blong, R.J., Graham, O.P. and Veness, J.A. 1982. The role of sidewall processes in gully development: some NSW examples. *Earth Surface Processes and Landforms*, 7, 381–385.

Bocco, G. 1991. Gully erosion: processes and models. *Progress in Physical Geography*, 15, 392–406.

Bocco, G. 1993. Gully initiation in Quaternary volcanic environments under temperate sub-humid seasonal climates. *Catena*, 20, 495–513.

Botha, G.A., Wintle, A.G. and Vogel, J.C. 1994. Episodic late Quaternary palaeogully erosion in northern KwaZulu-Natal, South Africa. *Catena*, 23, 327–340.

Bowyer-Bower, T.A.S. and Bryan, R.B. 1986. Rill initiation: concepts and evaluation of badlands topography. *Zeitschrift für Geomorphologie Supplement Band*, 59, 161–175.

Bradford, J.M. and Piest, R.F. 1977. Gully wall stability in Loess derived alluvium. *Journal of the American Soil Science Society*, 41, 115–122.

Bradford, J.M. and Piest, R.F. 1978. Gully wall stability in Loess derived alluvium. *American Journal of Soil Science*, 42, 323–328.

Bradford, J.M. and Piest, R.F. 1985. Erosion development of valley bottom gullies in the upper Midwestern United States. In D.R. Coates and J.D. Vitek (Eds) *Thresholds in Geomorphology*. Allen & Unwin, London, 75–101.

Bryan, K. 1941. Pre-Columbian agriculture in the Southwest as conditioned by periods of alluviation. *Annals of the Association of American Geographers*, 4, 219–242.

Bryan, R.B. 1987. Process and significance of rill development. *Catena Supplement*, 8, 1–16.

Bryan, R.B. 1990. Knickpoint evolution in rillwash. *Catena Supplement*, 17, 111–132.

Bryan, R.B. 1996. Erosional response to variations in interstorm weathering conditions. In M.G. Anderson and S.M. Brooks (Eds) *Advances in Hillslope Processes*. Wiley, Chichester, 589–612.

Bryan, R.B., Campbell, I.A. and Sutherland, R.A. 1988. Fluvial geomorphic processes in semi-arid ephemeral catchments in Kenya and Canada. *Catena Supplement*, 13, 15–35.

Bryan, R.B., Hawke, R.M.and Rockwell, D.L. 1998. The influence of subsurface moisture on rill system evolution. *Earth Surface Processes and Landforms*, 23, 773–789.

Bryan, R.B. and Jones, J.A.A. 1997. The significance of soil piping processes: inventory and prospect. *Geomorphology*, 20, 209–218.

Bryan, R.B. and Oostwoud Wijdenes, D.J. 1992. Field and laboratory experiments on the evolution of microsteps and scour channels on low angle slopes. *Catena Supplement*, 23, 1–29.

Bryan, R.B. and Price, A.G. 1980. Recession of the Scarborough Bluffs, Ontario, Canada. *Zeitschrift für Geomorphologie Supplement Band*, 34, 48–62.

Bryan, R.B., Yair, A. and Hodges, W.K. 1978. Factors controlling the initiation of runoff and piping in Dinosaur Provincial Park Badlands, Alberta. *Zeitschrift für Geomorphologie Supplement Band*, 29, 151–168.

Buckham, A.F. and Cockfield, W.E. 1950. Gullies formed by sinking of the ground. *American Journal of Science*, 248, 137–141.

Burkham, D.E. 1972. Channel changes of the Gila River in Safford Valley, Arizona 1846–1970. *US Geological Survey Professional Paper*, 655-G, 24pp.

Bull, W.B. 1997. Discontinuous ephemeral streams. *Geomorphology*, 19, 227–276.

Bull, L.J. and Kirkby, M.J. 1997. Gully processes and modelling. *Progress in Physical Geography*, 21, 354–374.

Bull, L.J., Kirkby, M.J., Shannon, J. and Hooke, J.M. 2000. The variation in estimated discharge in relation to the location of storm cells in SE Spain. *Catena*, 38(3), 191–209.

Bunting, B.T. 1961. The role of seepage moisture in soil formation, slope development, and stream initiation. *American Journal of Science*, 259, 503–518.

Butcher, G.C. and Thornes, J.B., 1974. Spatial variability in runoff processes in an ephemeral channel. *Zeitschrift für Geomorphologie Supplement Band*, 29, 83–92.

Campbell, I.A. 1974. Measurements of erosion on badlands surfaces. *Zeitschrift für Geomorphologie Supplement Band*, 21, 122–137.

Campbell, I.A. 1982. Surface morphology and rates of change during a ten year period in the Alberta badlands. In R.B. Bryan and A. Yair (Eds) *Badland Geomorphology and Piping* GeoBooks, Norwich, 221–238.

Carson, M.A. and Kirkby, M.J. 1972. *Hillslope Form and Process*. University Press, Cambridge, 475pp.

Champion, A.M. 1933. Soil erosion in Africa. *Geographical Journal*, 82, 130–139.

Colclough, J.D. 1965. Tunnel erosion. *Tasmanian Journal of Agriculture*, 34, 7–12.

Cooke, R.U. 1974. The rainfall context of arroyo initiation in southern Arizona. *Zeitschrift für Geomorphologie Supplement Band*, 21, 63–75.

Cooke, R.U. and Reeves, R.W. 1976. *Arroyos and Environmental Change in the Southwest*. Clarendon Press, Oxford.

Crouch, R.J. 1976. Field tunnel erosion; a review. *Soil Conservation Journal*, 32, 98–111.

Crouch, R.J. 1990. Rates and mechanisms of discontinuous gully erosion in a red-brown earth catchment, New South Wales, Australia. *Earth Surface Processes and Landforms*, 15, 277–282.

Crouch, R.J. and Blong, R.J. 1989. Gully sidewall classification: methods and applications. *Zeitschrift für Geomorphologie Supplement Band*, 33, 291–305.

Cumberland, K.B. 1944. *Soil erosion in New Zealand*. Soil Erosion and River Control Council, Wellington, 288pp.

Dalrymple, J.B., Blong, R.J. and Conacher, A.J. 1968. A hypothetical nine unit land surface model. *Zeitschrift für Geomorphologie*, 12, 60–76.

Dardis, G.F. 1989. Quaternary erosion and sedimentation in Badlands areas of southern Africa. *Catena*, 14, 1–10.

Dardis, G.F. and Beckedahl, H.R. 1988. Drainage evolution in an ephemeral soil pipe-gully system, Transkei, Southern Africa. In G.F. Dardis and B.P. Moon (Eds) *Geomorphological Studies in Southern Africa*. Balkema, Rotterdam, 247–265.

De Ploey, J. 1971. Liquefaction and rainwash erosion. *Zeitschrift für Geomorphologie*, 15, 491–496.

De Ploey, J. 1973. Ruissellement diffus, rauinement et badlands dans le bassin de Kasserine (Tunisie steppiqua), Livre jubilaire Solignac. *Annales Mines and Geology, Tunisie*, 26, 583–593.

De Ploey, J. 1974. Mechanical properties of hillslopes and their relation to gullying in central, semi-arid Tunisia. *Zeitschrift für Geomorphologie Supplement Band*, 21, 177–190.

De Ploey, J. 1989. A model for headcut retreat in rills and gullies. *Catena Supplement*, 14, 81–86.

De Ploey, J. 1990. Threshold conditions for thalweg gullying with special reference to Loess areas. *Catena Supplement*, 17, 147–151.

De Ploey, J. 1992. Gullying and the age of badlands; an application of the erosional susceptibility model Es. *Catena Supplement*, 23, 31–46.

Dietrich, W.E. and Dunne, T. 1993. The channel head. In K. Beven and M.J. Kirkby (Eds) *Channel Network Hydrology*. John Wiley & Sons Ltd, Chichester, 175–219.

Dietrich, W.E., Wilson, C.J. and Reneau, S.L. 1986. Hollows, colluvium, and landslides in soil mantled landscapes. In A.D. Abrahams (Ed.) *Hillslope Processes*. Allen & Unwin, London, 361–388.

Dietrich, W.E., Wilson, C.J., Montgomery, D.R and Romy Bauer, J.M. 1992. Erosion thresholds and land surface morphology. *Geology*, 20, 675–679.

Downes, R.G. 1949. A soil, landuse and erosion survey of parts of the counties of Moira and Dalatite, Victoria. *Bulletin of the CSIRO, Australia*, 243, 89pp.

Dunne, T. 1980. Formation and controls of channel networks. *Progress in Physical Geography*, 4, 211–239.

Dunne, T. 1990. Hydrology, mechanics and geomorphic implications of erosion by subsurface flow. In C.G. Higgins and D.R. Coates (Eds) *Ground Water Geomorphology: The Role of Subsurface Water in Earth Surface Processes and Landforms*. Geological Society of America Special Paper, 252.

Ebisemiju, F.S. 1989. A geomorphological approach to land use planning and soil conservation. *Journal of Environmental Management*, 28, 327–336.

Elliott, J.G., Gellis, A.C. and Aby, S.B. 1999. Evolution of Arroyos: incised channels of the Southwestern United States. In S.E. Darby and A. Simon (Eds) *Incised River Channels: Processes, Forms, Engineering and Management*. John Wiley & Sons, Chichester, 153–185.

Engelen, G.B. 1973. Runoff processes and slope development in Badlands National Monument, South Dakota. *Journal of Hydrology*, 18, 55–79.

Fenneman, N.M. 1922. Physiographic provinces and sections in Western Oklahoma and adjacent parts of Texas. *United States Geological Survey Bulletin*, 730, 126–129.

Gerits, J. 1986. *Implications of chemical thresholds and physico-chemical analysis for modelling erosion in southeast Spain*. Paper to Commission on Measurement, Theory and application in Geomorphology, Granada.

Gerits, J., Imeson, A.C. and Verstraeten, J.M. 1986. *Chemical thresholds and erosion in saline and sodic materials*. Estudios sobre Geomorfologia del sur Espana, Murcia, Universidad de Murcia.

Gerits, J., Imeson, A.C., Verstraeten, J.M. and Bryan, R.B. 1987. Rill development and badlands regolith properties. *Catena Supplement*, 8, 141–160.

Gutiérrez, M., Benito, G. and Rodriguez, J. 1988. Piping in badlands areas of the middle Ebro basin, Spain. *Catena Supplement*, 13, 49–60.

Gutiérrez, M., Sancho, C., Benito, G., Sirvent, J. and Desir, G. 1997. Quantitative study of piping processes in badland areas of the Ebro Basin, NE Spain. *Geomorphology*, 20, 237–253.

Guy, B.T. and Dickinson, W.T. 1990. Inception of sediment transport in shallow overland flow. *Catena Supplement* 17, 91–109.

Hack, J.T. 1942. *The changing physical environment of the Hopi Indians of Arizona*. Peabody Museum Papers of American Archaeology and Ethnology, University of Harvard, Cambridge, Massachusetts, 35.

Harvey, A. 1982. The role of piping in the development of badlands and gully systems in southeast Spain. In R.B. Bryan and A. Yair (Eds) *Badlands Geomorphology and Piping*. GeoBooks, Norwich, 317–336.

Haworth, E. 1897. *Physiography of Western Kansas. University Geological Survey, Kansas*, 2, 17–21.

Heede, B.H. 1971. Characteristics and processes of soil piping in gullies. *US Department of Agriculture Forest Service Research Paper*, RM-68, 1–15.

Heede, B.H. 1974. Stages of development of gullies in Western United States of America. *Zeitschrift für Geomorphologie*, 18, 260–271.

Higgins, C.G. 1990. Gully development. In C.G. Higgins and D.R. Coates (Eds) *Groundwater Geomorphology*. Geological Society of America, Special Paper, 252, 139–155.

Holmgren, G.G.S. and Flanagan, C.P. 1977. Factors affecting spontaneous dispersion of soil materials as evidenced by the crumb test. In J.L. Sherard and R.S. Decker (Eds) *Dispersive Clays and Related Piping and Erosion in Geotechnical Projects*. American Society of Testing Materials, Special Technical Paper, 623, 218–239.

Horton, R.E., 1945. Erosional development of streams and their drainage basins; hydrophysical approach to quantitative morphology. *Bulletin of the American Geological Society*, 56, 275–370.

Howard, A.D. 1988. Groundwater sapping experiments and modelling. In A.D. Howard, R.C. Kochel and H.E. Holt (Eds) *Sapping Features of the Colorado Plateau, a Comparative Planetary Field Guide*. NASA Scientific and Technical Information Division, 108pp.

Howard, A.D. 1994. A detachment-limited model of drainage basin evolution. *Water Resources Research*, 30, 2261–2285.

Howard, A.D. 1997. Badland morphology and evolution: interpretation using a simulation model. *Earth Surface Processes and Landforms*, 22, 211–227.

Howard, A.D. and MacLane, C.G. 1988. Erosion of cohesionless sediment by groundwater seepage. *Water Resources Research*, 24, 1659–1674.

Imeson, A.C., Kwaad, F.J.P.M. and Verstraeten, J.M. 1982. The relationship of soil chemical and physical properties to the development of badlands in Morocco. In R.B. Bryan and A. Yair (Eds) *Badlands Geomorphology and Piping*. GeoBooks, Norwich, 47–70.

Imeson, A.C. and Verstraeten, J.M. 1988. Rills on badlands slopes; a physico-chemically controlled phenomenon. *Catena Supplement*, 12, 139–150.

Imeson, A.C. and Verstraeten, J.M. 1989. The microaggradation and erodibility of some semi arid and Mediterranean soils. *Catena Supplement*, 14, 11–24.

Ingles, O.G. and Aitchinson, G.D. 1969. Soil water disequilibrium as a cause of subsidence in natural soils and earth embankments. *Proceedings of the Symposium on Land Subsidence, Tokyo*, 342–353.

Johnson, W.D. 1901, The High Plains and their utilisation. *US Geological Survey 21st American Report*, 139, 702–711.

Jones, J.A.A. 1971. Soil piping and stream initiation. *Water Resources Research*, 7, 602–610.

Jones, J.A.A. 1981. *The Nature of Soil Piping; A Review of Research*. GeoBooks, Norwich, 301pp.

Jones, J.A.A. 1982. Experimental studies of pipe hydrology. In R.B. Bryan and A. Yair (Eds) *Badlands Geomorphology and Piping*. GeoBooks, Norwich, 355–370.

Kilinc, M. and Richardson, E.V. 1973. *Mechanics of soil erosion from overland flow generated by simulated rainfall*. Colorado State University Hydrology Paper, 63, 82pp.

Kirkby, M.J. 1971. Hillslope process response models based on the continuity equation. In D. Brunsden (Ed.) *Slopes: Form and Process*. Institute of British Geographers, Special Publication 3.

Kirkby, M.J. 1980. The stream head as a significant geomorphic threshold. In D.R. Coates and A.D. Vitek (Eds) *Thresholds in Geomorphology*. Allen & Unwin, London, 53–73.

Kirkby, M.J. 1986. A two-dimensional simulation model for slope and stream evolution. In A.D. Abrahams (Ed.) *Hillslope Processes*. Allen & Unwin, London, 53–73.

Kirkby, M.J. 1994. Thresholds and instability in stream head hollows: a model of magnitude and frequency for wash processes. In M.J. Kirkby (Ed.) *Process Models and Theoretical Geomorphology*. John Wiley & Sons Ltd, Chichester, 295–352.

Kirkby, M.J. and Bull, L.J. 2000. Factors controlling gully growth in fine-grained sediments: A model applied to southeast Spain. *Catena*, 40, 127–146.

Kirkby, M.J. and Chorley, R.J. 1967. Throughflow, overland flow and erosion. *Bulletin of the International Association for Scientific Hydrology*, 12, 5–21.

Kochel, R.C., Howard, A.D. and MacLane, C.F. 1982. Channel networks developed by groundwater sapping in fine-grained sediments: analogs in some Martian valleys. In M.J. Woldenberg (Ed.) *Models in Geomorphology*. Allen & Unwin, Boston, 313–341.

La Roca Cervignon, N. and Calvo-Cases, A. 1988. Slope evolution by mass movements and surface wash. *Catena Supplement*, 12, 95–102.

Leopold, L.B. 1951. Rainfall frequency: an aspect of climatic variation. *American Geophysical Union Transactions*, 32, 347–357.

Leopold, L.B., Wolman, M.G. and Miller, J.P. 1964. *Fluvial Processes in Geomorphology*. Freeman & Company, San Francisco.

Leopold, L.B., Emmett, W.W. and Myrick, R.W. 1966. Channel and hillslope processes in a semi-arid area, New Mexico. *US Geological Survey Professional Papers*, 352-G, 1–23.

Leopold, L.B. and Snyder, C.I. 1951. Alluvial fills near Gallup, New Mexico. *US Geological Survey Water Supply Paper*, 1261.

Liggitt, B. and Fincham, R.J. 1989. Gully erosion: the neglected dimension in soil erosion research. *South African Journal of Science*, 18–20.

Llasat, M.C. and Puigcerver, M. 1994. Meteorological factors associated with floods in the northeastern part of the Iberian Peninsula. *Natural Hazards*, 9, 81–93.

McGee, W.J. 1897. Sheetflood erosion. *Geological Society of America Bulletin*, 8, 87–112.

Macklin, M.G., Lewin, J. and Woodward, J.C. 1992. Quaternary fluvial systems in the Mediterranean basin. In J. Lewin, M.G. Macklin and J.C. Woodward (Eds) *Mediterranean Quaternary River Environments.* A.A. Balkema, Rotterdam, 1–24.

Martin-Penela, A.J. 1994. Pipe and gully systems development in the Almanzora basin, southeast Spain. *Zeitschrift für Geomorphologie,* 38(2), 207–222.

Miller, J.P. 1958. Problems of the Pleistocene in cordilleran North America as related to reconstruction of environmental changes that affected early man. In T.L. Smiley (Ed.) *Climate and Man in the Southwest.* University of Arizona Press, Tucson, Arizona, 19–41.

Montgomery, D.R. and Dietrich, W.E. 1989. Source areas, drainage density and channel initiation. *Water Resources Research,* 25, 1907–1918.

Montgomery, D.R. and Dietrich, W.E. 1994. Landscape dissection and drainage area-slope thresholds. In M.J. Kirkby (Ed.) *Process Models and theoretical Geomorphology.* John Wiley & Sons, 221–246.

Morgan, A.L. 1976. *An investigation into the location, geometry and hydraulics of ephemeral soil pipes on Plynlimon, Mid-Wales.* BSc Dissertation, University of Manchester.

Moss, A.J., Walker, P.H. and Hutka, J. 1979. Raindrop simulated transportation in shallow water flows: an experimental study. *Sedimentary Geology,* 22, 165–184.

Nir, D. and Klein, M. 1974. Gully erosion induced by land use in semi arid terrain. *Zeitschrift für Geomorphologie Supplement Band,* 21, 191–201.

Oostwoud Wijdenes, D.J. and Bryan, R.B. 1991. Gully development on the Njemps Flats, Baringo, Kenya. *Catena Supplement,* 19, 71–90.

Oostwoud Wijdenes, D.J. and Bryan, R.B. 1994. Gully headcuts as sediment sources on the Njemps Flats and initial low-cost gully control measures. *Advances in Geoecology,* 27, 205–229.

Oostwoud Wijdenes, D.J. and Gerits, J. 1994. Runoff and sediment transport on intensely gullied, low-angled slopes in Baringo District. *Advances in Geoecology,* 27, 121–141.

Parker, G.C. 1963. Piping: a geomorphic agent in landform development of the drylands. *IAHS,* 65, 103–113.

Patton, P.C. and Schumm, S.A. 1975. Gully erosion in Colorado: a threshold performance phenomenon. *Geology,* 3, 88–90.

Piest, R.F., Bradford, J.M. and Spomer, R.G. 1975. Mechanisms of erosion and sediment movement from gullies. *Present and Prospective Technology for Predicting Sediment Yields and Sources,* ARS-USDA, 162–176.

Poesen, J.W.A. and Hooke, J.M. 1997. Erosion, flooding and channel management in Mediterranean environments of southern Europe. *Progress in Physical Geography,* 21, 157–199.

Prosser, I.P. 1991. A comparison of past and present episodes of gully erosion at Wangrah Creek, Southern Tablelands, New South Wales. *Australian Geographical Studies,* 29, 139–154.

Prosser, I.P. and Winchester, S.J. 1996. History and processes of gully initiation and development in eastern Australia. *Zeitschrift für Geomorphologie Supplement Band,* 105, 91–109.

Robinson, K.M. and Hanson, G.J. 1995. Large scale headcut erosion testing. *Transactions of the American Society of Agricultural Engineers,* 38, 429–434.

Rubey, W.W. 1928. Gullies in the Great Plains formed by sinking of the ground. *American Journal of Science,* 15, 417–422.

Sargunam, A., Riley, P., Arulanandan, K. and Krone, R.B. 1973. Physico-chemical factors in erosion of cohesive soils. *Journal of the Hydraulics Division, American Society of Civil Engineers,* 99, HY3, 555–558.

Savat, J. 1979. Laboratory experiments on erosion and deposition of loess by laminar sheetflow and turbulent rill flow. *Proceedings of the Seminar Agricultural Soil Erosion in Temperate Non-Mediterranean Climate,* 20–23, 139–143.

Savat, J. and Poesen, J. 1981. Detachment and transportation of loose sediments by raindrop splash. Part 1. The calculation of absolute data on detachability and transportability. *Catena,* 8, 1–17.

Schumm, S.A. 1956a. Evolution of drainage systems and slopes in badlands at Perth Amboy, New Jersey. *Bulletin of the American Geological Society,* 67, 597–646.

Schumm, S.A. 1956b. The role of creep and rainsplash on the retreat of badlands slopes. *American Journal of Science,* 254, 693–706.

Schumm, S.A. 1964. Seasonal variations of erosion rates and processes on hillslopes in Western Colorado. *Zeitschrift für Geomorphologie,* 5, 215–238.

Schumm, S.A. 1973. Geomorphic thresholds and complex response of drainage systems. In M. Morisawa (Ed.) *Fluvial Geomorphology.* State University of New York Publications in Geomorphology, Binghampton, 299–310.

Schumm, S.A. and Hadley, R.F. 1957. Arroyos and the semiarid cycle of erosion. *American Journal of Science*, 255, 161–174.

Schumm, S.A. and Lichty, R.W. 1963. Channel widening and floodplain construction along Cimarron River in Southwestern Kansas. *US Geological Survey Professional Paper*, 352D, 71–88.

Schumm, S.A. and Lusby, G.C. 1963. Seasonal variations in infiltration capacity and runoff on hillslopes of Western Colorado. *Journal of Geophysical Research*, 63, 3655–3666.

Seginer, I. 1966. Gully development and sediment yield. *Journal of Hydrology*, 4, 236–253.

Sharma, H.S. 1982. Morphology of ravines of the Morel Basin Rajasthan, India. In H.S. Sharma (Ed.) *Perspectives in Geomorphology*. Concept Publishing, New Delhi, 35–48.

Sherard, J.L., Ryker, N.L. and Decker, R.S. 1972, Piping in earth dams of dispersive clay. *Proceedings of the Special Conference on the Performance of Earth and Earth Supported Structures*, ASCE, 150–161.

Smith, T.R. and Bretherton, F.P. 1972. Stability and the conservation of mass in drainage basin evolution. *Water Resources Research*, 8, 1506–1529.

Smith, T.R., Birnir, B. and Merchant, G.E. 1997a. Towards an elementary theory of drainage basin evolution: I. The theoretical basis. *Computers and Geosciences*, 23, 811–822.

Smith, T.R., Merchant, G.E. and Birnir, B. 1997b. Towards an elementary theory of drainage basin evolution: II. Computational evaluation. *Computers and Geosciences*, 23, 823–849.

Tanaka, T., Yasuhara, M. and Marui, A. 1982. Pulsating flow phenomenon in soil pipes. *Annual Report of the Institute of Geoscience, University of Tsukuba*, 8, 33–36.

Temple, D.M. 1992. Estimating flood damage to vegetated deep soil spillways. *Applied Engineering in Agriculture*, 8(2), 237–242.

Terzaghi, K. and Peck, R.B. 1948. *Soil Mechanics and Engineering Practice*, John Wiley & Sons, New York, 566pp.

Torri, D. 1987. A theoretical study of soil detachability. *Catena Supplement*, 10, 15–20.

Torri, D. and Bryan, R. 1997. Micropiping processes and biancana evolution in southeast Tuscany, Italy. *Geomorphology*, 20, 219–235.

Torri, D., Sfalanga, M. and Chisci, G. 1987. Threshold conditions for incipient rilling. *Catena Supplement*, 8, 94–106.

Tsukamoto, Y., Ohta, T. and Noguchi, H. 1982. Hydrological and geomorphological studies of debris slides on forested hillslopes in Japan. *IAHS*, 137, 89–98.

Veness, J.A. 1980. The role of fluting in gully extension. *Journal of Soil Conservation*, 36, 100–108.

Waters, M.R. 1985. Late Quaternary alluvial stratigraphy of Whitewater Draw, Arizona: implications for regional correlation of fluvial deposits in the American Southwest. *Geology*, 13, 705–708.

Wells, N.A., Andriamihaja, B. and Rakotovololona, H.F.S. 1991. Patterns of lavaka, Madagascar's unusual gullies. *Earth Surface Processes and Landforms*, 16, 189–206.

Willgoose, G., Bras, I. and Rodriguez-Iturbe, I. 1991a. A coupled network growth and hillslope evolution model 2: Nondimensionalization and applications. *Water Resources Research*, 27, 1685–1696.

Willgoose, G., Bras, I. and Rodriguez-Iturbe, I. 1991b. A coupled channel network growth and hillslope evolution model 1: Theory. *Water Resources Research*, 27, 1671–1684.

Willgoose, G., Bras, I. and Rodriguez-Iturbe, I. 1991c. Results from a new model of river basin evolution. *Earth Surface Processes and Landforms*, 16, 237–254.

Willgoose, G., Bras, I., and Rodriguez-Iturbe, I. 1994. Hydrogeomorphology modelling with a physically based river basin evolution model. In M.J. Kirkby (Ed.) *Process Models and Theoretical Geomorphology*. Wiley, Chichester, 3–22.

Wise, S.M., Thornes, J.B. and Gilman, A. 1982. How old are the badlands? A case study from southeast Spain. In R.B. Bryan and A. Yair (Eds) *Badlands Geomorphology and Piping*. GeoBooks, Norwich, 259–278.

Yair, A., Bryan, R.B., Lavee, H. and Adar, E. 1980. Runoff and erosion processes and rates in the Zin Valley northern Negev, Israel. *Earth Surface Processes and Landforms*, 5, 205–225.

Yair, A. and Lavee, H. 1974. Aerial contribution to runoff on scree slopes in an extreme arid environment. *Zeitschrift für Geomorphologie Supplement Band*, 21, 106–121.

Yair, A., Sharon, D., and Lavee, H. 1978. An instrumented watershed for the study of partial area contribution of runoff in the arid zone. *Zeitschrift für Geomorphologie Supplement Band*, 29, 71–82.

Yalin, M.S. 1971. *Mechanics of Sediment Transport*. Pergamon Press.

10 Badland Systems in the Mediterranean

FRANCESC GALLART, ALBERT SOLÉ, JOAN PUIGDEFÀBREGAS
AND ROBERTO LÁZARO
Estación Experimental de Zonas Aridas (CSIC), Almeria, Spain

10.1 INTRODUCTION

The term *badlands* is currently used for areas of unconsolidated sediments or poorly consolidated bedrock, with little or no vegetation, that are useless for agriculture because of their intensely dissected landscape. Drainage density by V-shaped valleys is usually very high, and some of the degraded landscapes relate to piping erosion, mass movements or the outcrop of shallow saline groundwater (not necessarily characteristic of dissected landscapes) (Bryan and Yair, 1982b). Badlands are different from gullies in the sense that they are not only linear erosive forms normally cut in loose sediments, but also include hillslopes and divides that are usually carved in soft bedrock. Nevertheless, both forms may be closely related, as badland areas may be initiated or reactivated by gully development (Nogueras et al., 2000).

Badlands are frequently considered to be landscapes that are characteristic of dryland areas. Nevertheless, they also occur in wetter areas where high topographic gradients, bedrock weakness and high-intensity rainstorms, which are rather frequent in Mediterranean environments, coexist. This chapter, therefore, seeks to analyse badland dynamics for a range of precipitation which includes subhumid areas, explicitly incorporating the role of vegetation, not discussed in former reviews of badlands (Bryan and Yair, 1982a; Campbell, 1989). Examples are mainly from Spain, where a variety of environmental conditions are found, but also from other Mediterranean areas. The emphasis is mostly on surface processes, whilst the role of tectonics, relief and base level in influencing badland occurrence, although important within a geological context (Harvey, 1987; Alexander et al., 1994), has not been considered in this chapter.

10.2 GEOLOGICAL CONTROLS OF BADLAND OCCURRENCE AND FORMS

The main factor controlling badland formation is the particular character of the rocks or other materials which form the base for the interaction of weathering and erosion processes (Campbell, 1989). However, the existence of highly eroded slopes means the previous or simultaneous development of a high relief where a protective caprock has been removed and/or stream downcutting has occurred (Howard, 1994). In most Mediterranean regions Quaternary tectonics have been quite active, resulting in past and/or present uplifting in most badland

Dryland Rivers: Hydrology and Geomorphology of Semi-arid Channels. Edited by L.J. Bull and M.J. Kirkby.
© 2002 John Wiley & Sons, Ltd.

areas. Badlands form on soft or unconsolidated geological materials, mostly *soils*, or some *sediments or sedimentary rocks* (aeolian, glacial, colluvial and alluvial deposits).

Soils in badlands deserve special attention, because soils are the inter-phase between the lithosphere and the atmosphere, and so constitute one of the key elements either favouring or restricting the initiation of badland formation. When soils are resilient against erosion processes, gullies do not form; however, when soils, either because of their particular ground cover, i.e. sparse vegetation, and/or intrinsic properties, cannot withstand erosive forces, the topsoil is eroded and deep gullies develop, which may give rise to badlands if the underlying material is also erosion-sensitive.

Consequently, the characteristics of the materials underlying soils are crucial for the development of true badlands. However, not only does the degree of consolidation define a badland-prone material, but cementing agents and particle size range and distribution are also crucial.

Lithology is a major factor for badland production, and is probably of greater importance than tectonics, climate, topography or land use (Campbell, 1989; Gerits et al., 1987; Imeson and Verstraeten, 1988; Calvo et al., 1991a,1991b). The general characteristics of a soil, regolith or geological formation that favours badland relief are the unconsolidated or very poorly cemented material of clay and silt, sometimes with soluble minerals such as gypsum or halite (Scheidegger et al., 1968). Specific characteristics, like structure, mineralogy, physical and chemical properties, may play either a primary or secondary role in material disintegration and badland development. Fourteen badland areas, mainly from the western Mediterranean, are examined as examples (Table 10.1).

10.2.1   Structure–Microstructure; Morphology–Micromorphology

Both shales and mudstones have a considerable network of fissures and cracks, mainly due to unloading stresses when the bedrock goes from deep burial to near Earth–surface conditions, either because of tectonic activity or erosion dismantling the Earth's upper crust. This network of cracks and fissures, which can be seen through an optical microscope (Figure 10.1), is the predominant entry for atmospheric fluids coming into close contact with the rock and starting weathering processes. In Tabernas gypsiferous mudstones, gypsum-filled cracks are responsible for mudstone breakdown once the gypsum dissolves (Cantón et al., 2001). In Vallcebre mudstones, smectite aggregates start to swell when a network of cracks and fissures connects with atmospheric solutions (Solé et al., 1992).

Individual particles and/or clay aggregates, when observed through the scanning electron microscope (SEM), are seen to have important intergrain pore spaces (Figure 10.2), which may conduct weathering fluids by capillarity.

From fresh mudrock to the weathered state, shales and mudrocks have been monitored for temporal changes in their surface morphology by means of sequential photography (Farres, 1978; Harvey, 1982, 1987; Regüés et al., 1993, 1995; Pardini et al., 1995; Cantón et al., 2001) or from the differences between wet and dry bulk densities (Imeson, 1986; Bouma and Imeson, 2000). In all cases, changes were very fast and usually related to porosity enhancement.

10.2.2   Mineralogy

Certain minerals play an essential role in the breakdown of some rocks at near surface conditions and can be divided into two great groups: (a) those which may become soluble, like all soluble salts (halite), but also moderately soluble like sulphates (gypsum) or carbonates (calcite

Table 10.1 Characteristics of badland areas (mainly from the western Mediterranean)

Reference	Location	Type of rock	Rock age	Sand	Silt	Clay	Gypsum	CaCO₃ eq.	Smectite	SARp	WCT
Cantón, 1999	Tabernas (SE Spain)	Calcaric-gypsiferous mudstone	Upper Miocene	10	80	10	10–30	30	(+)	1–25	2
Berrad et al., 1994	Albox (SE Spain)	Calcaric-gypsiferous claystone	Upper Miocene	0	50	50	3–15	30	++	1–16	1
Gerits et al., 1987	Guadix (SE Spain)	Marine marl	Eocene–Miocene	9	62	29	3	50–60	n.a.	20–80	n.a.
Unpublished data	Los Guillermos (SE Spain)	Claystone	Neogene	15	75	10	0–4	40	+++	1–7	1
Martín-Penela, 1994	Vera (SE Spain)	Gypsiferous mudstone	Upper Miocene	n.a.	>50	>40	+++	Marls	(+++)	n.a.	n.a.
Unpublished data	Abanilla (SE Spain)	Mudstone	Neogene	9	76	15	7–12	22	(+)	1–40	2
Harvey and Calvo, 1989	Petrer (SE Spain)	Marl	Cretaceous	13	68	19	Traces	60	++	1–50	1
Benito et al., 1992	Huesca (NE Spain)	Shale	Miocene	n.a.	n.a.	n.a.	0	+	(+)	13–42	n.a.
Solé et al., 1992	Vallcebre (NE Spain)	Mudstone	Late Cretaceous	10	55	35	0	30–50	+++	<1	1
Meunier et al., 1987	Draix (SE France)	Black marls	Middle Jurassic	n.a.	n.a.	n.a.	0	30–60	(+)	n.a.	n.a.
Torri et al., 1994	Volterra (Central Italy)	Silty clay sediments	Pliocene	2	50	48	0	16	(+)	21	n.a.
Yair et al., 1980	Negev (S. Israel)	Sediments	Cretaceous–Neogene	n.a.	n.a.	30–80	+	+	+++	n.a.	n.a.
Gomer, D., 1995	Oued Mina (N. Algeria)	Marls	Triassic–Jurassic	Silty to silty clay			n.a.	+	n.a.	n.a.	n.a.
Imeson et al., 1982	Beni Boufrah (N. Morocco)	Silty colluvial sediments	n.a.	4–80	11–46	8–54	0	1–34	–	20–34	n.a.

n.a. = not available.   SARp = practical Sodium Absorption Ratio.   WCT = Water Coherence Test.

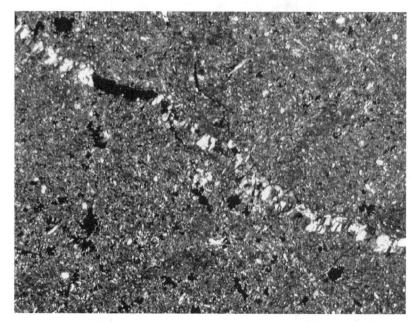

Figure 10.1   Photomicrograph (under cross-polarised light) of a fresh gypsiferous mudstone showing a crack filled with gypsum (from upper left to bottom right)

Figure 10.2   SEM image of a calcium–gypsiferous mudstone from Tabernas, showing a phantom of a typical twin gypsum crystal

and dolomite), especially when they can be dissolved because of the small size of their constitutive particles and/or solvent characteristics (Cantón et al., 2001); and (b) those swelling upon wetting, like clays, some of which, like smectite, can absorb water in amounts several times their dry weight, with consequent volume increases. Almost all lithologies in Table 10.1 reveal the presence of smectite.

10.2.3 Texture

Badlands usually develop on fine-grained clayey or silty sediments which come under the general generic heading of 'shale' (consisting of bedded silts, clay mud, siltstone, mudstone, mudshale, clayshale, claystone) (Fairbridge, 1968). Those developed around the Mediterranean are not an exception. Most parent materials are essentially silt-dominant, with clay as the second particle size, while sand is in general very poorly represented (Table 10.1).

Texture depends on four factors: particle-size distribution, grain shape, degree of crystallinity and relationship among grains (Terzaghi and Peck, 1967). Of these, particle-size distribution plays the key role in susceptibility for material disintegration and erosion: the larger the range of particle sizes, the higher the degree of packing, and hence the greater resistance to breakdown processes. This is especially true of materials that underwent minimum burial (Taylor and Smith, 1986). Conversely, the narrower the particle-size distribution, the higher the susceptibility for material disintegration, piping and, consequently, for badland development (Terzaghi and Peck, 1967). Particle-size analyses, by means of laser diffraction (Cantón, 1999; Pardini, 1996) from several badland sites around the Mediterranean (Table 10.1) show quite uniform fine textures: D_{60} ranges from 2.3 to 24 µm and D_{10} ranges from 1 to 3.7 µm. Uniformity coefficients (D_{60}/D_{10}) range from 2.3 to 9 µm, with a median of 3.5 µm, and indicate quite a uniform particle size.

10.2.4 Physical Properties

Besides textural properties, porosity is the second most important physical property. While being a considerably compacted rock in the fresh state, the overall porosity of mudstones determined by Hg-intrusion porosimetry is relatively high, around 10%, and steadily increases upon weathering up to the range of upper soil values, commonly 40–60% (Figure 10.3) (Solé et al., 1992; Bouma and Imeson, 2000; Cantón et al., 2001). The initial porosity of the fresh mudstone seems to enhance or restrict further weathering, leading to badland formation; in addition, the higher the macroporosity, the more unstable the badland regolith type (Imeson, 1986; Solé et al., 1992).

Geomechanical properties provide another important control for erosion: Atterberg limits (for consistency), swelling, and slaking behaviour are considered in many badland studies.

Consistency limits (Atterberg plastic and liquid limits) are good indices of material reactivity in relation to water; the higher the difference between them (known as the plasticity index I_p), the more stable the material. Shales, mudrocks and their weathering products usually have very low I_p, in a range from 4 to 20. Those from Spain in Table 10.1 have a median around 14, indicating their high susceptibility to fluidification.

Material coherence – or its opposite, slaking behaviour – can be evaluated by means of either the water coherence test (WCT; Emerson, 1967) or a modified version (Gerits et al., 1987). On a scale from 1 to 7 (least to most coherent) most badland-prone shales and mudrocks score 1 or 2, which is high slaking behaviour (Table 10.1).

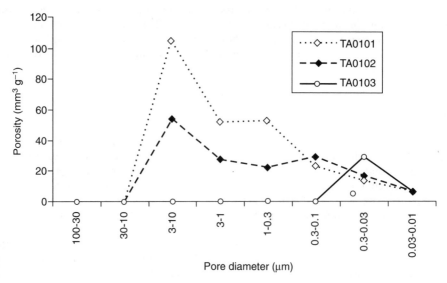

Figure 10.3 Hg-intrusion porosimetry of samples from a weathering profile. Identification and total porosity (mm³ g⁻¹): TA0103 = third deepest layer, fresh mudstone (33.72); TA0102 = second layer (131.15) , TA0101 = upmost, surface layer (250.25)

Swelling indices, while common in geotechnical studies, have been little used in badland studies. The Lambe (1951) swelling test was used by Berrad et al. (1994) to characterise erosion behaviour of soils in southeast Spain. The COLE index was also used: this evaluated three-dimensional and linear expansion of soft materials, like soils developed from marls at two sites near Guadix (southeast Spain) and in northeast Morocco (Gerits et al., 1987; Imeson et al., 1982).

10.2.5 Chemical Properties

The cohesion of mudrock is commonly due to thin films of slightly soluble cementing material. When mudrocks are permanently located above the water table, which is the case in most semi-arid Mediterranean regions, they are quite stable. However, submerged mudrocks, even for short periods, are likely to be very unstable because of their relatively high porosity and because of the leaching effects of submergence. Leaching removes the cementing substances and trans-forms the weathered mudstone into an almost cohesionless material that is no longer stable (Scheidig, 1934). In semi-arid regions, the stability of mudrocks is illusory because there are very short periods of local wetting followed by complete drying-out periods which destabilise the mudstone either through wetting–drying or solution–crystallisation processes (Goudie, 1990; Cantón et al., 2001).

Clay dispersion is a physico-chemical process relevant to erosion processes, particularly to the development of pipes, as discussed below. Materials (soils, regoliths or rocks) with a potential to disperse are those which contain a high exchangeable sodium percentage (ESP), saturating part of the exchangeable cations of their clays. This percentage is considered to be critical when higher than 13. As this parameter is relatively complicated to obtain, a well-correlated value has been designed by soil salinity specialists (USSLS, 1954), known as the sodium absorption ratio (SAR), which is easily calculated from the soluble cations extracted from the soil (rock or other material) saturated paste according to:

$$SAR = \frac{Na^+}{\sqrt{(Ca^{2+} + Mg^{2+})/2}}$$

where Na^+, Ca^{2+} and Mg^{2+} represent the concentrations of these cations in milli-equivalents per litre (meq l^{-1}). Lithologies, regoliths and soils with ESP or SAR values higher than 13 (SSSA, 1997) are susceptible to chemical dispersion upon wetting, meaning that they are destructured materials which cannot withstand the erosive impact of water.

10.3 GEOMORPHIC PROCESSES

10.3.1 Weathering

As most badlands are developed on bedrock, some weathering process is needed before erosion processes can act. Indeed, badlands are usually carpeted with the product of weathering called *regolith*, defined as the entire mantling cover of unconsolidated material on the surface of the Earth's crust regardless of its origin, though mostly formed by weathering of unaltered rocks (Fairbridge, 1968). Regolith is very vulnerable to erosion because it is usually composed of unbound mineral particles.

Regoliths may be considered as a special kind of soil, characterised mainly by very low organic matter content and mineralogical and chemical characteristics close to those of the parent rock, although with higher porosity because of physical weathering processes. Nevertheless, where parent rocks bear significant contents of gypsum or soluble salts, regoliths may become impoverished in these soluble materials because of leaching processes working alongside the physical weathering (Cantón et al., 2001). In badland areas with moderate erosion rates, the evolution of regolith may be complex, through the formation of shallow horizons with more compact structure and lower porosity, usually named *crusts*, that may be purely physical or may incorporate algae or lichens (Alexander and Calvo, 1990; Solé et al., 1997).

The literature on Mediterranean badlands indicates that a few weathering processes seem to contribute to their formation: wetting–drying (including swelling–shrinking, slaking and salt solubilisation–crystallisation) and freezing–thawing.

The action of wetting–drying cycles is the weathering process most commonly claimed for regolith formation. Indeed, water content variations mean changes in capillary forces able to perform significant physical work (Regüés et al., 2000), leading to the progressive disintegration of soft rock as observed in experimental conditions (Pardini et al., 1996; Cantón et al., 2001). Only a few wetting–drying cycles contribute to enlarge total porosity (Figure 10.4), which is assessed by the significant increase in the water absorption capacity of the mudstone (Cantón et al., 2001). Actually, in Tabernas badlands, a combination of three factors is responsible for mudstone weathering: repeated cycles of wetting–drying, existence of some primary porosity (intergrain pores and cracks and/or fissures from both uplifting and tectonic activity), and solubilisation–crystallisation of relatively soluble minerals, with gypsum the most abundant within this category. Also, a few wetting–drying cycles have been sufficient to reveal ion migration (especially Na^+, Ca^{2+}, Mg^{2+}, SO_4^-, HCO^- and Cl^-) within the mudstone, which leads to mineral dissolution (Cantón, 1999; Cantón et al., 2001).

Swelling–shrinking processes imply the presence of swelling clays, like smectite. In addition to mineralogical characterisation, the magnitude of this process can be assessed by the Lambe

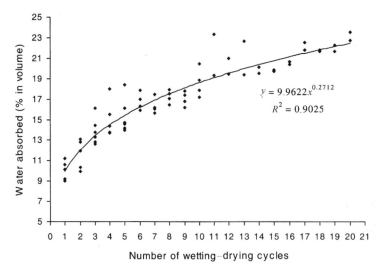

Figure 10.4   Water absorbed in a series of replicated mudstone blocks during 20 wetting–drying cycles and best-fitting model

swelling test or by the COLE index, as discussed above. Wetting–drying alternations in materials with expansive clays (smectite) cause the formation of nets of deep cracks and may also lead to the formation of a shallow layer of loose expanded regolith fragments, usually called *popcorn* (Hodges and Bryan, 1982).

Rapid wetting of air-dry fine materials can lead to aggregates bursting because of the compression of air trapped between the wetting fronts driven by capillary forces – a process called *slaking* (Yoder, 1936) that is relevant to erosion (Imeson and Verstraeten, 1989; Solé et al., 1992). Physico-chemical slaking has been found to be the main detachment mechanism in highly dispersive sodium-rich clays (Gerits et al., 1987) to the point that the detachment rate is controlled by the rate of advancement of the wetting front (Torri et al., 1994).

In areas subjected to freezing cycles, as in the Pyrenees and the Marnes Noires area in France, the weathering role of gelivation may be much more important than wetting–drying (Clotet et al., 1988, Regüés et al., 1995, Oostwoud Wijdenes and Ergenzinger, 1998). The growth of ice crystals on the freezing front, along with the cryosuction of the water retained in the matrix, can cause the breakdown of the rock and lead to the formation of a layer of popcorn without the existence of actual wetting–drying cycles, as observed in laboratory experiments (Pardini et al., 1996). In these environments, harsh thermal conditions and high regolith instability in shady aspects may provide a more important check on vegetation spreading than dryness on sunny aspects, leading to the increased occurrence of badland surfaces on these wetter aspects (Regüés et al., 2000).

10.3.2   Soil Formation and Resilience in Badland Environments

In badland areas, more- or less-developed soils do not form a continuous three-dimensional body covering the entire landscape. On the contrary, soils are restricted to those landscape patches where geomorphic agents allowed their accumulation or have not been able to destroy them. Usually badlands consist of mosaics of physiographic units where all the stages can be found between the bare parent material at the bottom and/or sides of gullies, and soils with some kind of vegetation cover in more stable positions. These mosaics are especially complex in semi-arid Mediterranean landscapes. In Tabernas, southeast Spain, strong interrelationships have been found between soil development, ground cover type, hydrological and geomorphological behaviour, and topographical attributes (Alexander et al., 1994; Cantón, 1999).

In a badland area it is important to distinguish between a developed soil and a regolith, which is the initial stage of soil formation. Steep slopes and gullies do not allow the formation of a developed soil because erosion processes are either frequent and/or intense. Soils are usually found on relatively stable surfaces because they result from either the evolution of the in-situ parent material or any accumulation of sediments, as in pediments or fluvial terraces.

Despite the similarity in parent material characteristics in a given badland area, differences exist in the relative importance of the soil properties, due in part to environmental or historical factors (Imeson et al., 1982; Solé et al., 1992; Torri et al., 1994). Almost all five recognised factors of pedogenesis (parent material, climate, topography, living beings, and time) differ within short distances in badland environments, depending on which physiographic unit is examined. Developed soils are found on flat or moderate slopes where climate and topography have acted together for enough time to leach weathering agents or to change particle-size distribution from the parent material. The development of some horizons may even decrease soil erodibility, while colonising plants provide protective ground cover and increase organic matter content and, thus, aggregate stability. Bare regoliths are mostly found on steep slopes where runoff and erosion predominate; colonising plants cannot provide enough cover or are not able to change regolith properties for a steady soil development. Intermediate states are found on moderately steep slopes where incipient soil development is favoured by a continuous or discontinuous cover of colonising species such as lichens, mosses or microphytic crusts. However, not all pedogenic processes lead to more resilient soils. In some instances, leached salts from upper parts of the landscape can accumulate downslope and produce either gypsic, saline or sodic horizons that restrict vegetation growth and/or decrease the soil resistance to erosive agents.

In general, the greater the soil resistance to erosion, the better protected will be the badland-prone material underneath. Soil resistance is related to both soil cover and intrinsic soil characteristics, like aggregate stability, which is strongly dependent on the amount and type of organic matter and clay. Organic matter content is related to both climate and vegetation. However, erosion is influenced not only by soil erodibility, but also by the impact of climatic erosivity, which can in some circumstances overcome the intrinsic resistance of a soil.

10.3.3 Infiltration and Runoff

Soil surface characteristics control infiltration, soil moisture and temperature regimes, runoff, and hence soil and landscape evolution. For this reason, soil surfaces are important in the understanding of badland evolution and geomorphological behaviour. Weather simulation has been used by several authors to study the response of badland surfaces to rainfall (Scoging, 1982; Imeson and Verstraeten, 1988; Calvo et al., 1991b; Solé-Benet et al., 1997; Bouma and Imeson, 2000; among others). On most badland surfaces water infiltrates with difficulty due to the presence of surface crusts or seals. Actually, all authors agree on the complexity of the response because of the high spatial and temporal variability of regolith properties. Runoff response is usually fast, with a very short 'time to runoff', less than 4 minutes in a variety of experiments (Imeson et al., 1982; Calvo et al., 1991b; Solé-Benet et al., 1997) and infiltration fronts reduce to a few centimetres. Micro-relief patterns due to micro-rills, crustose lichens, pedestals or pinnacles cause particular infiltration and runoff responses because the flow follows the micro-channels left by the micro-relief (Imeson and Verstraeten, 1988; Solé et al., 1997).

Nevertheless, badland regoliths may allow high infiltration rates when there are open cracks or highly porous popcorn structures. On badland surfaces with deep cracks, true overland flow may be rare, and crack flow can feed rills and main channels with water and sediment during storms that are not of sufficient duration to lead to crack closure (Yair and Lavee, 1985). In areas with shallower cracks or with longer rainfall events, crack closure usually means a strong reduction in permeability, allowing overland flow to occur. Therefore, the formation of runoff during storms may be a complex phenomenon that depends on lithology, antecedent conditions or a sufficient duration of the storm (Hodges and Bryan, 1982; Regüés et al., 1995). In areas subjected to freezing during winter, infiltration rates may show a clear seasonal pattern, being high in winter because of the formation of a deep highly porous regolith, and lower at the end of the non-freezing season because of regolith compaction and depletion (Regüés et al., 1995).

10.3.4 Erosion Processes

Rainsplash

During rainstorms, the impact of raindrops contributes to the destruction of popcorn layers and to the sealing of cracks, through the clogging role of detached regolith particles. In badlands with highly dispersive clays, the fine, detached particles may flow through cracks and micro-pipes from the beginning of the storm (Torri et al., 1994; Torri and Bryan, 1997), whereas in less dispersive materials, detached particles are transported by overland flow after the permeability and roughness of the regolith surface have been attenuated. Rainsplash is usually a very active process of detachment, as demonstrated through rainfall experiments and by the frequent existence of regolith pedestals below rock fragments and small plants (Regüés et al., 1995).

Piping

In badland areas developed on clay materials which are highly expansive or with high exchangeable sodium content, flow and particle detachment on a crack network can evolve into the extensive development of a net of pipes (Hodges and Bryan, 1982; Alexander, 1982; Torri et al., 1994), although dispersivity seems to be more relevant than expansivity (Gutiérrez et al., 1988). Piping phenomena may also cause the expansion of channel headcuts or gully margins by the collapse of macro-pipes or tunnel expansion (Harvey, 1982; Gutiérrez et al., 1988, López-Bermúdez and Romero-Díaz, 1989), which may also be developed in sediment fills of older dissected forms (Gallart, 1992).

Rills

These are very common micro-forms on badland surfaces, although they are usually non-permanent because of the diffusive role of drying or frost weathering. Nevertheless, rills usually reappear during rainfall events and are significant in sediment production and water and sediment conveyance. In badlands with thick cracked regoliths, rills may be fed by water and sediment, not through overland flow but through crack and micro-pipe flow (Bryan et al., 1978; Gerits et al., 1987). In badland surfaces with expansive materials where rills are fed only by crack and micro-pipe flow, it has been suggested that rills may not be formed by the classical role of overland flow, but by the collapse of areas close to the base of the regolith where moisture due to the formation of a perched water table reaches a critical value for liquefaction (Imeson and Verstraeten, 1988). This hypothesis may explain why rills on steep badland surfaces are frequently parallel (with few junctions) and start from the top of the divides without a band that is free of erosion. Following this hypothesis, space between rills would decrease as the slope increases, which reduces the threshold of moisture content for collapse, and would increase with the development of the crack system, since the higher efficiency of the subsurface drainage of the regolith would prevent local saturation.

Shallow Mass Movements

Regoliths in steep badland hillslopes are frequently affected by shallow mass movements during rainfall events or rainfall experiments. The unstabilised regolith mass may flow towards the valley bottom in the form of small mud or debris flows (Hodges and Bryan, 1982; Oostwoud Wijdenes and Ergenzinger, 1998). Poorly cracked regoliths are more prone to shallow mass movements due to easier saturation (Gerits et al., 1986) and pieces of coarser regolith may collapse or suffer some transfer downslope without reaching the channel system. Steep hillslopes can also be subject to the fall of regolith fragments, especially during dry periods (Gallart, 1992). In mountain areas with badlands developed on relatively cohesive marls, mass movements in the form of debris falls may be the main transport process of sediments from hillslopes to the channel network. Because of the coarseness of the materials eroded from hillslopes, the contribution of bedload transport to the total sediment exported from these areas ranges between 40 and 80%, decreasing with the increasing size of the catchment because of the increased attrition with increasing catchment size (Richard and Mathys, 1999). Frost creep may also be an active process of sediment transport along badland hillslopes in mountain areas, moving individual particles or pieces of frozen regolith.

Deeper Mass Movements

These are sometimes related to badland initiation and evolution, but their activity disorganises the characteristic fluvial landscape. Slumps or rapid wasting of soils may lead to the outcrop of bare soft rock triggering the formation of badlands (Clotet et al., 1988). On the other hand, the dissection of badlands may increase the topographic gradients and lead to the destabilisation of hillslopes (Alexander, 1982).

10.3.5 Erosion Rates

Badland areas look as if they have been caused by very rapid erosion. Erosion rates in badlands have been measured by three main methods, classified according to the temporal scale: (1) *long-term* methods (10^3–10^5 years) consist of the estimation of incision depths below some landscape element of known age (Yair et al., 1982; Bryan and Yair, 1982b); (2) *short-term* methods (1–10 years) usually form part of monitoring programmes and include the measurement of ground lowering using erosion pins (Alexander, 1982; Campbell, 1982; Benito et al., 1992; Clotet and Gallart, 1986; Lecompte et al., 1996), frames or profiles for detailed micro-topographic description of plots or sections (Campbell, 1974; Benito et al., 1992) and repeated topographic surveying (Schumm, 1956; Egels et al., 1989), as well as the monitoring of sediment production from small areas (Clotet and Gallart, 1986; Gutiérrez et al., 1995; Castelltort, 1995; Sirvent et al., 1996 and 1997) or small catchments (Richard and Mathys, 1999); and (3) *medium-term* methods ($10-10^2$ years) consist of the measurement of the volume of sediments retained by some trap that acted during a period of time that was too long to monitor (Clotet et al., 1988).

Nevertheless, erosion rates that are estimated through the different methods may show inconsistencies. For example, local short-term methods usually give higher rates than whole-landscape long-term methods (Yair et al., 1982), but this can be explained by the fact that erosion rates in a badland system vary from one point to another – an element that is very active at one moment may be fairly inactive a few decades later. Local topographic measure-ments (erosion pins) must be treated with special caution as they rarely cover entire functional units and can be prone to error as a result of ground expansion during the weathering processes. Short-term measurements through erosion pins may lead to completely inconsistent rates (or even negative ones) because periods of major regolith depletion alternate with periods of regolith formation, through bedrock expansion.

An example of the comparison between different methods is shown in Figure 10.5, which represents erosion rates in the badlands of the Vallcebre area (eastern Pyrenees, Spain) that were estimated through monitoring by erosion pins and two small plots (4 and 37.5 m^2 respec-tively), and the measurement of the volume of sediments in a 40-year-old natural trap that dammed a catchment of 3.1 ha (Clotet et al., 1988). Micro-catchment rates were obtained through the monitoring of sediments from an elementary catchment of 0.17 ha (Castelltort, 1995). The high temporal variation in rates obtained through erosion pins is caused by the expansion of regolith during freezing periods in winter. The uncertainty of badland erosion

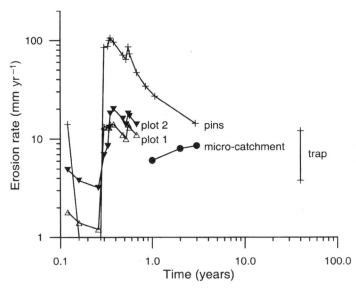

Figure 10.5 Erosion rates obtained through different methods and different time spans in badlands at Vallcebre (eastern Pyrenees, Catalonia, Spain). Original data from Clotet et al. (1988) and Castelltort (1995)

rates measured in the natural trap is due to the fact that badlands cover only 12% of the contributing catchment, and semi-degraded areas with visible erosion features cover 26%. Sirvent et al. (1997) compared erosion rates obtained through erosion pins, profilometers and sediment collection during a one-year period in an area free from freezing, and found differences of up to 20–30%, similar to differences with the same method at various points of the same badland unit.

The rates of badland erosion in dry areas are much more moderate than the appearance of the landscape leads one to expect (Yair et al., 1982; Wise et al., 1982). This paradox is only apparent because it is known that dry areas have limited potential for erosion (Langbein and Schumm, 1958), yet also have little potential of landscape recovery after infrequent geomorphic events (Wolman and Gerson, 1978). Badland areas are not protected by vegetation, and under the general scheme proposed by Langbein and Schumm (1958), erosion rates should be expected to grow rapidly with increased precipitation.

Erosion rates estimated for badland areas with different annual rainfall totals show indeed an increasing trend (Figure 10.6). Point Z represents long-term erosion rates estimated for the Zin-Havarim badlands in Israel using topographic lowering (Yair et al., 1982). Light point C represents a one-year erosion rate measured through monitoring of an elementary badland unit, whereas bold C represents the estimate of long-term channel incision assessed by the weathering rate, both at El Cautivo, Tabernas, southeast Spain (Cantón, 1999). Point A represents the long-term erosion rate in the Alberta badlands in Canada, estimated with the same method as for point Z (Bryan and Yair, 1982b). Point B represents a one-year erosion rate obtained through sediment monitoring from an elementary badland unit at Las Bardenas, Ebro depression, Spain (Gutiérrez et al., 1995). Light points V were obtained through the same method in the Vallcebre area, and bold V represents the three-year average (Castelltort, 1995). Finally, bold points **L**, **M** and **R**

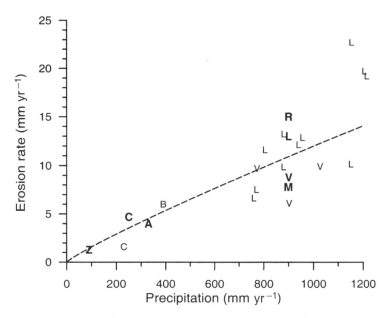

Figure 10.6   Erosion rates in badland areas across a wide range of precipitation. See text for explanation of the symbols and references

represent 11-year averages obtained through monitoring sediment discharge from small catchments of 0.13 ha (**R**), 8 ha (**M**) and 86 ha (**L**) in the experimental area at Draix, southeast France (Richard and Mathys, 1999), whereas light L points represent annual rates. Nevertheless, this graph does not necessarily mean that erosion rates in a badland area increase if the yearly precipitation increases in a geographic location; this may happen for annual values, but increased precipitation can also lead to the stabilisation of badlands because of increased vegetation cover (Nogueras et al., 2000). The relationships between vegetation, erosion rates and rainfall amounts are somewhat intricate, as will be discussed below.

On the other hand, in most kinds of soft bedrock where badlands are developed, erosion rates may become limited by weathering rates, both of which are controlled by lithology (Lecompte et al., 1996). This has been suggested in semi-arid areas where fresh bedrock outcrops in channel beds (Cantón, 1999) as well as in semi-humid areas where regolith is depleted almost every year (Clotet et al., 1988; Regüés et al., 1995). Nevertheless, some adjustment between erosion rates and weathering may be suggested, as regolith protects bedrock from weathering (Regüés et al., 1995, 2000), erosion rates in badlands usually being considered transport-limited (Campbell, 1989).

Finally, it is worth emphasising that the sediments coming from badland areas may be deposited on alluvial fans or plains after short transport. The drier the climate, the more ephemeral and local are runoff events, providing few chances for distant transport. In more humid climates, intense showers during summer are the main erosive circumstance, whereas long-distance transport is effected by long-lasting runoff events fed by autumn or winter rains of moderate intensity (Gallart et al., 1998).

10.4 VEGETATION AND GEOMORPHIC EVOLUTION OF BADLANDS

Both high erosion rates and low vegetation cover are related by feedback relationships that are at work in badland areas. Low vegetation cover means that the ground surface is unprotected against rainsplash and overland flow, whereas high erosion rates impoverish the soil's capacity to bear vegetation and make seedling survival difficult because of the physical instability of the ground (Guàrdia and Ninot, 1996; Guàrdia et al., 2000).

The implications of these features on the evolution of badlands are approached in two complementary ways in this section. The first is a computer experiment that tries to explore the occurrence of badlands along a precipitation range; as such, it works on a broad scale, bulking all the spatial differentiation of processes and rates, and does not attempt to explain their evolution through time. The second approach reviews the characteristics and role of the vegetation that grows in badlands on a narrow scale with emphasis on the issue of spatial patterns. Finally, a discussion integrates the outcomes of the two approaches.

10.4.1 Relationships between Erosion Rates and Vegetation through a Precipitation Range

In arid areas, both the potential for vegetation colonisation and the erosion rates are low; vegetation therefore plays a negligible role and badland dynamics are slow (Yair et al., 1982) and fully controlled by physical processes (Yair and Lavee, 1985). Nevertheless, badlands also occur in more humid areas where vegetation can control erosion rates; in these areas badland landscapes are poorly vegetated areas where high erosion rates and impoverished vegetation cover are related by feedback loops.

A simplified computer simulation model has been developed for analysing the interaction between vegetation and geomorphic processes for a wide range of annual precipitation. The model is built up with one independent variable (annual rainfall), two state variables (vegetation cover, regolith thickness) and two geomorphic processes (weathering, erosion). Weathering and erosion rates have been adjusted for soft rocks, characteristic of badlands. The annual rainfall positively controls vegetation cover, weathering rate and erosion rate. Weathering increases regolith thickness, which favours vegetation cover but decreases weathering rate itself. Erosion rate is diminished by vegetation cover and restricts vegetation cover itself and depletes regolith thickness. Vegetation growth rates are not directly simulated but the effect of erosion on vegetation increases for decreasing rainfall. A relevant limitation of this model is that it does not take into account the feedback role of increasing local topographic gradients when a badland area has already developed.

Mathematical expressions for the former relationships were established from previous studies when possible (Kirkby, 1976; Ahnert, 1987; Stocking, 1988) or postulated from field observations; they are all continuous functions. This approach represents a crude simplification of the system. The aim is to provide a tool for fundamental discussion but not for application, and attention is therefore paid to the qualitative description of the behaviour of the system rather than to exact rainfall depth, erosion rate or vegetation cover values.

Different values within the full range of state variables (vegetation cover, regolith thickness) for every constant annual rainfall amount within the gradient (from 50 to 1000 mm) were used as starting conditions for the model, which was run by iterations and stopped when relative rates of change were smaller than one per thousand in both state variables (stability condition). The degree of stability of the system at every value of vegetation cover was then obtained by averaging the inverses of rates of change of the system variables for this vegetation cover. (System stability is the degree of permanence of a given condition in a dynamic equilibrium, independent of its geomorphic activity or its state of degradation.) Consistent results were obtained from the beginning, but a few trials with slight changes in equations or parameters were necessary to avoid some spurious irregularities.

Erosion rates at system stability obtained for the different rainfall values are plotted in Figure 10.7. The lower line (dots) represents conditions near the vegetation optimum and shows the same trend as results previously obtained from actual data (Langbein and Schumm, 1958) and through simulation (Kirkby, 1976): erosion rates increase with the increasing rainfall amounts for dry climates, but decline when vegetation cover is sufficiently developed to control erosion. The upper line (triangles) represents another dynamic equilibrium condition at maximum geomorphic activity and gives the first original result: it is very close to the former line for dry climates, but shows a rapid increase for annual rainfall amounts higher than 250 mm because of the increase of available energy without the vegetation control, and finally becomes limited by the weathering rate when it exceeds 500 mm.

System stability estimates for different annual rainfall and vegetation cover values are represented in Figure 10.8. The first remarkable fact shown in this figure is that the model predicts the lack of areas devoid of vegetation for annual rainfall of less than 300 mm; this fact is congruent with the field evidence that badlands in dry areas usually have higher vegetation cover than badlands in more humid areas, sometimes in the form of seasonal vegetation or lichens (Alexander and Calvo, 1990; Solé et al., 1997).

For rainfall values below 350 mm, the model shows a large plateau or a very flat stability field, suggesting that, as in these dry conditions the gradients for system change are low, local degradations may remain for a long time as the natural tendency for recovery is also low.

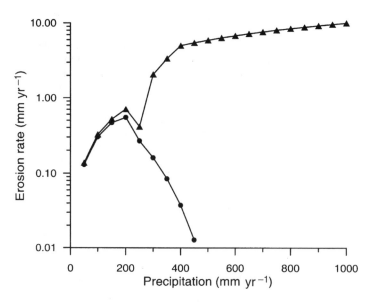

Figure 10.7 Erosion rates reached at system stability for different annual rainfall depths, obtained through a simulation model that analyses relationships between precipitation, weathering, erosion and vegetation cover. The lower line (dots) represents the more vegetated conditions whereas the upper line (triangles) represents the poorer vegetation cover

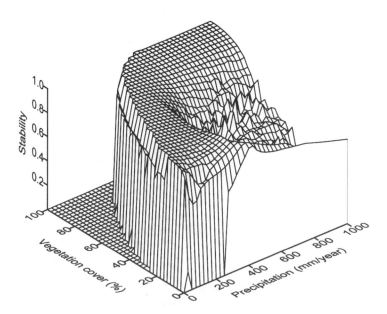

Figure 10.8 System stability estimates obtained with the same model as the former figure, plotted for the respective annual rainfall depths and vegetation cover percentages

Nevertheless, the lack of a distinct high-stability area for badlands in this low precipitation range is due to the fact that badland development is poorly related to vegetation controls and cannot be adequately handled by this approach. As discussed before, this approach does not consider the feedback mechanism created by badland topography.

Between 300 and 500 mm, a relative peak of system stability, situated near 10–20% of vegetation cover, suggests a significant presence of badland areas in those semi-arid climates. As shown in Figure 10.7, the development of vegetation may be sufficient in this precipitation range to control erosion rates, but moderate erosion rates are enough to maintain low vegetation covers, which conversely are not very effective in preventing erosion, thus suggesting that the areas within this range of precipitation may be highly sensitive to badland development because of vegetation disturbances.

Finally, for annual rainfalls higher than 500 mm, the model predicts a clear dichotomy between fully vegetated and completely bare conditions, but the general tendency is towards full vegetation. The range of system stability for bare areas is very narrow, and is maintained by high erosion rates. This suggests that they are triggered by special conditions (human disturbance, mass movements), and have a natural tendency towards recovery, which gives some hope for revegetation work (see also Alexander et al., 1994).

10.4.2 Characteristics of the Vegetation Growing in Badlands

Geomorphic processes responsible for badland evolution operate at different rates over a single landform, and, in consequence, badlands show a high degree of spatial heterogeneity. This feature has implications for the spatial distribution of vegetation, its establishment, and its relationships with runoff and sediment movement at the small scale.

Spatial Distribution of Vegetation in Badland Systems

One of the most conspicuous characteristics of vegetation in badland areas is its small-grain spatial heterogeneity, with sharp contrasted patches that match landforms fairly well. This feature is reported by several authors (Butler and Goetz, 1986; Alexander et al., 1994; Guàrdia and Ninot, 1996; Calzolari et al., 1993; Chiarucci et al., 1995; Guàrdia, 1995) and has been explicitly investigated by Lázaro and Puigdefàbregas (1994), Lázaro (1994), and Lázaro et al. (2000a). The main regularities reported by these authors are summarised as follows.

In the divides between gullies, less eroded strips support remains of soils and vegetation that are similar to the regional types on non-gullied hillslopes. On the upper hillslope sectors of gullies, plant communities show slight differentiation, with a higher proportion of pioneer species and changes of composition according to exposure. Backslopes show the strongest differentiation, with very low vegetation density and a prevalence of specialists on moving substrate. Footslopes and pediments are rich in runon-adapted communities that are equipped to withstand sediment deposition and salt accumulation. Finally, stabilised channels and swales host mesic and denser plant communities that are often spatially discontinuous according to the flow conditions of each channel sector.

Most badlands have a long history of activation and stabilisation phases that leave behind successive remains from different ages. This feature offers the opportunity to trace back long-term changes in plant communities, by looking at the botanical composition of samples taken from different-aged remains of the same landform. This was actually performed in the Tabernas badlands, in southeast Spain, by Alexander et al. (1994), who distinguished up to five erosion-sedimentation phases that left remnants in the badland landscape. After a botanical survey of the area, they were able to show that relative age was the main factor associated with vegetation differences in the three youngest phases of pediments. The trend with increasing age was a larger total plant cover, a drop in halophytic shrubs, an increase in annual plants, and a smaller but more diverse lichen cover. The factors underlying these changes were the differentiation and leaching of the top soil layers. In the two older classes of pediments, soil is already differentiated and their vegetation was associated with local factors rather than with time.

These findings show that the distribution of badland vegetation is far from random, and is more associated with landforms and time. These two factors determine its complex and often intricate pattern, with sharp boundaries and discontinuities.

More detailed studies of the floristic structure and dominant plant architectures of badland vegetation (Lázaro, 1994; Lázaro et al., 2000a; Guàrdia, 1995; Guerrero, 1998) enabled the species to be grouped into two main classes that were called 'endurers' and 'builders' by Lázaro (1994). Most perennial endurers are able to survive on moving substrates, but as they grow at very low densities their effects on soil evolution and sediment deposition are minimal. They often show prostrate habits and long roots that grow laterally and are able to sprout. Other endurers showing annual behaviour rely on their ability to reproduce themselves in the badland environment. Builders can only be established in fairly stable substrates, but they reach very high densities that help sediment deposition and soil evolution. They often show graminoid morphology, with shallow, fasciculate and very dense root systems. Crustose lichens that cover large proportions of the available surface constitute a group close to builders, which is also associated with quite stable substrates (Lázaro et al., 2000a).

Endurers typically are the most common plants on backslopes undergoing active erosion, while builders grow on more stable or depositional landforms such as pediments, gentle hillslopes or remnants of both on hillslopes that suffer from reactivated erosion. While the builder communities, even when they are few in number, include a large representation of the local flora, endurer communities include only a small number of species drawn from the same local stock.

Lichen crusts are widespread in arid climates (Lázaro et al., 2000a; Yair and Lavee, 1985) in fairly stable upper hillslope convex sectors. In such conditions, soil water storage is very low, because runoff exceeds run-on, and lichens can out-compete annual plants.

These spatial patterns of badland vegetation at the local scale change along climatic gradients. Specific literature on this subject has not been found, but the information available from a transect of field sites along the Mediterranean façade of Spain offers the opportunity of highlighting some regularities.

In the Pyrenean foothills, at the cold-humid end of the Mediterranean transect mentioned above, perennial plants, small shrubs and tussocks constitute the dominant life forms of badland vegetation (Guàrdia, 1995; Guerrero, 1998). In such conditions, soil moisture is high and north-exposed backslopes are less stable than those exposed to the south, because of the geomorphic effect of freezing. Plants living on northern aspects are adapted to both ground instability and harsh winter conditions (Regüés et al., 2000).

In south-east Spain, at the warm-dry end of the transect, crustose terricolous lichens and annual plants are widespread (Alexander and Calvo, 1990; Lázaro, 1994). Soil moisture is low and southerly slopes are the most unstable because of wetting and drying changes and deposition of salts close to the soil surface (Solé et al., 1997). In the Tabernas area, bare soil and eroded soil supporting only very sparse 'endurers' accounts for approximately 33% of the total area. Lichens and living crust can account for 32%, and the rest is more or less densely covered by higher plants, in mosaics of annual and perennial plant patches which act as builders (Cantón, 1999).

Factors Affecting Plant Regeneration and Survival

A review of the constraints on successful establishment of plants in the badland environment will help to explain the patterns of its spatial distribution. These constraints may be grouped in three classes: those that affect soil seed banks, seed germination and seedling survival.

Soil seed banks are replenished by seed yield and dissemination, while they are emptied by germination, seed mortality and seed wash. In most cases, seed yield is fairly irregular over time, because of particular rainfall patterns, or internal physiological factors. Both these factors apply in *mast* years (years in which fruits or seeds are produced largely over average values), although particular rainfall patterns are more common in dry climates and internal physiological factors prevail in humid conditions.

Examples of rainfall-driven mast years in the arid sector of badland distribution have been reported for alpha grass (*Stipa tenacissima*) and annual plants. Mast years for alpha grass correlate with heavy autumn rainfall (Haase et al., 1995) while spring rainfall determines the success or failure of seed yield for annual plants (Espigares Pinilla, 1994).

An example of internally driven mast years may be found in *Retama sphaerocarpa*, a legume bush common in swales and non-incised channels of the badland areas in southeast Spain. This species has a very deep root system (Haase et al., 1996) that ensures a regular water supply through the year. Mast years are therefore determined by the cycles of exhaustion and replenishment of resources triggered by abundant seed yields, rather than by particular rainfall sequences (Gutiérrez, personal communication).

Seed bank depletion by erosion and seed wash has been reported (García-Fayos et al., 1995; García-Fayos and Cerdà, 1997) to be relatively low, between one- and two-thirds of the total seed rain and 10% of the soil seed bank. In general, seed wash cannot be considered a cause of failure of seed supply, although it may affect some species with small or very irregular seed yield. This is the case of species with prevailing vegetative regeneration, such as the tussock grass *Achnatherum calamagrostis* (L.) (Guàrdia, 1995). In contrast, seed wash may cause local increases of the seed bank in depositional landforms in channels or pediments. Seed concentration of the sediments delivered at gully outlets may be larger than that recorded for the hillslopes by a factor of 40 or 50 (García-Fayos et al., 1995).

Information on soil seed bank size in badlands is still scarce, but available figures range between 250 and 1400 seeds m^{-2} (García-Fayos et al., 1995; Guàrdia, 1995; Guàrdia et al., 2000). This reserve is not large, but is sufficient to support germination, which in the cases reported does not reach 10% of the seed bank (Guàrdia, 1995; Guàrdia et al., 2000).

A number of factors affect germination in badland plants, including seed dormancy (Guàrdia, 1995), previous rainfall and temperature sequences (Guàrdia, 1995; Espigares Pinilla, 1994). Soil crusts are claimed to affect germination but no field information is available, and soil salinity has been reported to delay germination in badlands (Maccherini et al., 1996). Most of these factors work through topsoil properties, such as temperature, moisture and salt concentration, which vary spatially over the badland system and so may contribute to the observed spatial pattern of vegetation.

The first months after germination are often critical to the establishment of seedlings. At the Vallcebre field site in the Pyrenean foothills (in the humid climate sector of badland distribution), dissemination and emergence mainly occur in autumn and spring respectively. In such conditions, the first two summers after germination are crucial to seedling establishment: if summer drought occurs in any of these two years, mortality is greater on the more eroded back slopes than on the upper slopes (Guàrdia et al., 1996). Seedling mortality occurs nevertheless all the year round, which implies that mortality is the result of a combination of several factors, such as ground instability, dryness and low winter temperatures (Guàrdia et al., 2000).

In the warm-dry climate sector, annual plants that make up the bulk of badland vegetation germinate with the first autumn storms. However, the time sequence of dry and rainy spells during this season is a strong selection factor that determines the final species composition of the community (Espigares Pinilla, 1994).

Many of the species that make up the vegetation of badlands are mycorrhiza-dependent. This feature has been reported for a number of species in the arid climate sector of southeast Spain, such as *Stipa tenacissima*, *Anthyllis cytisoides*, *Retama sphaerocarpa*, *Stipa capensis*, *Lavandula spica* (Requena et al., 1996). Mycorrhizal infection is crucial for helping nutrient uptake and seedling establishment in poor substrates. However, it is known that soil erosion generally leads to the loss or reduction of mycorrhizal propagules present in the soil (Jasper et al., 1991; Requena et al., 1996); therefore, the lack of inoculum potential for mycorrhiza formation may contribute to the failure of seedling survival on the eroded or recently stabilised badland slopes.

The factors that help germination and seedling establishment also have spatial structure and show most favourable combinations in particular small areas or 'safe sites'. Therefore, it can be anticipated that on eroded badland slopes, mostly colonised by 'endurers', safe-site density is very small and widespread seedling establishment will occur only in particularly favourable years. On the contrary, 'builders' create their own safe site, and therefore increase the probability of success for plant establishment.

This hypothesis has been confirmed by field observations in the Tabernas badland area (Lázaro, 1994; Lázaro et al., 2000b). The botanical composition and cover of annual plant communities at the same set of plots was recorded in a humid and in a dry year. The largest and smallest between-year differences were found in 'endurer' and 'builder' communities, respectively.

Relationships of Vegetation with Runoff and Sediment Movement at the Micro-scale

The information available from Mediterranean badlands enables some general rules on the effects of vegetation on runoff and sediment discharge to be established. Results come mostly from field experiments with micro-plots in different plant cover types using either simulated or natural rainfall (Cantón, 1999; Solé et al., 1997; Calvo et al., 1991a).

'Endurer' populations on backslopes are very sparse and have no significant effect on runoff and sediment movement, which are controlled by other factors such as physical crusts, stone cover, gradient, micro-topography, etc. (Solé et al., 1997; Cantón, 1999).

Results from lichen-covered plots indicate that runoff coefficients are similar to those found on bare ground, while sediment outputs are dramatically reduced. Lichen crusts increase surface rugosity and surface detention of water (Yair et al., 1980). These effects favour initial infiltration and delay the time to runoff (Calvo et al., 1991b; Solé et al., 1997). The final outcome of these features is a desynchronising effect on runoff generation over large areas; and because of this, lichen cover may give high runoff coefficients on the small scale, but these drop abruptly when the contributing areas are bigger (Yair and Lavee, 1985).

This behaviour of lichen stands is very sensitive to cover, and lichen degradation, once it starts, is accelerated by increasing gaps through micro-piping and splash erosion (Yair et al., 1980; Cantón, 1999). Dense patches of 'builders' on hillslopes reflect an increase in infiltration and a decrease in runoff and sediment discharge (Cantón, 1999). The relations of these fluxes with plant cover are negative and often asymptotically non-linear (Solé et al., 1997). Sparse plant cover on upper and less steep slope sectors, more similar to nearby vegetation on non-gullied hillslopes, shows complex interaction with runoff and sediment movement. The effect of vegetation depends on both its density and its spatial structure, and increasing vegetation density causes an exponential decrease of runoff and sediment output. The effect of spatial structure is dynamic and species-specific, and is particularly significant in tussock grasses because they are able to intercept a significant percentage of downhill runoff and sediment flow. Field observations point to interception values of 50% for *Stipa tenacissima* tussocks (Puigdefàbregas and Sánchez, 1996). This lateral interception of sediments interacts with plant growth and helps the formation of banded structures along contour lines that increase the interception effect through a positive feedback. Beyond a certain threshold of downhill sediment flow, spatial structures of vegetation are disrupted, rills are initiated and sediment output rises dramatically (Puigdefàbregas and Sánchez, 1996).

Vegetation in stabilised drainage ways is usually relatively dense, or very dense in some points, and shows high roughness coefficients (Prosser, 1996) that substantially reduce runoff erosivity while increasing infiltration in the channel. This vegetation is very dependent on a water supply from hillslopes, and it has been shown that evapotranspiration from channel stands of *Retama sphaerocarpa* amounts to 130% of rainfall (Domingo et al., in press). For these reasons channel vegetation is prone to local collapses either by piping and mass failures, or by long drought spells, both of which may occur simultaneously and lead to the reactivation of channel incision and thus to erosion in the entire badland system (Nogueras et al., 2000).

A Synopsis on the Functions of Badland Vegetation

Badlands are primarily geomorphic landscapes triggered and maintained by linear incision. Nevertheless, when they develop in areas where climate allows some permanent vegetation, the interaction between vegetation development and hydrological and geomorphic processes becomes the main driver of the system.

When they grow beyond simple gully forms, badlands become complex landscapes, with fine compartmentation between small contiguous areas with diverse soil thickness, erosion and deposition rates, runoff and run-on, as well as sunshine and temperature regimes. This complexity interacts with vegetation, providing a wide range of habitats with different environmental attributes, most of which are characterised by feedback loops between vegetation and geomorphic processes.

The main colonisation trend of vegetation in badlands is similar to the general trends in plant colonisation of barren lands, including biological crusts, herbaceous plants, annuals or perennials depending on climate, shrubs and trees. Changes along this trend occur only if species-specific thresholds of soil stability and water availability have been attained. For example, lichen crusts can only establish themselves if soil movement is small enough, in which case they may persist as a steady state, but poor water availability enables them to out-compete grasses and shrubs.

Density changes in vegetation are determined by such processes as establishment and collapse, which are discontinuous (step-like) and have different constraints. For example, establishment is determined mainly by the density of safe sites, while collapse may be driven by drought spells, sediment wash or mudflows and mass movements. As these processes work at different rates, aggradational and degradational changes are not symmetrical, and hysteresis effects are frequent.

Along a given catena, the most run-on-dependent plant cover types are also likely to be the most sensitive to rainfall change. As these plants are installed in areas close to waterways, they interact with the flow of water and sediments and their collapse may change the hydraulics of the flow, allowing the incision of gullies and reactivating the erosion upstream.

In dry climates the lower rates of the processes driving geomorphic and vegetation changes lead to the persistence of traces of several stabilisation/reactivation cycles. This feature results in heterogeneous and fine-grained spatial patterns, with many relict surfaces with their associated vegetation, and relatively large global plant cover values. On the contrary, in humid zones, processes driving both the evolution of badlands and vegetation recovery work so fast that they remove most of the remnants of previous cycles or phases, and leave a large amount of bare ground or vegetated areas, depending on the current stage of stability or reactivation of linear incision.

10.5 A CLIMATIC CLASSIFICATION OF BADLANDS IN THE MEDITERRANEAN

As a summary of the preceding analysis of badland characteristics and processes in the Mediterranean, a tentative classification of badland landscapes into three main types may be proposed.

1. Arid Badlands

These develop in areas with annual precipitation below 200 mm, where dryness impedes any effective control of vegetation on erosion. They are 'physical' badlands, as vegetation plays no relevant role and geomorphic processes are fully controlled by bedrock and regolith characteristics. Differences between hillslopes of different aspect may exist, but these are related to the direct control of regolith moisture on weathering and erosion processes (Yair and Lavee, 1985). Arid badlands are very old, and their initiation is related to geological controls and drainage net evolution, but not to vegetation degradation by human action. The Zin-Havarim badlands (northern Negev, Israel) provide the best-known example of this type of badlands in the Mediterranean (Yair et al., 1980, 1982; Yair and Lavee, 1985).

2. Semi-arid Badlands

These develop in areas with annual precipitation within the range 200–700 mm, characterised by discontinuous permanent vegetation covers or annuals. Vegetation is able to exert some effective control on geomorphic processes, and the primary control of badland development on vegetation is due to the limitations of water availability imposed by thin regoliths, especially in sunny aspects. Badland landscapes within this range of precipitation are typically asymmetrical (Figure 10.9). The sunny aspects show impoverished or null vegetation cover because of the strong control on water availability effected by radiation, whereas the shady aspects may bear a vegetation cover close to 100% (Kirkby et al., 1990; Solé et al., 1997; Cantón, 1999). Semi-arid badlands are usually old (Wise et al., 1982) and may show the relicts of different phases of stabilisation and reactivation of erosion during their history (Calvo et al., 1991a; Alexander et al., 1994; Nogueras et al., 2000). The initiation of large badland areas is usually due to natural processes, but this precipitation range is very sensitive to gully and badland development as a result of human-induced degradation, especially where traditional agricultural practices concentrate runoff into ditches (Gallart, 1979, 1992). Examples of these badlands are frequent throughout the Mediterranean, the better-known examples being located in various parts of Spain and southern Italy.

Figure 10.9   Oblique aerial photograph of badlands at Tabernas (Almeria, SE Spain), taken from the SW. Observe the conspicuous asymmetry of the landscape

3.   Humid Badlands

These develop in areas, usually mountainous in character, with annual precipitation exceeding 700 mm and with frequent rainstorms during the summer. There is enough water to allow dense vegetation cover, including continuous permanent pastures. The growth of vegetation on bad-land surfaces is checked more by the high erosion rates than by dryness, reclamation with vegetation being feasible especially with the help of some structure used for fixing the regolith (Ballesteros, 1994; Crosaz and Dinger, 1999; Richard and Mathys, 1999). Therefore, the greater dryness of sunny aspects may not be relevant to badland formation, and badlands may be not clearly asymmetrical or may even show reversed asymmetry as freezing, rather than dryness, may control vegetation (Regüés et al., 2000). Humid badlands are usually much younger than arid or semi-arid types, their initiation being related to mass movements (Clotet et al., 1988) or to the degradation of vegetation (Ballais, 1999). The best-known examples of these badlands in Mediterranean regions may be found in the southern Alps (Draix, Haute Provence, France) and in the Pyrenees (Vallcebre, Catalonia, Spain).

REFERENCES

Ahnert, F. 1987. Process-response models of denudation at different spatial scales. *Catena Supplement*, 10, 31–50.

Alexander, D. 1982. Differences between calanchi and biancane badlands in Italy. In R.B. Bryan and A. Yair (Eds) *Badland Geomorphology and Piping*. GeoBooks, Norwich, 71–87.

Alexander, R.W. and Calvo, A. 1990. The influence of lichens on slope processes in some Spanish badlands. In J.B. Thornes (Ed.) *Vegetation and Erosion*. Wiley, Chichester, 385–398.

Alexander, R.W., Harvey, A.M., Calvo, A., James, P.A. and Cerda, A. 1994. Natural stabilization mechanisms on badland slopes: Tabernas, Almería, Spain. In A.C. Millington and K. Pye (Eds) *Environmental Change in Drylands: Biogeographical and Geomorphological Perspectives*. John Wiley and Sons, 85–111.

Ballais, J.L. 1999. Apparition et évolution de roubines à Draix. In N. Mathys (Ed.) *Les bassins versants expérimentaux de Draix, laboratoire d'étude de l'érosion en montagne*. Cemagref, Antony, 235–245.

Ballesteros, R. 1994. *Efecte dels factors locals y de la fertilitat en una experiència de revegetació en un àrea de badlands de la conca de l'Alt Llobregat*. Treball Final de Carrera, Escola Superior d'Agricultura de Barcelona (unpublished).

Benito, G., Gutiérrez, M. and Sancho, C. 1992. Erosion rates in badlands areas of the Central Ebro Basin (NE Spain). *Catena*, 19, 262–286.

Berrad, F., García-Rossell, L. and Martín-Vellejo, M. 1994. Les propriétés géomécaniques, un facteur de contrôle de l'érosion: Cas d'une zone aride su sudeste espagnol. In Oliverira, Rodrigues, Coelho and Cunha (Eds) *7th International IAEG Congress*, Balkema, Rotterdam, 331–338.

Bouma, N.A. and Imeson A.C. 2000. Investigation of relationships between measured field indicators and erosion processes on badland surfaces at Petrer, Spain. *Catena*, 40, 147–171.

Bryan, R.B. and Yair, A. (Eds) 1982a. *Badland Geomorphology and Piping*. GeoBooks, Norwich, 408pp.

Bryan, R.B. and Yair, A. 1982b. Perspectives on studies of badland geomorphology. In R.B. Bryan and A. Yair (Eds) *Badland Geomorphology and Piping*. GeoBooks, Norwich, 1–12.

Bryan, R.B., Yair, A. and Hodges, W.K. 1978. Factors controlling the initiation of runoff and piping in Dinosaur Provincial Park badlands, Alberta, Canada. *Zeitschrift für Geomorphologie Supplement Band*, 29, 151–168.

Butler, J. and Goetz, H. 1986. Vegetation and soil-landscape relationships in the North Dakota badlands. *The American Midland Naturalist*, 116(2), 378–386.

Calvo, A., Harvey, A.M. and Paya-Serrano, J. 1991a. Processes interactions and badland development in SE Spain. In M. Sala, J.L. Rubio and J.M. García-Ruiz (Eds) *Soil Erosion Studies in Spain*, pp. 75–90, Geoforma Ed., Logroño.

Calvo, A., Harvey, A.M., Paya-Serrano, J. and Alexander, R.W. 1991b. Response of badland surfaces in SE Spain to simulated rainfall. *Cuaternario y geomorfología*, 5, 3–14.

Calzolari, C., Ristori, J., Sparvoli, E., Chiarucci, A. and Soriano, M.D. 1993. Soils of biancana badlands: distribution, characteristics and genesis in Beccanello Farm (Tuscany, Italy). *Quaderni di Scienza del Suolo*, 5, 119–141.

Campbell, I.A. 1974. Measurement of erosion on badlands surfaces. *Zeitschrift für Geomorphologie Supplement Band*, 21, 122–137.

Campbell, I.A. 1982. Surface morphology and rates of change during a ten-year period in the Alberta badlands. In R.B. Bryan and A. Yair (Eds) *Badland Geomorphology and Piping*. GeoBooks, Norwich, England, 221–237.

Campbell, I.A. 1989. Badlands and badland gullies. In D.S.G. Thomas (Ed.) *Arid Zone Geomorphology*. Belhaven/Halsted Press, London, 159–183.

Cantón, Y. 1999. *Efectos hidrológicos y geomorfológicos de la cubierta y propiedades del suelo en paisaje de cárcavas*. Unpublished PhD thesis, Universidad de Almeria, Spain, 394pp.

Cantón,Y., Solé-Benet, A., Queralt, I. and Pini, R. 2001.Weathering of a gypsum-calcareous mudstone under semi-arid environment at Tabernas, SE Spain: laboratory and field-based experimental approaches. *Catena* (in press).

Castelltort, X. 1995. *Erosió, transport y sedimentació fluvial com a integració dels processos geomorfològics d'una conca. (Conca de Cal Rodó, Alt Llobregat.)* PhD Thesis, U. Barcelona, 235pp.

Clotet, N. and Gallart, F. 1986. Sediment yield in a mountainous basin under high Mediterranean climate. *Zeitschrift für Geomorphologie Supplement Band*, 60, 205–216.

Clotet, N., Gallart, F. and Balasch, J. 1988. Medium term erosion rates in a small scarcely vegetated catchment in the Pyrenees. *Catena Supplement*, 13, 37–47.

Crosaz, Y. and Dinger, F. 1999 Mesure de l'érosion sur ravines élémentaires et essais de végétalisation. Bassin versant expérimental de Draix. In N. Mathys (Ed.) *Les bassins versants expérimentaux de Draix, laboratoire d'étude de l'érosion en montagne*. Cemagref, Antony, 103–118.

Chiarucci, A., De Dominicis, V., Ristori, J. and Calzolari, C. 1995. Biancana badland vegetation in relation to morphology and soil in Orcia calley, central Italy. *Phytocoenologia*, 25(1), 65–87.

Domingo, F., VillaGarcía, L., Puigdefàbregas, J., Boer, M. and Alados-Arboledas, L. (in press). Assessing lateral water inputs to vegetation of an ephemeral river in semi-arid climate through rainfall-evapotranspiration relationships. *Agricultural Forest Meteorology*.

Egels, Y., Kasser, M., Meunier, M., Muxart, T. and Guet, C. 1989. Utilisation des mésures topographiques (Photogrammétrie terrestre et télémetrie sans réflecteur) pour la mesure de l'érosion de petits bassins versants expérimentaux et comparaison avec les mesures de transport solide à l'émissaire. *Bassins Versants expérimentaux de Draix, Compte rendu de Recherche no 2 en érosion et hydraulique torrentielle*. CEMAGREF, ONF-RTM, 3-25.

Emerson, V.V. 1967. A classification of soil aggregates based on their consistency in water. *Australian Journal of Soil Research*, 5, 47–57.

Espigares Pinilla, T. 1994. *Fluctuaciones en la dinamica de pastizales anuales mediterraneos: El papel de los factores meteorologicos en el momento de la regeneracion*. Tesis Doctoral, Universidad Autónoma de Madrid.

Fairbridge, R.W. 1968. *Encyclopedia of Geomorphology*. Dowden, Hutchinson and Ross, Inc. Pennsylvania, USA.

Farres, P. 1978. The role of time and aggregate size in the crusting process. *Earth Surface Processes and Landforms*, 3, 243–254.

Gallart, F. 1979. Observaciones sobre la geomorfología dinámica actual en la Conca d'Òdena (Alrededores de Igualada, provincia de Barcelona). In J. Muñoz, T. Aleixandre and J. Gallardo (Eds) *Actas de la III Reunión del Grupo Español de Trabajo del Cuaternario*. CSIC, Madrid, 123–134.

Gallart, F. 1992. Estudi Geomorfòlgic de la Conca d'Odena. *L'Estrat*, 3 (monogràfic), 9–45, Igualada (Barcelona).

Gallart, F., Latron, J. and Regüés, D. 1998. Hydrological and sediment transport processes in the research catchments of Vallcebre (Pyrenees). In J. Boardman and D. Favis-Mortlock (Eds) *Modelling Erosion by Water*. Springer-Verlag NATO-ASI Series I-55, Berlin, 503–511.

García-Fayos, P. and Cerdà, A. 1997. Seed losses by surface wash in degraded Mediterranean environments. *Catena*, 29, 73–83.

García-Fayos, P., Recatala, T.M., Cerdà, A. and Calvo, A. 1995. Seed population dynamics on badland slopes in southeastern Spain. *Journal of Vegetation Science*, 6, 691–696.

Gerits, J.J.P., Imeson, A.C. and Verstraeten, J.M. 1986. Chemical thresholds and erosion in saline and sodic materials. In F. López-Bermúdez and J. Thornes (Eds) *Estudios Sobre la Geomorfología del Sur de España*. Murcia, 71–74.

Gerits, J., Imeson, A.C., Verstraeten J.M. and Bryan, R.B. 1987. Rill development and badland regolith properties. *Catena Supplement*, 8, 141–160.

Gomer, D. 1995. *Écoulements et érosion dans des petits bassins-versants à sols marneux sous climat semi-aride méditerranéen*. Published by: 'Projet pilote d'aménagement integré du bassin versant de l'oued Mina, c/o Deutsche Gesellschaft für Technische Zusammenarbeit (GTZ) GmbH, D-65726 Eschborn'. 207 pp + annexes.

Goudie, A.S. 1990. Weathering processes. In D.S.G. Thomas (Ed.) *Arid Zone Geomorphology*. Belhaven/ Halsted Press, London, 11–24.

Guàrdia R. 1995. *La colonitzacio vegetal de les arees erosionades de la conca de La Baells (Alt Llobregat)*. Doctoral Thesis, University of Barcelona.

Guàrdia, R. and Ninot, J.M. 1996. Distribution of plant communities in the badlands of the upper Llobregat basin (southeastern Pyrenees). *Studia Geobotanica*, 12, 83–103.

Guàrdia, R., Alcantara, C. and Ninot, J.M. 1996. Dinàmica anual de l'emergència de plàntules a les àrees aixaragallades de la conca de Vallcebre (Alt Llobregat). *Folia Botanica Miscellanea*, 10, 211–229.

Guàrdia, R., Gallart, F. and Ninot, J.M. 2000. Soil seed bank and seedling dynamics in badlands of the upper Llobregat basin (Pyrenees). *Catena*, 40, 173–187.

Guerrero, J. 1998. *Respuestas de la vegetación y de la morfdología de las plantas a la erosion del suelo. Valle del Ebro y Prepirineo atagonés*. Publicaciones del Consejo de Protección de la Naturaleza de Aragón. 12. Zaragoza, 231pp.

Gutiérrez, M., Benito, G. and Rodríguez, J. 1988. Piping in badland areas on the middle Ebro basin, Spain. *Catena Supplement*, 13, 49–60.

Gutiérrez, M., Sancho, C., Desir, G., Sirvent, J., Benito, G. and Calvo, A. 1995. *Erosión hídrica en terrenos arcillosos y yesíferos de la Depresión del Ebro*. ICONA, Universidad de Zaragoza, 389pp.

Haase, P., Pugnaire, F.I. and Incoll, L.D. 1995. Seed production and dispersal in the semi-arid tussock grass *Stipa tenacissima* (L.) during masting. *Journal of Arid Environments*, 31, 55–65.

Haase, P., Pugnaire, F.I., Fernández, E.M., Puigdefàbregas, J., Clark, S.C. and Incoll, L.D. 1996. An investigation of rooting depth of the semi-arid shrub *Retama sphaerocarpa* (L.) Boiss. by labelling of ground water with a chemical tracer. *Journal of Hydrology*, 177, 23–31.

Harvey, A. 1982. The role of piping in the development of badlands and gully systems in south-east Spain. In R.B. Bryan and A. Yair (Eds) *Badland Geomorphology and Piping*. GeoBooks, Norwich, England, 317–335.

Harvey, A. 1987. Patterns of Quaternary aggradational and dissectional landform development in the Almeria region, southeast Spain: a dry region tectonically-active landscape. *Die Erde*, 118, 193–215.

Harvey, A.M. and Calvo, A. 1989. Distribution of badlands in SE Spain: implications of climatic change. In A.C. Imeson and De Groot (Eds) *Landscape-ecological impact of climatic change*, discussion report on Mediterranean region.

Hodges, W.K. and Bryan, R.B. 1982. The influence of material behavior on runoff initiation in the Dinosaur Badlands, Canada. In R.B. Bryan and A. Yair (Eds) *Badland Geomorphology and Piping*. GeoBooks, Norwich, England, 13–46.

Howard, A.D. 1994. Badlands. In A.D. Abrahams and A.J. Parsons (Eds) *Geomorphology of Desert Environments*. Chapman & Hall, London, 213–242.

Imeson, A. 1986. Investigating volumetric changes in clayey soils related to subsurface water movement and piping. *Zeitschrift für Geomorphologie Supplement Band*, 60, 115–130.

Imeson, A.C. and Verstraeten, J.M. 1988. Rills on badland slopes: a physico-chemically controlled phenomenon. *Catena Supplement*, 12, 139–150.

Imeson, A.C. and Verstraeten, J.M. 1989. The microaggregation and erodibility of some semi-arid mediterranean soils. *Catena Supplement*, 14, 11–24.

Imeson, A.C., Kwaad, F.J.P.M. and Verstraeten, J.M. 1982. The relationship of soil physical and chemical properties to the development of badlands in Morocco. In R.B. Bryan and A. Yair (Eds) *Badland Geomorphology and Piping*. GeoBooks, Norwich, England, 47–70.

Jasper, D.A., Abbot, L.K. and Robson, A.D. 1991. The effect of soil disturbance on vesicular-arbuscular mycorrhizal fungi in soils from different vegetation types. *New Phytologist*, 118, 471–476.

Kirkby, M.J. 1976. Hydrological slope models. The influence of climate. In E. Derbyshire (Ed.) *Geomorphology and Climate*. John Wiley and Sons, Chichester, 247–267.

Kirkby, M.J., Atkinson, K. and Lockwood, J. 1990. Aspect, vegetation cover and erosion on semi-arid hillslopes. In J. Thornes (Ed.) *Vegetation and Erosion*. Wiley, Chichester, 25–39.

Lambe, T.W. 1951. *Soil Testing for Engineers*. John Wiley & Sons, New York, 165pp.

Langbein, W.B. and Schumm, S.A. 1958. Yield of sediment in relation to mean annual precipitation. *Transactions of the American Geophysical Union*, 39, 1076–1084.

Lázaro, R. 1994. *Relaciones entre vegetación y geomorfología en el área acarcavada del Desierto de Tabernas*. Unpublished PhD thesis. Universidad de Valencia.

Lázaro, R. and Puigdefàbregas, J. 1994. Distribución de la vegetación terofítica en relacion con la geomorfologia en areas acarcavadas cerca de Tabernas (Almeria). *Monografias de Flora y Vegetación Béticas*, 7-8, 127–154.

Lázaro, R., Alexander, R. and Puigdefàbregas, J. 2000a. Cover distribution patterns of lichens, annuals and shrubs in the Tabernas Desert, Almería, Spain. In R.W. Alexander and A.C. Millington (Eds) *Vegetation Mapping: From Patch to Planet*. Wiley & Sons, Chichester, 19–40.

Lázaro, R., Rodrigo, F.S., Gutiérrez, L., Domingo, F. and Puigdefàbregas, J. 2000b. Analysis of a thirty year rainfall record (1967–1997) from semi-arid SE Spain: a plant ecological perspective. *Journal of Arid Environments* (in press).

Lecompte, M., Lhenaff, R. and Marre, A. 1996. Premier bilan de six années de mesures sur l'ablation des roubines des Baronnies méridionales (Préalpes françaises du sud). *Revue de Géographie Alpine*, 1996(2), 11–16.

López-Bermúdez, F. and Romero Díaz, M.A. 1989. Piping erosion and badland development in SE Spain. *Catena Supplement*, 14, 59–74.

Maccherini, S., Chiarucci, A., Torri, D., Ristori, J. and De Dominicis, V. 1996. Influence of salt content of pliocene clay soil on the emergence of six grasses. *Israel Journal of Plant Sciences*, 44, 29–36.

Martín-Penela, A.J. 1994. Pipe and gully systems development in the Almanzora Basin (SE Spain). *Z. Geomorph*. N.F., 38(2), 207–222.

Meunier, M., Cambon, Olivier, Mathys, N., Combes, 1987. *Bassins versants expérimentaux de Draix*. Compte rendu de recherche no. 1 en érosion et hydraulique torrentielle. CEMAGREF-ONF-RTM.

Nogueras, P., Burjachs, F., Gallart, F. and Puigdefàbregas, J. 2000. Recent gully erosion in the El Cautivo badlands (Tabernas, SE Spain). *Catena*, 40, 203–215.

Oostwoud Wijdenes, D.J. and Ergenzinger, P. 1998. Erosion and sediment transport on steep marly hillslopes, Draix, Haute-Provence, France. An experimental field study. *Catena*, 33, 179–200.

Pardini, G. 1996. *Evoluzione temporale della microtopografia superficiale, della micromorfologia e della struttura in relazione ai processi di meteorizzazione nelle marne smectitiche di Vallcebre*. Ph.Dr Thesis. Facultat de Geologia, Universitat de Barcelona (unpublished).

Pardini, G., Pini, R., Barbini, R., Regüés, D., Plana, F. and Gallart, F. 1995. Laser elevation measurements of a smectite-rich mudrock following freeze-thawing and wet-drying cycles. *Soil Technology*, 8, 161–175.

Pardini, G., Vigna Guidi, G., Pini, R., Regüés, D. and Gallart, F. 1996. Structural changes of smectite-rich mudrocks experimentally induced by freeze–thawing and wetting–drying cycles. *Catena*, 27, 149–165.

Puigdefàbregas, J. and Sánchez, G. 1996. Geomorphological implications of vegetation patchiness on semiarid slopes. In M.G. Anderson and S.M. Brooks (Eds) *Advances in Hillslope Processes*, vol. 2. Wiley & Sons, Chichester, 1027–1060.

Prosser, I.P. 1996. Thresholds of channel initiation in Historical and Holocene Times, Southeastern Australia. In M.G. Anderson and S.M. Brooks (Eds) Advances in Hillslope Processes, vol. 2. Wiley & Sons, Chichester, 687–708.

Regüés, D., Guàrdia, R. and Gallart, F. 2000. Geomorphic agents versus vegetation spreading as causes of badland occurrence in a Mediterranean subhumid mountainous area. *Catena*, 40, 173–187.

Regüés, D., Pardini, G. and Gallart, F. 1995. Regolith behaviour and physical weathering of clayey mudrock as dependent on seasonal weather conditions in a badland area at Vallcebre, Eastern Pyrenees. *Catena*, 25, 199–212.

Regüés, D., Llorens, P., Pardini, G., Pini, R. and Gallart, F. 1993. Physical weathering and regolith behaviour in a high erosion rate badland area at the Pyrenees: research design and first results. *Pirineos*, 141/142, 63–84.

Requena, N., Jeffries, P. and Barea, J.M. 1996. Assessment of natural mycorrhizal potential in a desertified semiarid ecosystem. *Applied and Environmental Microbiology*, 62(3), 842–847.

Richard, D. and Mathys, S. 1999. Historique, contexte technique et scientifique des BVRE de Draix. Caractéristiques, données disponibles et principaux résultats acquis au cours de dix ans de suivi. In N. Mathys (Ed.) *Les bassins versants expérimentaux de Draix, laboratoire d'étude de l'érosion en montagne*. Cemagref, Antony, 11–28.

Scoging, H. 1982. Spatial variations in infiltration, runoff and erosion on hillslopes in semi-arid Spain. In R.B. Bryan and A. Yair (Eds) *Badland Geomorphology and Piping*. GeoBooks, Norwich, England, 89–112.

Scheidegger, A.E., Schumm, S.A. and Fairbridge, R.W. 1968. Badlands. In R. Fairbridge (Ed.) *Encyclopedia of Geomorphology*. Dowden, Hutchinson and Ross, Inc., USA, 43–48.

Scheidig, A. 1934. *Der Löss (Loess)*. Dresden, 233pp.

Schumm, S.A. 1956. Evolution of drainage systems and slopes in badlands at Perth Amboy, New Jersey. *Bulletin of the Geological Society of America*, 67, 597–646.

Sirvent, J., Gutiérrez, M. and Desir, G. 1996. Erosión e hidrología de áreas acarcavadas. In T. Lasanta and J.M. García-Ruiz (Eds) *Erosión y Recuperación de Tierras en Áreas Marginales*. Instituto de Estudios Riojanos – SEG, 109–135.

Sirvent, J., Desir, G., Gutiérrez, M. and Sancho, C. 1997. Erosion rates in badland areas recorded by collectors, erosion pins and profilometer techniques (Ebro basin, NE Spain). *Geomorphology*, 18(2), 61–75.

Solé, A., Josa, R., Pardini, G., Aringhieri, R., Plana, F. and Gallart, F. 1992. How mudrock and soil physical properties influence badland formation at Vallcebre Pre-Pyrenees (NE Spain). *Catena*, 19(3-4), 287–300.

Solé, A., Calvo, A., Cerdà, A., Lázaro, R., Pini, R. and Barbero, J. 1997. Influence of micro-relief patterns and plant cover on runoff related processes in badlands from Tabernas (SE Spain). *Catena*, 31, 23–38.

Stocking, M.A. 1988. Assessing vegetative cover. In R. Lal (Ed.) *Soil Erosion Research Methods*. Soil and Water Conservation Society, Ankeny, Iowa, 163–185.

Taylor, R.K. and Smith, T.J. 1986. The engineering geology of clay minerals: swelling, shrinking and mudrock breakdown. *Clay Minerals*, 21, 235–260.

Terzaghi,K. and Peck, R.B. 1967. *Soil Mechanics in Engineering Practice*, 2nd Edn. John Wiley & Sons, New York.

Torri, D. and Bryan, R. 1997. Micropiping processes and biancana evolution in southeast Tuscany, Italy. *Geomorphology*, 557.

Torri, D., Colica, A. and Rockwell, D. 1994. Preliminary study of the erosion mechanisms in a biancana badland (Tuscany, Italy). *Catena*, 23, 281–294.

Wise, S.M., Thornes, J.B. and Gilman, A. 1982. How old are the badlands? A case study from south-east Spain. In R.B. Bryan and A. Yair (Eds) *Badland Geomorphology and Piping*. GeoBooks, Norwich, England, 259–278.

Wolman, M.G. and Gerson, R. 1978. Relative scales of time and effectiveness of climate in watershed geomorphology. *Earth Surface Processes and Landforms*, 3, 189–208.

Yair, A., Goldberg, P. and Brimer, B. 1982. Long term denudation rates in the Zin-Havarim badlands, northern Negev, Israel. In R.B. Bryan and A. Yair (Eds) *Badland Geomorphology and Piping*, GeoBooks, Norwich, England, 279–291.

Yair, A., Lavee, H., Bryan, R.B. and Adar, E. 1980. Runoff and erosion processes and rates in the Zin valley badlands, northern Negev, Israel. *Earth Surface Processes and Landforms*, 5, 205–225.

Yair, A. and Lavee, H. 1985. Runoff generation in arid and semiarid zones. In M.G. Anderson and T.P. Burt (Eds) *Hydrological Forecasting*. Wiley, Chichester, 183–220.

Yoder, R.R. 1936. A direct method of aggregate analysis of soils and a study of the physical nature of erosion losses. *Journal of the American Society of Agronomy*, 28, 337–351.

Part IV   Flooding in Ephemeral Channels

11 Floods: Magnitude and Frequency in Ephemeral Streams of the Spanish Mediterranean Region

FRANCISCO LÓPEZ-BERMÚDEZ, CARMELO CONESA-GARCÍA
AND FRANCISCO ALONSO-SARRÍA
Department of Physical Geography, University of Murcia, Spain

11.1 INTRODUCTION

The landscape in Mediterranean semi-arid environments is mainly formed by the action of surface water runoff, often associated with extraordinary events of high magnitude. The hydro-morphological role of Mediterranean floods has been recognised for many years, however controversy still surrounds the subject and many unknown factors remain. The data required to evaluate flooding processes, especially in ephemeral channels ('*ramblas*'), are not readily available because it is difficult to record these events in the right place and at the precise time. The magnitude of torrential flows in Mediterranean regions can also damage or destroy gauging instruments.

'Ramblas-rivers' and 'ramblas' are hydrologic features of great geomorphological effectiveness in semi-arid Mediterranean environments (Roselló, 1985; Mateu Bellés, 1990; López-Bermúdez et. al., 1988). Surface runoff is almost exclusively controlled by rainfall, with subterranean flows having little influence. Mediterranean 'ramblas' commonly experience highly variable rainfall, soil properties and land uses, which results in randomly distributed flows.

This chapter focuses on the magnitude and frequency of torrential events causing floods in the Spanish Mediterranean region.

11.2 HISTORICAL PERSPECTIVE

Flooding (in the outlets of major Mediterranean rivers) and flash floods (in steep tributaries and ephemeral channels) are a remarkable natural hazard in Mediterranean environments because of the impact on the landscape and the implicit social and economic problems. They are also indicators of desertification processes. Countries' policies have been conditioned by the magnitude and frequency of events in order to attempt to solve the problem. The relationship between Mediterranean society, floods and flash floods has also been a main factor in evolution of the physical landscape. Despite the permanence of physical factors related to floods, flood hazard is not constant during history. Yet, the sensibility of human settlements and activities depends on the model of economic development of the society and the model of

Dryland Rivers: Hydrology and Geomorphology of Semi-arid Channels. Edited by L.J. Bull and M.J. Kirkby.
© 2002 John Wiley & Sons, Ltd.

land occupation. Major catastrophic events show, in most cases, that changes in social relations and these elements deeply influence the ways in which society has managed water.

The information on historic floods that have affected Mediterranean rivers and ephemeral channels is widely available. In Spain there are detailed references to some floods from the mid-thirteenth century (e.g. the Segura River in 1258, 1292, 1356; the Turia River in 1321, etc.). Since 1450 there have been almost 2400 recorded flood events in the Mediterranean area of the Iberian Peninsula. The historical archives have supplied useful information about floods, particularly after the sixteenth century (events in 1545, 1568, 1651, 1653, 1704, 1767, 1830, etc.). Most researchers (Couchoud and Sánchez, 1965; Molina Sempere et al., 1994; Benito et al., 1996) have consulted these records, and interpreted an increase in both flood magnitude and frequency related to climatic changes. Lemeunier and Pérez Picazo (1988) suggest a slight increase in precipitation at the end of the middle ages in southeast Spain. This trend has not been uniform, but has accelerated (during the second half of the sixteenth and eighteenth centuries) and then receded (with a maximum in 1760–1880). After this, there appears to have been a decrease in extreme rainfall events and an increase in aridity. Similar results have been obtained by Barriendos (1994) who studied the historical climate in Cataluña from the fifteenth to the nineteenth century by statistically analysing the frequency of *rogations* (crying out for rain or for the end of severe rainfall events). Barriendos (1994) showed that there were three marked rainfall periods: the first between the sixteenth and seventeenth centuries that coincides with the little ice age; another during the last third of the eighteenth century; and a final pulse in the middle of the nineteenth century.

There are many historical references to extreme floods along the Spanish Mediterranean coast. In Cataluña, floods produced by ramblas usually occur on the littoral and prelittoral plains. The El Vallés flood in September 1962 was one of the worst disasters to have occurred in Spain in recent years. Valencia has been at the mercy of the River Turia for centuries (Arenillas and Sáez, 1987) and from the sixteenth to the nineteenth century 30 large floods

Figure 11.1   Cartagena floods by ramblas of Benipila and Hondón, 30 September, 1919

have been recorded. In 1957 a flood of about $4000\,\text{m}^3\,\text{s}^{-1}$ (Mateu Bellés, 1988) caused severe economic and human losses that are remembered even today. For the Júcar river, more than 70 large events have been recorded since 1388, the most important of which were in September 1864 and October 1982, when discharges of 12 000 and $5000\,\text{m}^3\,\text{s}^{-1}$ were recorded respectively (Albentosa Sánchez, 1989). Alcira, located in an abandoned meander, was destroyed in 1472 and in 1916 water reached a height of 8.5 m (Lautensach, 1964).

Flash floods in the Segura basin have similar characteristics to those in the Júcar river, most of them being synchronous with those of Valencia. López-Bermúdez et al. (1979) compiled a chronology of major floods in this basin and analysed historical documentation about their impacts. This and other works show that the most extreme floods take place in the lower Segura and the Guadalentín. The confluence of both rivers was the cause of floods in October 1651, October 1776, November 1884 and April 1946. In the nineteenth century, the Guadalentín, one of the most energetic water courses in the western Mediterranean, experienced two spectacular floods: the first in April 1802, which broke the Puentes reservoir with 604 lives lost; and the second in October 1879 (flooding of 'Santa Teresa'), which resulted in total flow of 58M m^3 in just over eight hours in Lorca (Espejo Arevalo, 1963). The twentieth century has also had many floods, in terms of morphological and socio-economic effects. Twenty-two large events have taken place at different points of the Guadalentín basin. One especially remarkable event was the 'San Miguel flood' (September 1919) that caused great economic and human loss on the coast, especially in Cartagena (Figure 11.1). Another notable event was the flood in the 'Rambla de Nogalte' in October 1973, with discharges of more than $2000\,\text{m}^3\,\text{s}^{-1}$ recorded at Puerto Lumbreras.

11.3 FACTORS CONTROLLING THE FLOOD MAGNITUDE

The main factor causing floods in the Spanish Mediterranean region is the high rainfall intensity. Other factors include natural or artificial obstructions within the channel, accidental unretainment of water, dam failures (Puentes–Murcia in 1802, Ribadelago–Leon 1957, Tous–Valencia in 1982, Aznalcollar–Sevilla in 1998). Weather causes are usually emphasised in the development of prevention policies because of the use of straightforward statistical analysis of weather data. However, Mediterranean flash floods can be increased or decreased by other factors including drainage basin and network morphometry, lithology, and land use change.

11.3.1 Climatological Factors

In view of the marked variation in flood response to different rainfall conditions, it is perhaps inevitable that many attempts have been made to classify floods on the basis of the storm event itself. In the fluvial semi-arid systems the most important and disastrous floods are pluvial, especially flash floods. These events are often the result of convective storms or high-intensity rain cells associated with frontal storms and high cold drops. A third pluvial type is caused by orographically reinforced frontal storms, producing severe and long-duration floods. Cold drops in height do not produce rainfall by themselves, but contribute to the maintaining of unstable conditions reinforcing frontal elevation of low level air masses. Other elements, such as the arrival of warm and wet air from the Mediterranean and polar waves in the upper troposphere, increase the number of low-level cells at the surface. Rainfall magnitude can be extreme if orographic air elevation coincides with advective-frontal air elevation, as in

November 1955, October 1973, November 1982 and September 1989 (López-Gómez, 1983; García de Pedraza, 1983; Miró-Granada, 1983).

11.3.2 Drainage Basin Factors

Area is probably the most important basin characteristic that affects flooding because it affects the time of concentration as well as the total volume of stream flow generated by a given catchment-wide precipitation event. Relief patterns in small basins in the Spanish Mediterranean coast (Cordillera Costero Catalana, Sistema Ibérico, Cordilleras Béticas), mean that several small basins can be affected by a single storm, each producing a flash flood. In High Guadalentín, there are several channels (Corneros, Rambla Seca and Turrilla) that are very sensitive to floods as a result of these variables. Basins rounded in shape and with moderate bifurcation ratios (3.8) may contribute to the formation of high sharp floodpeaks. In Campo de Cartagena, Rambla del Albujon has the same shape parameters with a dendritic form (Conesa-García, 1990). Between high areas in the Meseta and inner mountains in Cataluña, Valencia, Murcia and eastern Andalucía, usually higher than 400 m, and its litoral planes, there is an intermediate band (about 60 km wide) with high slopes where most of the streams are incised.

This band is also characterised by a high lithological diversity (limestones, loam, sandstones, etc.), where infiltration capacity is lower, even during high-intensity events (Mateu Bellés, 1990). Limestones and dolomites have a low regulation capacity because they occur on steep relief (e.g. Sierras de Revolcadores, Mojantes, María, El Carche, Aitana, Mondúber, etc.) with a quick flow response to heavy precipitation. In certain catchments or headwaters in the Betics (Ramblas de Nogalte, Béjar, Moreras, Peñas Blancas, etc.) the most important rocks are metamorphic with low storage potential, often resulting in rapid and intensive flooding. However, the most important lithological control on the concentrated runoff appears in neogene basins, where loams have a very low infiltration capacity. Ramblas and gullies develop in this kind of material.

11.3.3 Drainage Network and Channel Morphometrics

Channel–network characteristics, especially Hortonian indices, have been included in hydrological prediction models such as the Geomorphological Unit Hydrograph (Rodríguez-Iturbe and Valdés, 1979; Gupta et al., 1980; Singh, 1983). These kinds of model have been applied to Mediterranean basins by Rosso (1984), García Bartual (1989), Camarasa (1995) and Alonso-Sarría (1995). Main Mediterranean channels have dense tributary systems, with high drainage densities and torrential morphometric coefficients. This situation reflects fragile, highly erodible soils and a scarce vegetation cover, particularly in badlands areas with drainage density values higher than $50 \, km \, km^{-2}$ (Los Guillermos–Rambla Salada, or Albudeite basin, both in the Murcia region). Channel shape and form also affect flood magnitude.

11.3.4 Human Actions

Vegetation cover being removed by human activity, particularly in southern ranges, is the most important human factor causing mountain torrents (Vázquez de Prada, 1978). Deforestation is a result of the need for wood for fuel and shipbuilding, etc., with the use of wood being greatest in the eighteenth century (Bauer Manderscheid, 1980). Since the eighteenth century, there has been concern about forest destruction and its effect on floods and several hydrologic and forestry works were developed as follows:

1. Committees were organised to study the origin of major flood events (4 November 1864; 14 October 1879)
2. Headwaters Afforestation Plan (3 February 1880)
3. National Forestry and Hydrology Service (1901)
4. Afforestation Law (24 June 1908)
5. National Afforestation Plan (1926)
6. National Plan on Hydraulic Works (1933).

However, the success of these policies has been insufficient, as was evident during flood events in 1946, 1948, 1951, 1973, 1982 and 1989 when discharges reached similar levels as those seen previously. Moreover, an increase in forest fires since the 1970s has meant higher runoff rates and flow concentration.

Other land uses also contribute to a reduction in infiltration rates. Urban growth changes rainfall–runoff processes, decreasing lag time and increasing flood peaks. Urban drainpipes usually flow at the same points out of the cities, producing a difficult hydraulic problem. The urbanisation of channels and bad agricultural practices increase the risk of floods.

Another factor that increases flood magnitude is the building of infrastructure works (dams and canals) since the end of the nineteenth century. The effect of these works, especially canals, can worsen flood magnitudes affecting larger areas (Machí i Felici, 1988). In some cases, an artificial, more gentle slope is created which reduces runoff velocity, thus increasing water height. For example, during the flood event of 15 October 1879, the Guadalentín flood water was retained by the railway embankment reaching a height of 2.7 m. The collapse of this made the flood worse. Similar examples appear on the Jucar basin.

For centuries and even millennia, Mediterranean societies have tried to reduce flood areas, flood peaks and flood volumes using several engineering projects such as canals, dams, deviations or dikes (Mateu Bellés, 1990). By combining several solutions (Roselló, 1983, 1985) it is possible to reduce the effects, but it may not be enough, or it may even increase the damage if the event magnitude is higher than that for which the structure is designed. In certain cases, these kinds of structural solutions (canalisation, meander cutting, dams) have been applied to ephemeral channels, but due to the torrential nature of the events, in most cases the solutions have failed. The usual problems are: (i) a canal section that is too narrow (Costa Mas, 1988); (ii) embankment failure; and (iii) fast reservoir infilling with geomorphic consequences downstream. Due to the geomorphic activity of these kinds of channels, it is better to use an alternative structure in order to prevent erosion in headwaters such as: (i) small dams (with smaller catchment areas); (ii) soil conservation by terracing; (iii) increase forest cover.

In most of the cities affected by ramblas with historically high geomorphological activity, canalisation is considered the most suitable structural solution, although after canal construction, which results in a narrower section, it is hardly possible to improve prevention by any further action. Drain capacity must be improved by several actions (i) removing the sediments from the canal, (ii) increasing the canal depth, or even (iii) increasing bank heights. The first solution is very common in order to maintain the channel cross-section and Manning's n coefficient for the required flow regime. The second solution is not very common and must not change canal slope. The third solution requires undesirable urbanistic changes such as high walls along the river sides (Martín Vide, 1997).

(a)

(b)

Figure 11.2   The Rambla de Nogalte crossing through Puerto Lumbreras village as (a) a torrential flow (7 September 1989) and (b) a dry channel occupied by human activity (26 January 1990)

Experience has shown that hydraulic works are not enough to achieve a significant decrease in flood hazard (Figure 11.2). It is also necessary to reduce the human occupation of hazardous areas in order to reduce the risk and increase warning and prevention systems (early alarm systems, contingency plans). In Spain, risk mapping of fluvial banks and floodplains (including those in ephemeral channels) have legal support (Water Law, 2 August 1985; Public Ownership of Water Regulations, 11 April 1986). However, this mapping is only based on floods with a 50-year return period. Recently the Ministry of Works and Transport has used the 'Floodway' concept developed by *Federal Emergency Management Agency (FEMA)* for the National Flood Insurance Program. This concept has been applied in a new study on flooding hazard areas and distinguishes between 'high runoff channel' situations for flooding with a 100-year return period, and 'dangerous flood areas' with a 500-year return period. The Spanish government has developed the SAIH Program (*Automatic System for Hydrologic Information*), organised as a real time hydrologic and hydraulic data transmission system. Its versatility allows incorporation of the latest technological innovations as well as the future capabilities that society demands.

Due to the physical and human factors mentioned above, Mediterranean basins of coastal Spain have many areas at risk from flooding. In Spain there are 1400 hazard points with regard to flood events, with almost 50% of these located in the Mediterranean fluvial systems (Berga Casafont, 1992), and 75% in the east and southeast of the Iberian Peninsula (Comisión Nacional de Protección Civil, 1984). These points generally correspond with the fast economic and social development in these areas.

11.4 PHYSICAL MAGNITUDE

11.4.1 Hydrologic Magnitude

Flood magnitudes can be expressed as a wide range of measurable variables (Ward, 1978): flooding water depth, discharge rate, shape of the flood hydrograph and discharge volume. Undoubtedly the oldest and still probably the most common index of flood magnitude is water depth or stage. There are available data on water height for several historic floods in ephemeral streams. During the Segura and Guadalentín flash flood (October 1879), water height reached 0.6–1.8 m in the irrigated area of Murcia city (Couchoud and Sánchez, 1965); 2.5–3.2 m in Cartagena main streets due to the Benipila and Hondón ramblas overflowing in September 1919 ('El Eco' del Cartagena, 29 October 1919), and 0.8–2.6 m in Rambla de las Moreras in Mazarrón (Murcia) in October 1989 (Rodríguez Estrella et al., 1992). However, direct measurement of water height and discharge in ephemeral channels has been difficult due to the massive flow during events and the vulnerability of measurement equipment. That is the reason why, despite its value for analysis and flood prevention, available data is scarce and rarely continuous (Mateu Bellés, 1989; Segura, 1990; López-Bermúdez et al., 1998). Interest is normally focused on peak stage or discharge since this is the moment of greatest danger and maximum inundation, and many deterministic approaches exist to relate precipitation and catchment variables to maximum flood discharge.

Due to the lack of gauges in most Mediterranean ephemeral channels, empirical models for runoff estimation are widely used. Equations developed by Greager, Scimeni, Gauguillet, Kuickling, Gutmann and Hoffman to calculate peak floods were used to study the event of October 1973 (Heras, 1973). Maximum estimated discharges were 1100–1200 $m^3 s^{-1}$ in Guadalfeo and Albuñol, 2000 $m^3 s^{-1}$ in Adra, 3000 $m^3 s^{-1}$ in Guadalentín and 3500 $m^3 s^{-1}$ in Almanzora. Myers's (1969) equation has also been used in the torrential channels of Ovejas,

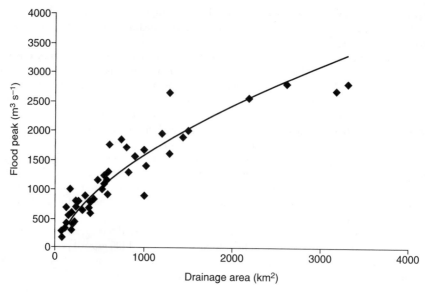

Figure 11.3 Event flood peaks versus catchment area for ephemeral stream basins in the Spanish Mediterranean region (based on various sources)

Maldo and Agua Amarga (Alicante province) during the intense rainfall event in October 1982, giving values of between 200 and 400 m^3 s^{-1} (Instituto Universitario de Geografía de Alicante, 1986). Some of these equations were fitted to larger basins, despite the fact that there is often a better relation between flow peak and basin area in small basins (Figure 11.3).

Also of interest is the relationship between magnitude and time, which defines the shape of the flood hydrograph and the discharge volume, that is, the total quantity of water measured on the hydrograph in the basal time defined by the flood level. In general, the flood hydrograph obtained from ephemeral streams consists almost entirely of quickflow, that is, water which reaches the stream channel very quickly during and immediately after precipitation. There are three kind of floods in relation to hydrograph shape: (i) flash floods, also described as 'walls of water' (Marshall, 1952), (ii) single peak floods and (iii) multiple peak floods.

Flash Floods

These hydrographs correspond to sharp peaks with the floodwaters rising and falling almost equally rapidly, and are frequent in small drainage areas and headwater streams within the Mediterranean coastal basins. These flood-producing processes result in more severe floods according to certain flood-intensifying conditions, particularly those basin and network variables which together affect the size and disposition of the quickflow-generating source areas (IASH, 1974). Catchment environmental characteristics in southeastern Spain contribute to steep hydrographs showing a concentrated discharge curve and an accented flood peak. In extreme hydrological conditions associated with areas of very high rainfall intensities that are maintained over several hours, infiltration capacities are exceeded and runoff hydrographs exhibit very sharp peaks. This hydrograph shape may be particularly elongated when the storm affects the whole watershed, such as occurs occasionally over numerous gully catchments – e.g. Bocaoria gully catchment on the Carthagene coast during the floods of 17 October 1972

Figure 11.4   Morphological results of Biescas flash flood

and 21 February 1985 (Conesa-García, 1985); or Zarza gully catchment draining to the Salada rambla (Murcia) on 30 September 1997. Some of these hydrographs show two or three almost vertical rising limbs separated by moderate to steep increasing rates, depending on the variations of rain intensity. Apart from the semi-arid Mediterranean environments, and mountainous areas of the general Mediterranean region, there are several examples of flash-flood basins completely affected by convective storms that have produced violent flash floods. The most important case, and one of the most intense small-area short-duration storms of this type, occurred in Biescas (Huesca, Pyrenees) in August 1996 (Figure 11.4), producing more than 250 mm in only one hour and a peak flow of $500 \, \mathrm{m^3 \, s^{-1}}$ near the outlet (García Ruiz et al., 1996).

Single Peak Floods

This is the most common type of flooding in most Mediterranean ephemeral channels. It is characterised by a time base duration of several days, (usually less than 6), and an overhanging flood peak. The hydrograph peak occurs very quickly, usually within a day in small and middle size basins, and thereafter discharge is mainly determined by the surface runoff. Particularly in the semi-arid fluvial systems the saturation and groundwater flows have little influence on runoff generation (Schick, 1988), and Hortonian overland flow is almost the exclusive runoff input (Yair and Lavee, 1985; Yair, 1990; Camarasa, 1995). The rate of exhaustion of storage is reflected in the recession limb of this hydrograph type (see Figure 11.5). The falling limb quickly declines in accordance with catchment conditions, explaining why most of the single event floods recorded in the Segura basin show symmetrical and sharp hydrographs. There is a slight smoothing in the late recession stage, which is related to subsurface flows through alluvial materials in the low sectors of the channel. One of the best examples of this type of flash

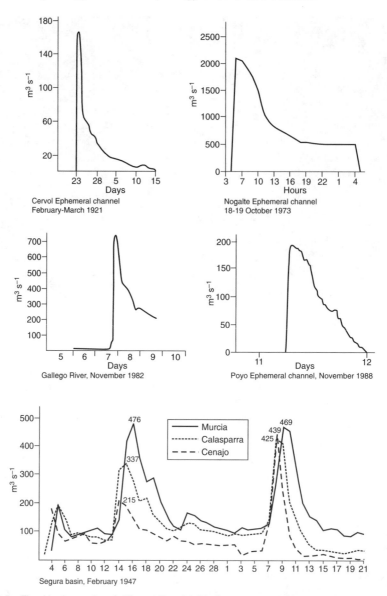

Figure 11.5 Flood hydrographs of different Spanish Mediterranean catchments

flood took place in Rambla de Nogalte in October 1973 (flood peak: 2100 m³ s⁻¹ in Lorca). This was one of the largest flood events in the Guadalentín–Viznaga system over the past 100 years; and was used as a flood model by CHS (Segura Water Authority) to develop the Flood Prevention Plan (Confederación Hidrográfica del Segura, 1985). It was the product of a severe storm related to an intense cold low-pressure cell. Rainfall was general in diverse basins in the southeastern Mediterranean. Single peak hydrographs were produced with very steep rising limbs, high magnitude peaks and base durations of less than two days.

Multiple Peak Floods

Successive flood peaks are caused by complex weather situations, associated with multiple convective cells, and cyclogenetic processes in the western Mediterranean, or to different response patterns in the tributary basins to the main channel. In such cases, although the individual flood peaks may not exceed, or even closely approach, those of single peak events, record peak discharges combine with flooding of an extended duration to produce multiple peak floods with disastrous results. This situation occurred in the Segura main channel during February–March 1947 with several high-magnitude stage increases. The flood hydrograph was, in this case, two main peaks (476 and 469 m^3 s^{-1}) and had a time base (6–8 days) longer than in single peak floods. Antecedent precipitation conditions played an extremely important part and precipitation before the flood event led to a substantial wetting of the catchment enabling runoff soon after commencement of the storm.

11.4.2 Morphological Impacts

The geomorphological effects of flood events depend on magnitude–frequency relationships. However these relations are not adequately understood in Spanish ephemeral streams. There are several studies on the geomorphological effects of isolated events such as flooding in October 1973 in Rambla de Nogalte (Navarro Hervás, 1985) (Figure 11.6); October 1982 in Segura tributaries (López-Bermúdez and Gutiérrez Escudero, 1983); ephemeral channels in Alicante (Instituto Universitario de Geografía de Alicante, 1986); September 1992 in Rambla las Arenas (Martín Vide et al., 1993); and August 1996 in Arás Gully (García Ruiz et al., 1996). Morphological impacts reported in these events include modifications of the channel where banks were overtopped, bank erosion, and floodplain sedimentation. An exceptional event

Figure 11.6 Morphological effects of the Rambla de Nogalte flash flood (October 1973)

took place in the Arás Gully (Biescas) with substantial changes in the whole torrential moun-tain system, including channel widening, blocks accumulating to form a boulder fan, and the accumulation of differentiate sediments in sieve pattern. Several authors have analysed extreme events in semi-arid systems (Thornes, 1976; Harvey, 1984; Conesa-García, 1995) finding marked spatial variability in the geometric impacts of floods, especially between the constrained channel along upper reaches and the non-trenched unconfined part of the channel on the downstream fan surface. Headwater reaches experienced incision processes and lower reaches experienced sedimentation processes and the movement of bars. These systems, with very variable flow regimes, erosion and sedimentation processes, are dominated by events of high magnitude and low frequency.

About 40% of sediment transport is by events of more than a 10-year return period (Richards, 1982). These flood events are characterised by large bedload and suspended sedi-ment concentrations. Sediment loads of 40% of the volume of flow have been recorded in events such as the extreme flood in 1973 on the Nogalte in southeast Spain (Heras, 1973). With bedload being the main component of the load, there is a large difference between the main channel and the floodplain bedloads. In various sections of Rambla Salada (Murcia), hydraulic radius equals 1 m by the Meyer–Peter and Müller equation, estimated unit bedload discharges vary between 0.01 and $3 \times 10^{-3}\,m^2\,s^{-1}$ along the channel and between 0.73 and $2.5 \times 10^{-3}\,m^2\,s^{-1}$ in floodplain areas (Programme MEDALUS IV, Project 4).

Ephemeral channel morphology depends on the discharge regime in the range of competent discharges. Conesa-García (1995) classifies the floods according to their geomorphological effectiveness, using field data of three streams in southeast Spain. On the 'ramblas', the major thresholds, implying an overall modification of the fluvial system, have a return period of 3.5 to 6 years, depending on catchment characteristics. Moderate floods which tend to cause local adjustments occur with a return period of about 2 years. Conceptual frameworks for analysis of the impacts of floods are moving away from simple threshold models of metamor-phosis by large events, and towards realising the importance of sequences of events and positive feedbacks (Poesen and Hooke, 1997). It is difficult to develop a form–process equilibrium because of the large variability in runoff and solid discharges, long drought periods and short but intense flash floods.

11.5 SOCIO-ECONOMIC MAGNITUDE

Throughout history ramblas have been occupied by humans. Present settlements and those discovered by archaeologists show an intensive use of ephemeral channels for water deviations to irrigate by sluices, canals, wells in the channel bed, direct cultivation of the bed, use of ramblas as roads, and sand extraction. This extensive use also extends to floodplains with highways, profitable cultivation, houses and industrial settlement. These uses in a high flood hazard area result in a serious land use problem, especially because it is not usually taken into account in land allocation policies, even though the Ministry of Environment produces maps on hazard points and risk areas.

In some cases, nearby mountains increase the risk in highly occupied areas such as the surroundings of Valencia where the Rambla del Poyo and Rambla del Carraixet produce hydrographs faster and higher in magnitude that those produced by Turia, a major river that crosses the city (García Alandete, 1992). The city of Alcoy is located on five interfluves in the Serpis river valley, and the older and higher quarters as well as recently built areas have been affected historically by the flood risk, especially the geometric changes in the channel due to geomorphological processes such as mass movements or bank breaking (Costa Mas and

Table 11.1   Large floods in Cataluña during the period 1960–1990

Date	Site	Victims	Damage (million pesetas)
September 1962	Vallés	973	1 600
October 1970	Gerona	–	600
September 1971	Bajo Llobregat	24	6 000
September 1971	Gerona	–	800
November 1982	Prepirineo	14	45 000
November 1983	Bajo Llobregat	8	4 000
October 1987	Litoral	5	3 000
December 1987	Muga and Fluviá	–	600
November 1988	Bajo Llobregat, Maresme and Fluviá	8	4 700
August 1989	Litoral and Lérida	–	1 000
	Total	1032	67 300

Source: Berga Casafont (1992).

Matarredona Coll, 1988). Nowadays several examples of this type of risk are not only related to extreme runoff events, but also to the increase in human settlements and activities. These include tourism at the outlet of the Rambla de las Moreras in Murcia (Rodríguez Estrella et al., 1992), agriculture in Puerto Lumbreras with the hazard of Rambla de Nogalte (Conesa-García, 1995; Conesa-García et al., 1996), towns divided by ramblas, as in the case of Totana (Murcia) with Rambla de la Santa (Botía Pantoja, 1992), and Almería with Rambla de Belén (Calvo Álvarez, 1992).

The problem is extended to the subhumid environments in the Spanish Mediterranean area. In Cataluña the historically intensive occupation of steep ramblas has resulted in high economic losses. Table 11.1 shows the most important flash-flood events during 1960–1990, which caused losses of more than 70 000 million pesetas and more than 1000 deaths, most of them during the September 1962 event.

Economic and social problems related to floods and flash-flood events in the Mediterranean increase massively in urban areas (Calvo García-Tornel, 1998). In Cataluña there are 90 flood risk areas (Berga Casafont, 1991), 11 of which are high risk because of the urban and industrial development in their surroundings. Urban development increases land vulnerability because of the concentration of population and human activities (Mateu Bellés, 1990). But, as urban planning is also developed without taking into account fluvial networks, urban development becomes an aggressive practice against channels. Also protection works and infrastructures may produce a false sense of security that may be very dangerous during extreme events. Hence a massive settlement increase is produced in those areas that are, theoretically, well protected. Finally a positive feedback develops: protection works increase protection for new settlements, and new settlements develop due to the protection (Berga Casafont, 1992).

11.6   ESTIMATION OF DISCHARGE FREQUENCY

Flood frequency or statistical estimation of the probable occurrence of specific flood thresholds is commonly used for flood prediction, for design of hydraulic structures and for explaining morphological adjustments at different time-scales. Instantaneous peak discharges must be used in analyses of this type, but as floods in the Mediterranean are flashy and energetic, it is difficult to take measurements and the quantity and quality of data are very limited (Camarasa, 1995; Poesen and Hooke, 1997). Many attempts have been made in diverse Mediterranean environments to calculate flood frequency and to develop a method that would be applicable to ungauged catchments. One much used approach is the estimation of regional flood frequency curves (Mimikou, 1987; AMHY group of the UNESCO FRIEND project – Prudhomme, 1995). Other researchers in semi-arid environments have related flood frequency to basin characteristics (Ferrari et al., 1993) or to regime theory, an empirically based method of describing the geometrical properties of the natural watercourses in terms of the parameters which influence its formation (Nouh, 1988; Conesa-García and Álvarez-Rogel, 1996).

In Spain only a few ephemeral channels have runoff records, with short and fragmented series: Ramblas del Moro, Judío and Algeciras in the Segura basin, Rambla de les Ovelles in Alicante, Rambla de la Viuda, Rambla de Cervera, Barranc de Carraixet in El País Valenciano, etc. The development of the *Hydrological Information Automatic System (SAIH)* in the larger Mediterranean Spanish basins (Ebro, Júcar, Segura, Tajo, Guadalquivir, Pirineo Oriental and Sur) means a great improvement in the runoff knowledge of large basins and also on ramblas with high-magnitude flows. Flood frequency can be studied from old records and compiled chronologies are available for some basins (López-Bermúdez, 1973; López-Bermúdez et al., 1979; Muñoz Bravo, 1989; Molina Sempere et al., 1994; Benito et al., 1996) (Figure 11.7) although trying to resolve whether frequency is increasing or if land use or climate change is a factor is difficult. However, there is no worth while documentation before the nineteenth century, nor has there been any systematic research. Previous documents refer simply to extreme floods, and it is difficult to assess either the frequency or the magnitude–frequency ratio.

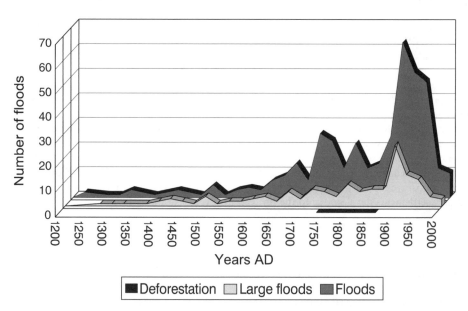

Figure 11.7 Evolution of flood frequency due to 'ramblas' and 'river-ramblas' in southeast Spain over the last eight centuries (Molina Sempere et al., 1994; Benito et al., 1996)

11.6.1   Flood Frequency Distributions

Some Mediterranean 'river-ramblas' (Guadalmedina, Andarax, Almanzora, Guadalentín, Vinalopó, Cervol, Palancia, etc.) have more extensive records of annual maximum discharges, which are useful for calculating probability distributions. Runoff estimation for specific return periods (Qt) can be achieved from a theoretical probability distribution which reflects the positive skewness of extreme event distributions. This can be obtained with original distributions (exponential, gamma, Gumbel-EV1, and Pearson-III) or transformed (log-normal, log-Gumbel, log-Pearson tipo III) (Benson, 1968). Because of the lack of theoretical reasons to develop a particular model, it is better to choose a log-normal model in order to take into account the multiplicative random factors over hydrologic regime in ephemeral channels. The Gumbel extreme value type 1 (EV1) is somewhat restrictive, but alternative probability functions fitted to annual series only diverge markedly in the tails of the distribution and at high return periods in particular. The log-Pearson III distribution can be well applied to a set of data containing zero values (Jennings and Benson, 1969), which occurs frequently in arid and semi-arid watercourses.

The Gumbel extreme value distribution (Gumbel, 1967) has been applied in several ephemeral channels in the Mediterranean (Heras, 1973) (Table 11.2). EV1 Gumbel distribution has been used by Conesa-García and Martínez Alcocer (1995) to estimate short return period floods in lower Segura before canalisation. Results show a $T = 6.25$ years for overflow events affecting local Murcia irrigated areas and $T = 4$ years for dominant discharge events. At present this reach is designed to $400 \, \text{m}^3 \, \text{s}^{-1}$ with steeper slope because of the meander cutting during the last decade. These works have eliminated the overflowing hazard in the short and middle terms.

11.6.2   Monthly and Seasonal Flood Frequency

Monthly and seasonal flood frequency in Mediterranean ephemeral streams is related to the daily rainfall regime. Martín Vide (1985) established, from statistical analyses of rainfall

Table 11.2   Peak flow discharges estimated for 'river-rambla' basins and different return periods in southern Spain

'Rambla'-river	Area (km²)	Mean runoff coefficient	25	50	100	500
			\multicolumn{4}{c}{Q (m³ s⁻¹)}			
Guadiaro	1504	0.69	1100	1350	1750	2000
Guadalhorce	3157	0.65	1350	1650	2150	2700
Guadalmedina	180	0.78	300	350	450	600
Vélez-Guaro	610	0.76	850	1100	1300	1750
Guadalfeo	1295	0.76	1400	1750	2100	2650
Albuñol	115	0.80	300	450	550	700
Adra	746	0.77	950	1350	1600	1850
Andarax	2188	0.71	1300	1600	2000	2550
Aguas	539	0.73	450	650	800	1000
Almanzora	2611	0.68	1350	1900	2200	2800
Guadalentín	3302	0.65	1150	1550	1850	2350

Source: Heras (1973).

records in five weather stations in the Spanish Mediterranean (Cap de Begur, Barcelona, Valencia, San Javier and Málaga), a larger amount of flood events and extreme rainfall days (greater than $50 \, mm \, d^{-1}$) during autumn especially in Murcia, Levante and Cataluña, superior by 50 mm (40–60% of total). October is the month with highest hazard, and September the second. Winter shows a higher number on the south Mediterranean coast (50%) with autumn in second place. Summer extreme rainfall events decrease when latitude decreases, with a maximum on the coast of Cataluña, because of the great number of connective storms and their intensity (especially in August).

Larger Mediterranean flood events in ephemeral channels show a similar regime. Most of these events took place in autumn: October 1919 in the Cardener, tributary of the Llobregat River; October 1923 in the Jucar basin; October 1973 in the Rambla de Nogalte (Segura River). The chronicle of flood events in southeast Spain (López-Bermúdez et al., 1979; Muñoz Bravo, 1989; Navarro Hervás, 1991) is very clear. Autumn is the season with a large amount of historical floods (41.3%); with summer only 25.4%. However, the distribution is not the same throughout the summer and most of the flood events are concentrated at the beginning (late June) and the end (early September). In spring, frequency is about 17.4% and in winter 15.9%. Flooding probability and its intensity during high hazard months are quite interesting parameters but they have not yet been sufficiently studied. Conesa-García (1988) has applied the Waylen (1985) method to estimate the monthly probability of floods in the Guadalentín River. This model is based on a Poisson frequency distribution:

$$p(m(t)) = \frac{(\exp(-\Lambda(t)) \cdot \Lambda(t)^{m(t)}}{m(t)}$$

where $p(m(t))$ is the probability of obtaining m flood events during time intervals similar to the selected (monthly or seasonal). m is the number of flood events during that interval along the analysis period, and can be estimated by moments as the excess frequency during the same period. The number of overtopping events is obtained from $m(t)$ events with an instantaneous peak higher than a critical threshold, Q_0 ('truncation level'), and a local maximum flood H_i. Results obtained with this method are similar to those previously shown, that is, October is the month with highest flood hazard inundation ($p(m(t)) = 0.29$ in the Lorca Valley).

11.6.3   Bankfull Discharge Frequency

The bankfull discharge, also described as channel-forming or dominant discharge, is the main geomorphic agent in the main channel, in terms of velocity v and shear stress τ. Higher discharges are less frequent and hardly increase the effects of v and τ. Cross-sections adjust to accommodate a uniform bankfull frequency, with an average return period, depending on the hydrological regime characteristics. Wolman and Leopold (1957) suggested a common return period for bankfull discharge of 1–2 years, a consistency implying mutual adjustment of flood-plain surface and channel-bed elevations. Dury et al. (1963) estimated the average bankfull return period as $T = 1.58$ years, by calculating bankfull flow from meander wavelength. In Spanish rivers, return periods of $T = 1.5$–7 years fit better, with higher values in the most irregular regimes. On various ephemeral streams studied by Conesa-García (1993) in southeast Spain this type of event controlling channel form occurs with a return period of about 2 to 6 years.

It is undoubtedly an oversimplification to assume a constant return period for dominant discharge. Harvey (1969) attributes the systematic variation in bankfull frequency to the hydrological regime. Hey (1975) relates the annual flood probability to the grain size of bed-forms, suggesting that bankfull conditions are more frequent in sand-bed streams than in gravel-bed streams. Kilpatrick and Barnes (1964) also identify variations in overbank frequency, but relate it to channel slope.

On ephemeral channels, it is difficult to estimate the bankfull frequency because of the lack of gauge data and information on channel dimensions. In some cases (larger ramblas with gauge records) it is possible to relate hydraulic geometry parameters with discharge, and define equations to estimate discharge from its return period. An often used approach in flood estimation from channel size is the channel-geometry method (Riggs, 1974; Osterkamp and Hedman, 1977; Wharton, 1992, 1995), which relate the flow dates of gauging stations and the channel dimensions. Recently Consesa-García and Álvarez-Rogel (1996) used it in several cross-sections in the River Segura, relating runoff for several return periods (Q_t) with the geometrics channel in two geomorphic levels: (i) level of functional channel or bankfull and (ii) level of flooding flat (overtopping limit). From that study, they found a bankfull width (W_b) and discharge of 2–5 year return period (100–200 m³ s⁻¹). Overtopping area (A_{ot}), including floodplain, is better correlated with extreme discharges (200–400 m³ s⁻¹ and 5–20 year return period). Best fitted equations are power relations between discharge and channel width, $Q_t = A \cdot W_b$ (Table 11.3).

The use of these equations in other reaches of the system has been useful in estimating the frequency of the channel formation discharges; on the other hand, it offers a model to discover if the relation between channel shape and discharge represents the equilibrium in the studied reach. Residuals indicate if the channel is infradimensional or overdimensional in relation to the water volume flowing during bankfull estimation, and determine its prospect of changing in the near future.

Table 11.3 Geometry-channel equations to estimate mean annual discharges (Q_{ma}) and discharges for different return periods (Q_t)

Segura Upper

$Q_{ma} = 0.296 \cdot W_b^{1.775}$	$Q_{ma} = 7.13\,(W_b - 30.2) + 125.9$
$Q_{10} = 25.11 \cdot W_b^{0.717}$	$Q_5 = 7.20 \cdot W_b^{0.997}$
$Q_{ma} = 0.062 \cdot A_b^{1.524}$	$Q_5 = 3.53 \cdot A_b^{0.823}$
$Q_{ma} = 1.66 \cdot W_{ot}^{0.776}$	$Q_{ma} = 58.44 \cdot e^{0.003 \cdot W(ot)}$
$Q_{ma} = 5.49 \cdot e^{0.004 \cdot A(ot)}$	$Q_5 = 0.002 \cdot A_{ot}^{1.792}$
$Q_{10} = 0.05 \cdot A_{ot}^{1.313}$	$Q_{20} = 0.347 \cdot A_{ot}^{1.05}$
$Q_{ma} = 0.046 \cdot \lambda^{1.314(ot)}$	$Q_5 = 3.08 \cdot \lambda^{0.705}$
$Q_{10} = 147.4 \cdot D_i^{0.285}$	$Q_{20} = 207.4 \cdot D_i^{0.232}$

Other environment semi-arid rivers

Ephemeral channel in California (Hedman, 1970):
$$*Q_r = 258 \cdot W_d^{1.54} \cdot D_d^{0.6}$$
Semi-arid regions of west USA (Osterkamp and Hedman, 1977):
$$*Q_{10} = 4.14 \cdot W_b^{1.63}$$
Ephemeral channels in south-east of Montana State (Omang et al., 1983):
$$*Q_2 = 10 \cdot W_b^{1.16}$$

* Discharge in cubic feet per second. Channel dimensions also in feet.
W_b = Bankfull level channel width (m); A_b = Bankfull level cross-section area (m²); W_{ot} = Overtopping level channel width (m); A_{ot} = Overtopping level cross section area (m²); λ = Meander wavelength (m); D_i = Depth from alluvial surface (m).

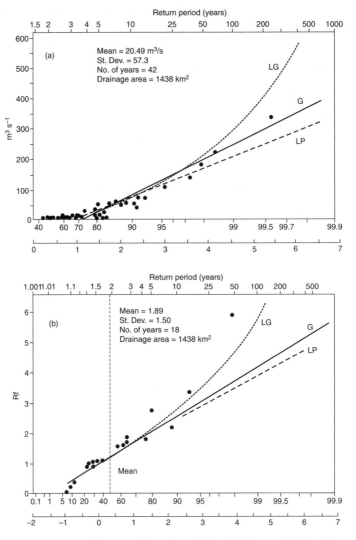

Figure 11.8 Return periods of Guadalentín maximum daily discharges and their ratios to mean flow in Lorca after Gumbel, log-Gumbel and log-Pearson III distributions (Conesa-García, 1988)

ACKNOWLEDGEMENTS

This work has been developed within the framework of Project MEDALUS III-4 (*Mediterranean Desertification and Land Use*): 'Ephemeral Channels and Rivers.' Contract No. ENV4-CT95-0118 (DG XII-DTEE) of the European Union.

REFERENCES

Albentosa Sánchez, L. 1989. *El clima y las aguas. Geografía de España.* Editorial Síntesis, Madrid, 170–179.

Alonso-Sarría, F. 1995. *Análisis de episodios lluviosos y sus consecuencias hidrológicas sobre una cuenca mediterránea semiárida no aforada* (*Cuenca de la Rambla Salada, Murcia*). Thesis. Universidad de Murcia, 501pp.

Arenillas, M. and Sáez, C. 1987. *Los ríos. Guía Física de España.* T. 3. Alianza Editorial, Madrid, 386pp.

Barriendos, M. 1994. *El clima histórico de Catalunya. Aproximación a sus características generales (SS. XV-XIX)*, Univ. Barcelona, Dep. Geografía Física y A.G.R., Tesis Doctoral.

Bauer Manderscheid, E. 1980. *Los montes de España en la Historia*, Serv. de Publ. Agrarias del Ministerio de Agricultura, Madrid.

Benito, G., Machado, M.G., Passmore, D.G., Brewer, P.A., Lewin, J., Branson, J. and Wintle, A.G., 1996. Climate change and flood sensitivity in Spain. In A.J. Brown and K.J. Gregory (Eds) *Global Continental Changes: The Context of Palaeohydrology.* Geological Society Special Publication 115. Geological Society, London, 85–98.

Benson, M.A. 1968. Uniform flood-frequency estimating methods for Federal Agencies. *Water Resources Research*, 4, 891–908.

Berga Casafont, L. 1991. La problemática de las inundaciones en España. *II Congreso Nacional de Ingeniería Civil.* Santander.

Berga Casafont, L. 1992. Avenidas con afecciones urbanas. In J. Dolz, M. Gómez and J.R. Martín (Eds) *Inundaciones y redes de drenaje urbano.* UPC Universitat Politecnica de Catalunya, Monografías 10, Barcelona.

Botía Pantoja, A. 1992. Problemática de las inundaciones con afecciones urbanas en la cuenca del Segura. In J. Dolz, M. Gómez and J.R. Martín (Eds) *Inundaciones y redes de drenaje urbano.* UPC Universitat Politecnica de Catalunya, Monografías 10, Barcelona.

Camarasa, A.M. 1995. *Génesis de crecidas en pequeñas cuencas semiáridas.* Ministerio de Obras Públicas, Transportes y Medio Ambiente. Confederación Hidrográfica del Júcar, València, 252pp.

Calvo García-Tornel, F. 1998. Risque naturel ou risque social. Espaces et Societés à la fin du XXe siècle. *Les Documents de la Maison de la Recherche en Sciences Humaines de Caen*, 7, 185–191.

Calvo Alvarez, J.M. 1992. Problemática de las inundaciones con afecciones urbanas en la cuenca Sur. In J. Dolz, M. Gómez and J.R. Martín (Eds) *Inundaciones y redes de drenaje urbano.* UPC Universitat Politecnica de Catalunya, Monografías 10, Barcelona, 395–410.

Comisión Nacional de Protección Civil. Comisión Técnica de Inundaciones 1984. *Las inundaciones en España.* Informe-Resumen.

Conesa-García, C. 1985. Procesos fluvio-torrenciales y morfología del barranco de Bocaoria (oeste de la sierra de Cartagena). *Actas IX Coloquio de Geografía*, T. I. Asociación de Geógrafos Españoles, Murcia, 334–343.

Conesa-García, C. 1988. Inundaciones en Lorca: riesgo y expectativa. *Cuadernos Informativos*, 5. Consejería de Política Territorial y Obras Públicas, Comunidad Autónoma de la Región de Murcia, 56–70.

Conesa-García, C. 1990. *El Campo de Cartagena: clima e hidrología de un medio semiárido.* Universidad de Murcia – Ayuntamiento de Cartagena, Murcia, 450pp.

Conesa-García, C. 1993. Identificación de umbrales y sucesos morfológicos asociados a corrientes torrenciales en el Sureste Peninsular. *Jornadas sobre El Estado Actual de la Investigación en la Ciencia Regional en Murcia.* Consejería de Fomento y Trabajo de la Comunidad Autónoma de Murcia, Murcia 194–212.

Conesa-García, C. 1995. Torrential flow frequency and morphological adjustments of ephemeral channels in southeast Spain. In E.J. Hickin (Ed.) *River Geomorphology.* John Wiley & Sons Ltd, Chichester, Ch. 9, 169–192.

Conesa-García, C. and Martínez Alcocer, P. 1995. Magnitud y frecuencia de sucesos hidromorfológicos del Bajo Segura anteriores a su encauzamiento. *Papeles de Geografía*, 22. Universidad de Murcia, 67–86.

Conesa-García, C., Álvarez-Rogel, Y., Belmonte Serrato, F., Vivero Martínez, M.A. and Rodríguez Tello. 1996. Simulación mediante SIG de áreas inundables en el tramo inferior de la Rambla de Nogalte (Cuenca del Segura). *VII Coloquio de Geografía Cuantitativa, Sistemas de Información Geográfica y Teledetección*, A.G.E., Vitoria.

Conesa-García, C. and Álvarez-Rogel, Y. 1996. El método de geometría de cauces aplicado a la estimación de caudales máximos de crecida en la Vega Alta del Segura. *Cadernos*, 21. Laboratorio Xeolóxico de Laxe, La Coruña, 469–482.

Confederación Hidrográfica del Segura 1985. *Plan Hidrológico Nacional. Estudio de fenómenos hidrológicos extremos*, T. IV. Ministerio de Obras Públicas y Urbanismo.

Costa Mas, J. 1988. Obras de defensa en los cursos alóctonos de la provincia de Alicante, Reunión científica internacional sobre *Avenidas fluviales e inundaciones en la Cuenca del Mediterráneo*. Centre Européen de Coordination et de Recherche et Documentation en Sciences Sociales de Viena – Instituto de Geografía de la Universidad de Alicante.

Costa Mas, J. and Matarredona Coll, E. 1988. Avenidas y problemas de tadules en la ciudad de Alcoy. Reunión científica internacional sobre. *Avenidas fluviales e inundaciones en la Cuenca del Mediterráneo*. Centre Européen de Coordination et de Recherche et Documentation en Sciences Sociales de Viena – Instituto de Geografía de la Universidad de Alicante.

Couchoud Sebastia, R. and Sánchez Ferlosio, R. 1965. *Hidrología histórica del Segura*. Madrid (reedición facsimil, Colegio de Ingenieros de Caminos, Canales y Puertos de Murcia, 1984).

Dury, G.H., Hails, J.R. and Robbie, M.B. 1963. Bankfull discharge and the magnitude–frequency series. *Aus. J. Science*, 26, 123–124.

Espejo Arévalo, M.D. 1963. *Lorca y la inundación de 1802*. Memoria de Licenciatura. Facultad de Filosofia y Letras. Universidad de Murcia, 137 pp.

FEMA (Federal Emergency Management Agency) 1990. *National Flood Insurance Program. Regulations for Floodplain Management and Flood Hazard Identification*. Revised. Oct.

Ferrari, E., Gabriele, S. and Villani, P. 1993. Combined regional frequency analysis of extreme rainfalls and floods. In Z. Kundzewicz, D. Resbjerg, S.P. Simonovic and K. Takeuchi (Eds) *Extreme Hydrological Events. Proceedings of an International Symposium*. IAHS Publication, 213. Wallingford, IAHS Press, 333–346.

García Alandete, J.I. 1992. Actuaciones de la Confederación Hidrográfica del Júcar en materia de defensa de poblaciones contra crecidas. In J. Dolz, M. Gómez and J.R. Martín (Eds) *Inundaciones y redes de drenaje urbano*. UPC Universitat Politecnica de Catalunya, Monografías 10, Barcelona, 349–360.

García Bartual, R. 1989. Estimación de la respuesta hidrológica de una cuenca sobre la base de la Teoría del Hidrograma Unitario Geomorfológico. *Cuaternario y Geomorfología*, 3, 1–7.

García de Pedraza, L. 1983. Situaciones atmosféricas tipo que provocan aguaceros torrenciales en comarcas del mediterráneo español. *Estudios Geográficos*, 170–171, 61–73.

García Ruiz, J.M., White, S.M., Martí, C., Valero, B., Errea, M.P. and Gómez Villar, A. 1996. *La catástrofe del barranco de Arás (Biescas, Pirineo Aragonés) y su contexto espacio-temporal*. Consejo Superior de Investigaciones Científicas, Instituto Pirenaico de Ecología, Zaragoza, 54pp.

Gumbel, E.J. 1967. Extreme value analysis of hydrologic data. *Statistical Methods in Hydrology*. National Research Council of Canada, Ottawa, 147–181.

Gupta, V.K., Waymire, E. and Wang, C.T. 1980. A representation of an instantaneous unit hydrograph from geomorphology. *Water Resources Research*, 16(5), 855–862.

Harvey, A. 1969. Channel capacity and the adjustment of streams to hydrologic regime. *Journal of Hydrology*, 8, 82–98.

Harvey, A. 1984. Geomorphological response to an extreme flood: a case from southeast Spain. *Earth Surface Processes and Landforms*, 9, 267–279.

Hedman, E.R. 1970. Mean annual runoff as related to channel geometry of selected streams in California, United States Geological Survey. *Water Supply Paper*, 199-E, Washington DC.

Heras, R. 1973. *Estudio de máximas crecidas de la Zona de Alicante–Almería–Málaga y de las lluvias torrenciales de octubre de 1973*. Madrid, Centro de Estudios Hidrográficos. Memoria diciembre, p. 12.

Hey, R.D. 1975. Design discharges for natural channels. In R.D. Hey and J.D. Davies (Eds) *Environmental Impacts Assessment*. Farnborough, Saxon House, 71–81.

IASH 1974. Flash floods. *Proceedings of the Paris Symposium*. Paris.

Instituto Universitario de Geografía de Alicante 1986. *Inundaciones en la ciudad y término de Alicante*. Universidad de Alicante – Ayuntamiento de Alicante, 177pp.

Jennings, M.E. and Benson, M.A. 1969. Frequency curves for annual flood series with some zero events or incomplete data. *Water Resources Research*, 5, 276–280.

Kilpatrick, F.A. and Barnes, H.H. 1964. Channel geometry of piedmont streams as related to frequency of floods. *United States Geological Survey Professional Paper*, 422E, 1–10.

Lautensach, H. 1964. *Geografía de España y Portugal*. Ed. Vicens Vives, Barcelona, 814pp.

Lemeunier, G. and Pérez Picazo, M.T. 1988. Avenidas y problemas de tadules en la ciudad de Alcoy. Reunión científica internacional sobre *Avenidas fluviales e inundaciones en la Cuenca del Mediterráneo*. Centre Européen de Coordination et de Recherche et Documentation en Sciences Sociales de Viena – Instituto de Geografía de la Universidad de Alicante.

López-Bermúdez, F. 1973. La Vega Alta del Segura. Clima, Hidrología y Geomorphología, Universidad de Murcia, 288 pp.

López-Bermúdez, F., Navarro Hervás, F. and Montaner Salas, E. 1979. Inundaciones catastróficas, precipitaciones torrenciales y erosión en la provincia de Murcia. *Papeles de Geografía*, 8, 49–92.

López-Bermúdez, F. and Gutiérrez Escudero, J.D. 1983. Descripción y experiencias de la avenida e inundaciones de octubre de 1982 en la Cuenca del Segura. *Estudios Geográficos*, XLIV. Consejo Superior de Investigaciones Científicas, Madrid, 87–99.

López-Bermúdez, F., Navarro Hervás, F., Romero Díaz, A., Conesa-García, C., Castillo Sánchez, V., Martínez Fernández, J. and García Alarcón, C. 1988. *Geometría de cuencas fluviales: las redes de drenaje del Alto Guadalentín*. Proyecto LUCDEME IV. Monografías 50. Instituto Nacional para la Conservación de la Naturaleza, Murcia, 227pp.

López-Bermúdez, F., Conesa-García, C. and Alonso-Sarría, F. 1998. Ramblas y barrancos mediterráneos: medio natural y respuesta humana. *Mediterráneo*, 12/13, 223–242.

López-Gómez, A.1983. Las lluvias catastróficas mediterráneas. *Estudios Geográficos*, 170–171, 11–29.

Machí i Felici, J. 1988. Influencia de las infraestructuras lineales en los llanos de inundación. Reunión científica internacional sobre *Avenidas fluviales e inundaciones en la Cuenca del Mediterráneo*. Centre Européen de Coordination et de Recherche et Documentation en Sciences Sociales de Viena – Instituto de Geografía de la Universidad de Alicante.

Marshall, W.A.L. 1952. The Lynmouth floods. *Weather*, 7, 338–342.

Martín Vide, J. 1985. *Pluges i inundacions a la Mediterrània*. Collecció Ventall, Ketres Editora, Barcelona, 58–64.

Martín Vide, J.P., Roselló Estelrich, R., Niñerola Chifoni, D. and Gómez Navarro, L. 1993. *La avenida del 9 de septiembre de 1992 en la riera de las Arenas*. Dep. De Ingeniería Hidráulica, Marítima y Ambiental, Universidad Politécnica de Cataluña, Barcelona.

Martín Vide, J.P. 1997. *Ingeniería fluvial*. Politext, Àrea d'Enginyeria Civil. Ediciones UPC, Universitat Politecnica de Catalunya, Barcelona, 82–83.

Mateu Bellés, J.F. 1988. Crecidas e inundaciones en el País Valenciano. En *Guía de la Naturaleza de la Comunidad Valenciana*. Edicions Alfons el Magnànim. Diputación Provincial de Valencia, 595–636.

Mateu Bellés, J.F. 1989. Ríos y ramblas mediterráneos. En *Avenidas fluviales e inundaciones en la cuenca del Mediterráneo*. Instituto Universitario de Geografía. Universidad de Alicante. Caja de Ahorros del Mediterráneo, Alicante, 133–150.

Mateu Bellés, J.F. 1990. Avenidas y riesgo de inundación en los sistemas fluviales mediterráneos de la Península Ibérica. *Boletín de la Asociación de Geógrafos Españoles*, 10, 45–86.

Mimikou, M. 1987. Regional treatment of flood data. In V.P. Singh (Ed.) *Regional Flood Frequency Analysis*. Reidel, Dordrecht, 91–101.

Miró-Granada, J. 1983. Consideraciones generales sobre la meteorología de las riadas en el Levante Español. *Estudios Geográficos*, 170–171, 31–53.

Molina Sempere, C.M., Vidal-Abarca, M. and Suárez, M.L. 1994. Floods in arid southeast Spanish areas: a historical and environmental review. In G. Rossi, N. Harmancioglu and V. Yevjevich (Eds) *Coping with Floods. Proceedings of the NATO Advanced Study Institute on Coping with Floods, Erice, Italy*, 3–15 November 1992. Kluwer Academic, Dordrecht, 271–278.

Muñoz Bravo, J. 1989. Enseñanza de las avenidas históricas en la cuenca del Segura. In *Avenidas fluviales e inundaciones en la cuenca del Mediterráneo*. Instituto Universitario de Geografía de Alicante. Caja de Ahorros del Mediterráneo, 459–467.

Myers, V.A. 1969. The estimation of extreme precipitation as the basis for desing floods; résumé of practice in the United States. *Floods and their Computation*. IASH/UNESCO/WMO, 84–104.

Navarro Hervás, F. 1985. Morfoestructura y comportamiento hídrico de la rambla de Nogalte. *Actas IX Coloquio de Geografía*, T. I. Asociación de Geógrafos Españoles. Murcia.

Navarro Hervás, F. 1991. *El Sistema Hidrográfico del Guadalentín*. Consejería de Política Territorial. Obras Públicas y Medio Ambiente, Murcia.

Nouh, M. 1988. Regime channels of an extremely arid zone. In W.R. White (Ed.) *International Conference on River Regime*. Hydraulics Research, Wallingford, 55–66.

Omang, R.J., Parret, C. and Hull, J.A. 1983. Mean annual runoff and peak flow estimates based on channel geometry of streams in South-eastern Montana, United States Geological Survey. *Water Resources Investigations*, 82, 4092, Washington DC.

Osborn, H.B. and Lane, L.J. 1969. Precipitation-runoff relation for very small semiarid watersheds. *Water Resources Research*, 5(2), 419–425.

Osterkamp, W.R. and Hedman, E.R. 1977. Variation of width and discharge for natural high-gradient stream channels. *Water Resources Research*, 13(2), 256–258.

Poesen, J.W.A. and Hooke, J.M. 1997. Erosion, flooding and channel management in Mediterranean environments of southern Europe. *Progress in Physical Geography*, 21(2), 157–199.

Prudhomme, C. 1995. *Modèles synthétiques des connaissances en hydrologie. Application à la regionalization des crues en Europe alpine et mediterranéen*. PhD thesis, Université de Montpellier.

Riggs, H.C. 1974. Flash flood potential from channel measurements. In *Flash Floods Symposium*. International Association Hydrological Sciences, Publication Num. 112, 52–56.

Richards, K. 1982: *Rivers, Forms and Processes in Alluvial Channels*. Methuen & Co. Ltd, London, 358pp.

Rodríguez-Estrella, T., López-Bermúdez, F., Navarro Hervás, F. and Albacete Carreira, M. 1992. El riesgo de inundabilidad y zonación para diferentes usos del llano de inundación de la rambla litoral de las moreras. La avenida de septiembre de 1989. In F. López-Bermúdez, C. Conesa-García and A. Romero Díaz (Eds) *Estudios de Geomorfología en España*, I. Sociedad Española de Geomorfología, Murcia, 353–364.

Rodríguez-Iturbe, I. and Valdés. 1979. The geomorphic structure of hydrologic response. *Water Resources Research*, 15(6), 1409–1420.

Rodríguez-Iturbe, I., Gonzalez Sanabria, M. and Bras, R.L. 1982. The geomorphoclimatic theory of the instantaneous unit hydrograph. *Water Resources Research*, 18(4), 877–886.

Roselló, V.M. 1983. La revinguda del Xúquer i el desastre de la Ribera (20–21 de octubre 1982). *Cuadernos de Geografía*, 32–33, 3–38.

Roselló, V.M. 1985. Ramblas y barrancos: un modelo de erosión mediterránea. *Actas, Discurso, Ponencias y Mesas Redondas. IX Coloquio de geógrafos Españoles*. Universidad de Murcia, Murcia, 177–184.

Rosso, R. 1984. Nash model relation to Horton order ratios. *Water Resources Research*, 20(7), 914–920.

Segura, F. 1990. *Las Ramblas Valencianas*. Departamento de Geografía. Universidad de Valencia. Valencia, 229pp.

Schick, A.P. 1988. Hydrologic aspects of floods in extreme arid environments. In *Floods Geomorphology*. Wiley, Chichester, 189–203.

Singh, V.P. 1983. *A Geomorphic Approach to Hydrograph Synthesis, with Potential for Application to Ungaged Watersheds*. Tech. Rep., Water Resources Research Institute, Louisiana State University, 101pp.

Thornes, J.B. 1976. *Semi-arid Erosional Systems*. Occasional Paper 7. London School of Economics, London.

Vázquez de Prada, V. 1978. *Historia económica y social de España*, vol. III. *Los siglos XVI y XVII*. Confederación española de Cajas de Ahorros, Madrid, 409pp.

Ward, R. 1978. *Floods. A Geographical Perspective*. The Macmillan Press Ltd, London, 244pp.

Waylen, P.R. 1985. Stochastic flood analysis in a region of mixed generating processes. *Institute of British Geographers. Transactions* (New Series), vol. 10, no. 1, 96.

Wharton, G. 1992. Flood estimation from channel size: guidelines for using the channel-geometry method. *Applied Geography*, 12, 339–359.

Wharton, G. 1995. The channel-geometry method: guidelines and applications. *Earth Surface Processes and Landforms*, 20, 649–660.

Wolman, M.G. and Leopold, L.B. 1957. River flood plains: some observations on their formation. *United States Geological Survey Professional Paper*, 282C, 87–107.

Yair, A. and Lavee, H. 1985. Runoff generation in arid and semi-arid zones. In Anderson and Burt (Eds) *Hydrological Forecasting*, Wiley, Chichester, 183–220.

Yair, A. 1990. Spatial variability in runoff in semi-arid and arid areas. *Seminario UIMP* (multicopiado).

12 Synoptic Conditions Producing Extreme Rainfall Events along the Mediterranean Coast of the Iberian Peninsula

FRANCISCO ALONSO-SARRÍA, FRANCISCO LÓPEZ-BERMÚDEZ AND CARMELO CONESA-GARCÍA

Area de Geografía Física, Universidad de Murcia, Spain

12.1 INTRODUCTION

Environmental risks associated with ephemeral channels in the Mediterranean basin are related to the occurrence of extreme rainfall events during very short time periods. These events produce high runoff discharges, flash floods, high erosion risk on hillslopes and in channels, and floods in the main basin. The risk associated with these events is due, not only to the high precipitation volumes, but also to the high intensity and consequently high kinetic energy of the rainfall. Extreme rainfall events have been studied for many years by Spanish climatologists and physical geographers (e.g. Albentosa, 1983; Camacho, 1994; Elias Castillo and Ruiz Beltrán, 1979; Font Tullot, 1983a; García de Pedraza, 1983; García Miralles and Carrasco Andreu, 1958; Gil Olcina, 1988; López Bermúdez and Navarro Hervás, 1979; López Gómez, 1983; López Bermúdez and Gutiérrez Escudero, 1983; López-Bermúdez and Romero Díaz, 1993; Medina, 1984; Riosalido, 1994, 1998; Rivera and Martínez, 1983; Segura, 1987), and are also found in other Mediterranean regions such as the south of France (Wainwright, 1996).

The dynamic origins of rainfall events over the Iberian coast are related to global atmospheric circulation modified by geographic factors. Due to its latitudinal position, the Mediterranean basin receives precipitation generated by frontal depressions related to western Atlantic flows, convective rainfall events associated with eastern flows and cyclogenetic processes generated in the Mediterranean basin itself. The latter event is more common in spring and in autumn.

In the past only daily rainfall data from very sparse networks of rain gauges were generally available, but now a huge amount of information is becoming available for the Mediterranean coast of the Iberian Peninsula, including SAIHs (Automatic Hydrologic Information Service) for main basins, weather radar and diverse rain gauges. The integration of all these systems contributes to the study of the spatial and temporal variability of rainfall events in the Mediterranean (Camarasa, 1988, 1991).

This chapter presents a review of synoptic conditions that have produced extreme rainfall events in the Iberian Peninsula, however it is also necessary to take into account features that

Dryland Rivers: Hydrology and Geomorphology of Semi-arid Channels. Edited by L.J. Bull and M.J. Kirkby.
© 2002 John Wiley & Sons, Ltd.

occur at a lower scale than the synoptic. These include mesoscale convective complexes, multi-cell storms and supercell storms that affect smaller areas but with higher average intensity rainfall. Finally some case studies of extreme rainfall events in different regions of the Spanish Mediterranean are analysed to highlight typical situations.

12.2 SYNOPTIC CONDITIONS

Previous work has identified the most important synoptic conditions that encourage the development of severe thunderstorms that result in extreme rainfall events (Albentosa, 1983; Camacho, 1994; García de Pedraza, 1983; García Miralles and Carrasco Andreu, 1958; Gil Olcina, 1988; López-Bermúdez and Navarro Hervás, 1979; López Gómez, 1983; López-Bermúdez and Gutiérrez Escudero, 1983; López-Bermúdez and Romero Díaz, 1993; Medina, 1984; Riosalido, 1994, 1998). These are as follows:

1. At 500 Hpa level (5500 m), low circulation indices of Rossby waves leads to polar low troughs or even cut-off lows to the Iberian Peninsula, and to the eastern Atlantic close to Portugal.
2. At the surface level, a low cell appears over Morocco or the Gulf of Cádiz. This low produces the incursion of warm air from North Africa to Spain from the south or southeast, crossing the Mediterranean Sea. Other surface conditions include easterly winds or cyclogenetic activity over the Mediterranean Sea.

12.2.1 Conditions at 500-Hpa Level

The most relevant element at 500 Hpa (5500 m) is westerly flow. This tends to follow a wave-like motion from west to east. These kinds of waves are called Rossby waves and can appear in two stable states: first, western straight and high-speed (more than 150 km h^{-1}) flow, called high-index flow; or, secondly, as high-amplitude, short wavelength and low-speed (70–150 km h^{-1}) waves, known as low-index flow (Haltiner and Martin, 1990).

When Rossby waves adopt a low-index pattern, meridian circulation is more important than zonal circulation, so there is a very important meridian heat interchange that produces penetrations of warm air from North Africa and polar or arctic cold air to the Mediterranean. On the other hand, low-index circulation introduces low-pressure troughs over the Iberian Peninsula and eastern Atlantic Ocean. These low troughs can contain low-pressure cells and may develop a cut-off low with cold air inside. These conditions result in a strong temperature gradient and the introduction of instability at the surface. Polar troughs usually develop a divergent flow in their eastern sector that encourages the development of a low-pressure cell on the surface beneath it.

Polar troughs can show a regressive behaviour because they have a lower relative speed with respect to the global westerly flow, caused by the conservation of its absolute speed. Hence the trough, and a cut-off low inside it, can remain over the Gulf of Cádiz for a long time, or may move slowly across the Peninsula or Strait of Gibraltar. A low cell often appears inside the polar trough, generally at the latitude of Galicia (Medina, 1984). This cell moves to the bottom of the trough. If the base of the trough is located over the Gulf of Cádiz, a clear diffluence over southeast Spain results. If the polar trough is centred over the Peninsula, the diffluence is not as clear.

Palmen (1949) and Palmen and Newton (1969) have shown that the deformation and deepening of polar troughs in the western circulation concludes with the formation of cut-off lows. The polar air mass within the trough loses contact with the original polar air mass and a closed low develops. This type of cut-off low tends to have an erratic behaviour with easterly and southerly movements that follow surface winds or the average wind between the surface and 500 Hpa.

Llasat (1987) has shown that the Palmen–Newton cut-off process produces most of the cut-off lows affecting the Iberian Peninsula. Llasat (1991) distinguishes between cut-off lows originating from meridian troughs (65%) and Z-shaped retrograde troughs in combination with high-pressure cells over Britain or the Gulf of Vizcaya. Finally, Gallego (1995) distinguishes between Z-shaped troughs, S-shaped troughs and V-shaped (meridian) troughs.

A wide range of Spanish meteorologists and climatologists (e.g. Capel Molina, 1981; Castillo Requena, 1978; Martin Vide, 1989; Llasat, 1991; Capel Molina, 1989a) have taken account of previous attempts to define cut-off lows and some (Fontaine and Portela, 1959; Reuter, 1954; Zimmerschied, 1954; Medina, 1976; Bessé et al., 1979; Hardman, 1983) have proposed that cut-off lows have the following characteristics:

- A cut-off low is a closed low cell that coincides with a minimum temperature in the upper and middle troposphere.
- A cut-off low may reflect surface conditions even though pressure at the sea surface is not well defined and an anticyclone may develop.
- Relative topography 500/100 Hpa must reflect a closed equiscalar line.
- The observed maximum temperature difference is 14°C and the minimum is 0°C.
- Cut-off lows can extend from the 300–100 Hpa level up to 700–800 Hpa and the related hazard grows with this thickness.
- The average diameter is about 3–8 degrees in latitude
- Sometimes the eastern sector of a cut-off low can be linked with a warm anticyclone that may be a blocking system.

Llasat (1991) described a model of cut-off low positions and affected areas. If a cut-off low appears over the Gulf of Cádiz, the areas affected are mainly to the south (Málaga-Almería), southeast (Murcia) and east (Valencia) of the Mediterranean coast. If the cut-off low is located over the Atlantic coast of Morocco, the affected area covers the south and east Mediterranean, and is hence further north than the previous situation. When the cut-off low is located over the Straits of Gibraltar or Alborán Sea, the affected area covers the east Mediterranean and the northeast (coast of Catalonia). Finally, if the cut-off low appears inside the Peninsula, the affected area covers Catalonia and the eastern Pyrenees.

12.2.2 Surface Conditions

The latitudinal location of the Iberian Peninsula and its position with respect to the Atmospheric Global Circulation centres means it is situated in a transition area between the temperate and humid climates of northern Europe. Western winds, polar cyclones, and the arid climates of North Africa dominate the area, which are in turn controlled by subtropical high-pressure cells. Due to the seasonal motion of these atmospheric circulation centres, the annual precipitation distribution is clearly seasonal. During winter, the Iberian Peninsula is dominated by a westerly circulation that produces rainfall volumes similar to those of temperate climates (at least in the western sector). However, during summer most of the territory, except the northern Spanish coast, is affected by subtropical anticyclones such as occur in arid zones. These conditions lead to a critical fall in precipitation during summer months.

The Mediterranean coast is less affected by this type of movements because the oceanic air masses that cross the Peninsula during wet seasons usually arrive at the Mediterranean after dropping rain, especially where peninsular ranges lead to vertical motion of the air masses. Sometimes, Atlantic low-pressure cells arrive at the western Mediterranean by crossing the Strait of Gibraltar from the Gulf of Cádiz and these produce rainfall events of some intensity on the seaboard. Air masses that produce precipitation on the Mediterranean seaboard usually originate over the Mediterranean or Gibraltar. This fact introduces differences between Atlantic areas with maximum precipitation in winter and Mediterranean areas with maximum precipitation in autumn (Font Tullot, 1983b; Martin Vide, 1987; Albentosa, 1991).

This means that most of the key surface situations that develop into high intensity rainfall events are those with a clear easterly wind that arrive in eastern or southeastern Spain. This has been highlighted by Spanish climatologists (e.g. García de Pedraza, 1990; Gallego, 1995). However rainfall events in northeastern Spain, and some in southeastern regions, are more related to cyclogenetic processes in the Mediterranean (Jansa-Clar, 1988; Gallego, 1995).

Eastern and Southeastern Conditions

Easterly wind conditions are mainly driven by an anticyclone over Europe and one or more thermal lows over North Africa. Southeasterly flows appear because of the combination of deep low-pressure cells over the Gulf of Cádiz that reflect a polar trough or cut-off low at 500 Hpa. Under these conditions, the 850 Hpa surface promotes a high-temperature gradient between the Atlas Range in North Africa and the Iberian Peninsula, with a massive penetration of warm air at the lower layers of the troposphere. This penetration is associated with low-pressure cells that developed in the north of Algeria and are related to the extension of easterly and south easterly winds to the Mediterranean coast of the Iberian Peninsula. At the coast, the temperature gradient is extremely strong and coastal ranges with a southeastern aspect tend to be affected by severe rainstorms.

Cyclogenetic Processes in the Mediterranean

Cyclogenetic processes in the Mediterranean are mainly a consequence of the modifications that specific geographic conditions introduce to the temperature distributions of air masses. The Meteorological Office (1962) stated that 91% of the Mediterranean cyclogenetic situations are developed in situ and are related to the leeward effect of the Alps. The Alps increase the pressure gradient of low cells, which affect the incoming colder air masses in comparison with the Mediterranean air mass. These cyclogenetic situations can increase the development of easterly flow from the Mediterranean Sea to the coast.

Following the work of Radinovic (1976), Jansa-Clar (1988) demonstrated that the western Mediterranean is a highly cyclogenetic area, where low cells originate or are deepened. Such processes increase during the cold season and can disappear during summer. The areas where this occurs most frequently are the gulfs of Genova-Baleares, León and Algeria.

For many years Spanish climatologists have debated the existence of a Mediterranean air mass with its own characteristics due to inputs of polar air. Hence, each new air mass will be colder than the pre-existing one with *Mediterranean characteristics*. The contact of two different air masses produces a frontogenesis process that leads to the presence of a Mediterranean front 52 days per year (Jansá-Guardiola, 1966). It is important to stress that the boundary of the air mass runs over the Mediterranean Sea, and the angle between its direction and the dominant orographic direction is relevant when explaining the distribution and intensity of rainfall along the coast. Gallego (1995) pointed out several south, southeast, east and northeast flow situations, when investigating the relationship with severe precipitation events.

Southeast conditions usually originate from a low cell located over the Gulf of Cádiz, the Alborán Sea or Morocco, with or without a high-pressure cell located over the northern Alps. Under these conditions, air masses that affect the Mediterranean sectors of the Iberian Peninsula come from North Africa or the Atlantic Ocean. Easterly conditions can be related to cyclogenetic processes over Algeria, with or without another low cell over Gibraltar or Sicily, or simply with a large thermal low located over North Africa with one or two high-pressure cells over Europe. Under these conditions air can arrive at the coast with cyclonic or anti-cyclonic curvature. In the first case air proceeds from North Africa and in the second from the Atlantic Ocean after a run over Europe. The length of the Mediterranean run of these air masses, and in consequence the amount of water vapour that they contain, is the key factor in explaining the amount of rainfall that they produce (Miró-Granada, 1976, 1983; Font Tullot, 1983a, 1983b; López Gómez, 1983; García de Pedraza, 1983; Gil Olcina, 1989). Northeasterly conditions appear with low-pressure cells over the Gulf of Geneva, Italy, or Tunis, and an extended high-pressure cell on the Atlantic Ocean extended to the Iberian Peninsula.

12.3 MESOSCALE SITUATIONS

Regardless of synoptic conditions, convective phenomena developed at a more local scale can produce severe storms. A storm can be considered as a spectacular and violent manifestation of atmospheric convection. It may be composed of one or more cells from short duration convective cells to highly organised multicell storms. Single-cell storms are rarely large (usually 5–10 km), are short-lived in time (less than an hour) and tend to be related to afternoon convection maximum. They have an evolution with three stages: cumulus stage, mature stage and dissipation stage (Sumner, 1988; Llasat, 1991).

The cumulus stage begins with the updraft of wet and warm air. This process develops some cumulus clouds that join to produce an isolated cell. The warm updrafts ascend all over the area below the cloud giving out latent heat to the air, which rises in temperature and increases the turbulent processes, but at the same time decreases surface pressure. If the cloud arrives at a negative temperature level, ice particles begin the Bergeron processes. The mature stage (15–30 minutes) commences when ice particles become heavier and updrafts can no longer keep them in the air. This precedes precipitation and downdrafts, which eventually results in a pressure increase and divergence at surface level. When rainfall reaches the surface, intensity values may be extremely high during the first 5–15 minutes, after which rainfall intensity stabilises at lower values. During the dissipation stage (about 30 minutes) downdrafts are the dominant air motion in the cloud and convective cells decrease in size. In addition rainfall evaporation at the surface causes a decrease in air temperature and the end of the turbulent convective processes.

More complex storms can be self-preserving, and hence become larger and longer-living. Austin and Houze (1972) established a hierarchical pattern of cloud systems by distinguishing between:

- synoptic rain areas
- large mesoscale precipitation areas
- small mesoscale precipitation areas (100–1000 km^2)
- convective cells (10–30 km^2) that are quickly generated and dissipated.

Sumner (1988) discusses systems with a positive feedback that result in long-duration severe local storms, and classifies them in four categories:

- Multicell storms
- Squall-line storms
- Supercell storms
- Mesoscale convective complexes.

The same author stated that there is a continuum between storm types, so it is difficult to draw boundaries.

12.3.1   Multicell and Supercell Storms

Static storms will rapidly dissipate by ending convective movement due to downdraft, which results from the precipitation process. If rain falls from a convective thermal area it can input wet air to other thermals, developing daughter cells on the periphery. Such storms may propagate subject to local winds. Often there is a cluster of four or more different cells with different evolution stages. Chisholm and Renick (1972) stated that multicell storms could include up to 30 or more independent cells. The lifetime of a single cell in a multicell storm is about 30 minutes (Browning, 1962). Multicell storms propagate in discrete jumps as new cells, moving to the right of the airflow in the northern hemisphere, are generated. The related shear keeps the downdraughts isolated with respect to the updrafts, ensuring continuation of the storm (Sumner, 1988).

Supercell storms were defined by Browning (1962) as very intense rainstorms characterised by a strong influence of shear in the upper layers. They usually appear at the mature stage of a multicell storm and are more complex with a highly organised inner circulation system. The conditions necessary for them to appear are:

- great instability
- average wind conditions below the cloud level
- considerable wind shear

Supercells appear as small cyclones with a general cyclonic rotation of updraft and divergence at the top of the cloud. They can have different shapes from circular to elliptical with dimensions of 20–30 km in the horizontal and 12–15 km of vertical thickness, with a long anvil extended 100–300 km in the wind direction. Updrafts are strong, up to 25–40 m s^{-1}. Finally, they are associated with significant hail production.

12.3.2   Mesoscale Convective Complexes

These are generally larger than other convective precipitation systems and can be considered as assemblages of amorphous clusters of storm cells covering areas up to thousands of square kilometres. Maddox (1980) classified them on the basis of scale, morphology and duration. They are more complex and less clearly understood than smaller cloud systems.

Sumner (1988) distinguished between amorphous clusters of cumulus form clouds often associated with small, synoptic scale, low-pressure areas (using the example of the Iberian Peninsula thermal low), and the true mesoscale convective complexes (MCCs) as defined by Maddox (1980). These are clusters of convectionally driven storms which begin as single thermally driven cumulonimbus, that are larger than individual thunderstorms by more than two orders of magnitude. The Maddox (1980) size criteria seem overly restrictive to both meteorologists and climatologists (Zipser, 1982; Browning and Hill, 1984; Riosalido et al., 1989) since they exclude certain mesoscale convective features that are similar in behaviour but smaller in size. Zipser (1982) proposed the term Mesoscale Convective Systems (MCSs), defining them as a structure of clouds within a mesoscale convective system.

MCCs and MCSs begin with a set of individual thunderstorms developed in response to localised convection, perhaps linked to topographic features. Thunderstorm growth leads to closer grouping and cell borders disappear. This way a large and relatively homogeneous area of intense precipitation is developed. The dissipation stage begins with the end of development of new convective elements due to surface cooling and upper or middle level dissipation.

12.4 LOCAL CONDITIONS

During the second half of the twentieth century, the synoptic situations that produce extreme rainfall events were studied extensively (Font Tullot, 1983a; Gil Olcina, 1988; López Gómez, 1983; Olcina-Cantos, 1994). However, recent studies (Millán et al., 1994) have demonstrated that local and regional processes may be more important than was previously thought. These processes can introduce relevant changes in the general precipitation patterns.

Recent studies have shown that the climate of the Mediterranean basin is not only the result of overlapping temperate and arid climate processes but is also due to the interaction of subtropical anticyclones and temperate western circulation that creates a climate with specific characteristics and processes. Research developed at the Centre for the Mediterranean Environmental Studies (CEAM) has proposed that the origin of extreme rainfall events is caused by (i) surface processes associated with the Mediterranean secondary air mass, (ii) the temperature gradients between the air and the sea surface and (iii) air masses entering the basin (Estrela and Millán, 1994). These may result in the development of a blocking system.

On the other hand, it is very important to account for local wind systems in coastal regions, and these can be important in the development of severe precipitation events. Local winds are mainly sea breezes, valley winds, and convective cells that originate in a heat accumulation by south and southeast aspect slopes. Jansá-Clar (1989) proposed that although sea breeze phenomena have a daily cycle, if the Sun-exposed surface is adequately large, a heat accumulation may produce a convective cell which will propagate with a vertical motion. This process means a pressure decrease and a suction effect on the air situated over the sea surface.

All these phenomena, acting together, especially under appropriate synoptic conditions, produce air instability and the development of large convective storms. In order to study these instability processes, it is necessary to take into account a large number of meteorological variables such as low troposphere humidity (Chalker, 1949); the temperature at different levels (Showalter, 1953); dew point (Sartor, 1962); and the shear between the wind at different levels (Llasat, 1991).

It is now clearly established that there is a strong connection between atmospheric processes and sea processes. Sea–atmosphere energy transfers are a key factor in understanding atmospheric thermodynamics (Coantic, 1979). The sea surface has a very low albedo, so it absorbs most of the solar radiation. Quereda (1982) established, by theoretical estimations, that the sea surface close to Castellón absorbs 0.25 cal cm^{-2} min^{-1} on average per year (Q_i). This energy is compensated by a sea–atmosphere transference (Q_a) in the form of latent heat flux (Q_e) and convective energy (Q_h). Figure 12.1 shows the monthly evolution of this balance. There is a net heat flux from sea to atmosphere from September to March, but October shows a large transfer

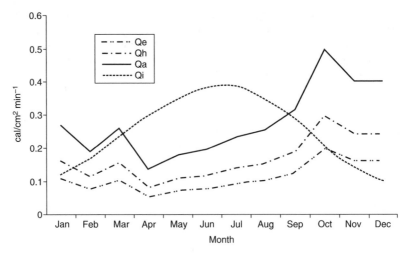

Figure 12.1   Energy balance at the interface sea–atmosphere close to Castellón (after Quereda, 1982)

to compensate the extreme temperature gradient between a warm sea after the heat accumulation during summer months and the first cold air masses that arrive at the Mediterranean.

Heat transfer and thermal convection reinforce cyclogenetic processes in the Mediterranean, developing an intense instability when the gradients are greater than 4–6 degrees. This situation produces convective cumulonimbus and could even have an influence on the synoptic scale processes by deepening the low cells that arrive in the basin. Quereda (1982, 1983) studied 20 blocking situations with cut-off lows in the western Mediterranean. These showed a significant heat accumulation in the western Mediterranean during the previous months. Moreover, important negative anomalies were registered in the northern Atlantic almost three months before. From these studies, it was stated that blocking situations originated by these cut-off lows can decelerate Rossby circulation, reinforcing the blocking.

In a study of rainfall events in the Murcia region from 1947 to 1993 (with a time resolution of 6 hours), Alonso-Sarría (1995) found that synoptic conditions significantly affect rainfall event volumes but not the rainfall intensity. Rainfall intensity seems to be more influenced by the time of the year when the rainfall event occurs (and the associated temperature gradient between sea and atmosphere).

Finally, it is also necessary to take into account the effects of relief and topography. A huge amount of work has been published on this topic, but it is especially important on the Mediterranean side of the Iberian Peninsula. All ranges parallel to the coast introduce an uplift effect on wet air masses driven from the sea by easterly winds. Air circulation around and over topographic barriers can be very complex. Air can move around areas of high relief or stay anchored, producing high-duration and high-intensity rainfall events if the air masses are saturated.

12.5   CASE STUDIES

Many examples of severe storms could be presented, but we have selected ten of the most severe ones, attempting to provide a representative sample for the Spanish Mediterranean coast. The

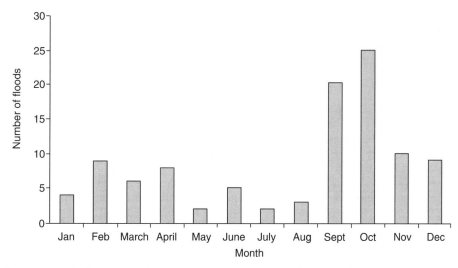

Figure 12.2 Big floods in the Segura basin since 1258 (after López-Bermúdez, 1973, and including recent data)

examples are in chronological order since there is more available information for the later storms. Figure 12.2 shows the number of big floods in each month since the thirteenth century, although care should be taken during analysis because of the low quality of ancient data. In the twentieth century, a large number of intense storm events have affected the Mediterranean coast of the Iberian Peninsula. Table 12.1 shows the frequency of extreme events per month in the same area. A clear movement of the maximum (from November to October) can be seen moving from Catalonia to Andalucía. These data should also be used as qualitative data rather than quantitative.

The list below shows the dates of the chosen events and the main areas affected; southeast Spain means Murcia and eastern Andalucía.

- September 1971   Catalonia
- October 1973   Southeast Spain

Table 12.1   Frequency of extreme events per month (after López-Bermúdez and Romero Díaz, 1993)

	Catalonia	Valencia	Baleares	Murcia	Andalucia
January					1
February					1
March			2		
April					
May		1			
June					
July				1	
August			1		
September		3	1	4	
October	4	8	5	5	12
November	14	10	3	5	3
December			2		

- October 1977   Valencia, Almería and Murcia
- October 1982   Valencia and north of Segura basin
- November 1982   Catalonia and south Spain
- November 1983   Catalonia
- July 1986   Murcia
- November 1987   Valencia and southeast Spain
- September 1989   Valencia and Murcia
- August 1996   Pyrenees

These events have been selected because of empirical relevance in terms of rainfall volume and intensity. The analysis is based on synoptic-scale evolution, on the development of thermo-convective cells, and on the resultant rainfall records. These are related to previous presented theoretical typology when possible. However such classification is mainly based on recent developments in mesoscalar analysis which is not available for most of the events recorded.

12.5.1   September 1971

A cut-off low began to develop on 18th September and moved southwestwards from Ireland, decreasing its latitude until 41°N, centred over Castilla on 20th and 21st and disappearing on 22nd near Galicia. This cut-off low was very deep, developing from the 850-Hpa level to the 100-Hpa level. At the surface, anticyclonic conditions caused a close contact between the Mediterranean air mass and the sea surface. This air mass therefore got warmer and wetter before being driven to the coast by an easterly airflow (caused by a surface low cell that appeared between the Mediterranean coast and Algeria). This low drove the Mediterranean air masses towards the Catalonia coast. The forced uplift of this air mass due to the coastal ranges of Catalonia, the presence over the continental area of a polar maritime air mass and the intense temperature vertical gradient led to high instability.

The associated rainfall event affected mainly Barcelona. It began on the 19th, and concluded on 22nd with a maximum intensity peak on the 20th. Massive rainfall amounts were recorded: 233 mm in Figueras and 308 mm in Esparraguera. On the final day of rainfall the Llobregat, Tordera and Güell rivers were in flood, causing major damage.

12.5.2   October 1973

This event was the cause of the catastrophic flash flood in the Rambla de Nogalte (Murcia), resulting in massive channel change, destruction and human loss of life (Heras, 1973; Capel-Molina, 1974). On 16th and 17th October the high-pressure wave that was a continuation of an Azores anticyclone (at 500 Hpa and located around 40°W), approached the Iberian Peninsula, increasing its amplitude as it moved. Between this wave and the Peninsula, a polar trough located over 20°W drove cold air to the Iberian latitude. Temperature at this time reached −24°C.

At the surface, a deep low over the Atlantic Ocean, southwest of Ireland, had a well-developed cold front that crossed the Iberian Peninsula driving a very unstable cold air mass. The instability was due to the cyclonic curvature of winds and to the temperature gradient due to basal warming. This instability resulted in some rainfall events and the arrival of a very unstable air mass at the Alboran Sea.

On the 18th, the trough suffered a cut-off, isolating a low over the Gulf of Cádiz with a high-pressure penetration over the Peninsula. This, combined with the fact that a thermal low was

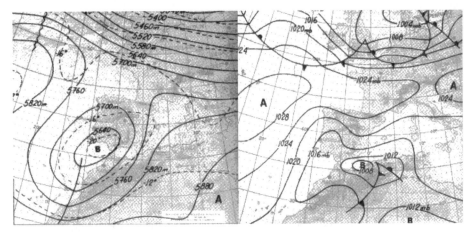

Figure 12.3   500-Hpa level and surface conditions on 18 October 1973 at 12:00 GMT

centred over North Africa, resulted in easterly and southeasterly flows that brought a tropical continental air mass over the Peninsula. This was wet and unstable because of its travel over the Mediterranean Sea. Convergence of this air mass with the polar cold air mass produced a strong cyclogenetic process over the Alborán Sea. The generated low produced updraft movements that were enhanced by the cold low at the 500-Hpa level and the coastal relief. This situation continued during the 19th with the deepening of the Alborán Sea low (Figure 12.3).

The updrafts were the cause of a strong cumulonimbus development and many dispersed cells appeared all over southeast Spain. Cold air at high levels contributed to the Bergeron–Findeisen process. The result was very intense storm events with total rain exceeding 100 mm. The most intense rainfall events were located on slopes with a south or southeast aspect, which faced the eastern winds. The unstable air masses increased their vertical motion and turbulence because of the presence of topography and due to their high relative humidity, a light uplift was enough to saturate the air.

12.5.3   October 1977

This event affected the southeast and east sector of the Peninsula (Almería and Valencia provinces, along with the Murcia region) and was described by Capel-Molina (1977). The event began with a very wide polar trough affecting the North Atlantic and western Europe. This trough developed and deepened while moving to the east, from the 21st to the 23rd. On the 22nd at 0 hours GMT, the 500-Hpa surface level showed a −15°C isotherm located over Madeira, Lisboa and San Sebastian. The polar trough appeared from Iceland to the Canary Islands and affected most of the northeastern Atlantic. At the surface a cold front crossed the Peninsula, producing low-intensity rainfall events.

On the 23rd the polar trough affected all of the Iberian Peninsula and the north of Europe, with its axis located at 10°W. Cold air continued its movement to the south, and Gibraltar registered a temperature of −19°C at the 500-Hpa level. At the surface, pressure gradients were very low. On the 24th, evolution occurred very rapidly at a high level. The polar trough underwent a retrogression process, eventually becoming a cut-off low centred over Casablanca. The western circulation changed to a two-branched pattern: one crossed north Europe and the other

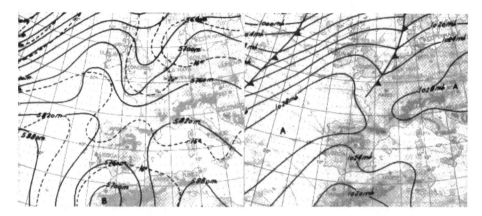

Figure 12.4   500-Hpa and surface situation on 24 October 1977 at 00:00 GMT

had a cyclonic movement around the low centred on the Iberian Peninsula. The presence of the upper level low initiated a surface cyclogenetic process. Convergent flows of warm air from the south or southeast increased the thermal gradient and marked the beginning of rainstorm activity (Figure 12.4).

On the 25th the cut-off low continued over North Africa but moved inland, hence completely affecting the southern Peninsula. At the surface the low continued driving southeasterly winds but with a higher maritime motion due to the low movement and the presence of a strong anticyclone in the north. The maximum rainfall occurred in María with 100 mm, Torrente with 140 mm and Villalonga with 180 mm.

On the 26th the low reduced in area but became deeper. Near the Alboran Sea, the temperature reached −20°C at 500 Hpa, increasing thermal gradients. At the same time surface pressure gradients were quite low, resulting in an air accumulation and uplift by the coastal ranges. All these factors led to a condition of instability and to the occurrence of extreme rainfall events in the southern and southeastern Iberian Peninsula (Tijola, 207 mm; Serón 224 mm; Uleila, 208 mm; Alcira, 215 mm), however these were not as extreme as the October 1973 events.

12.5.4   October 1982

This event was one of the most important rainfall events in Spain during the second half of the century because it resulted in the failure of the Tous dam in the Turia River basin (Valencia). As a consequence, much research was carried out (e.g. Gil Olcina, 1983a, 1983b; Morales Gil et al., 1983; Quereda, 1983; López-Bermúdez and Gutiérrez Escudero, 1983). From a physical point of view, this event stands out because of a high heat accumulation in the Mediterranean Sea in the previous months, reaching an anomaly of +8.1°C due to strong and long-lasting anticyclonic conditions (Quereda, 1983). During August and September there was some Arctic air expansion to correct the disequilibrium, however the most important anticyclone arrived on 16th October as a deep trough with a cold core over 10–15°W. At the same time a high-pressure area was located over the Mediterranean and a deep low appeared at the surface over Great Britain. All of the Iberian Peninsula was affected.

On 18th October there was a change at 500 Hpa and the polar trough took a NW–SE direction, reducing its circulation index. A clear cut-off low appeared over the centre of the Peninsula, bifurcating the western flow circulation. At the surface another low pressure cell with a Mediterranean front (partly due to a warm air intrusion from Africa), was over the Peninsula. On the 20th the cold low appeared over the Gulf of Cádiz and Gibraltar with a penetration of Saharan air from the southwest. A trough developed from North Africa to the eastern Peninsula, inside which a cyclogenetic event was occurring. On the 20th advection caused the penetration of the warm Saharan surface air and the development of a cyclogenetic process from North Africa to the eastern coast of the Peninsula. Saharan air suffered a föhn effect over the Atlas range, arriving at the Alborán Sea adiabatically heated and with a high evaporative capacity. Hence a warm and wet air mass arrived over eastern Spain collided with the Iberian mountains and reached the cold air mass located at high levels. These conditions resulted in a mesoscale convective system with a positive feedback because of the arrival of warm and wet air from the sea. The low maintained itself until the 22nd over the Algerian coast (Figure 12.5).

Rainfall started on the 19th over the eastern Peninsula, but the highest rainfall intensity values were recorded on the 20th. The event lasted until the 23rd over Catalonia. The total amount of rainfall during this event reached very high values: Cofrentes, 480 mm; Enguera, 545 mm; Bircop, 476 mm; Jalance, 476 mm; Ontur, 196 mm; and Hellín, 170 mm.

12.5.5   November 1982

This event affected Catalonia and the south Pyrenees area, along with the Gibraltar and Malaga coast and has been described by Olcina-Cantos (1994) and Llasat (1991). The main reason for this event was the crossing of a cold front and the penetration of warm and wet air from North Africa. Over the coastal ranges of Andalucia there was an important development of cumulonimbus and rainfall resulted due to multiple factors. An antecedent period of stability lasting several days had caused the formation of a warm and wet Mediterranean air mass, and sea surface proximity, coastal relief and aspect (southeast) all contributed to a strong uplift of air masses. A large number of relatively stationary isolated multicell rainstorms appeared all

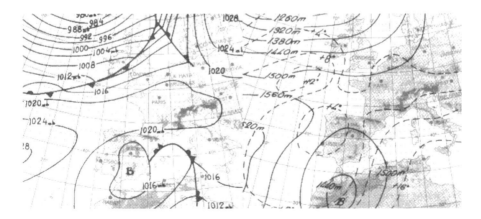

Figure 12.5   500 surface situation and 500-Hpa level on 19 October 1982 at 00:00 GMT

over the Mediterranean coast. This resulted in the most severe rainfall event in Catalonia and the south Pyrenees since 1940 during 6th, 7th and 8th November 1982 (340 mm in Pobla de Lillet, 556 mm in La Molina and 576 mm in Valcebollere). The intense rainfall and the massive wind velocity (reaching $170\,\mathrm{km\,h^{-1}}$) produced strong waves with catastrophic consequences to people and infrastructure. Rainfall affected all of the Peninsula for 48–60 hours.

12.5.6   November 1983

This event, described in Llasat (1991), began on 6th November with a low-pressure cell located over the Gulf of Cádiz, which finally moved to the north of Algeria. At the same time, a large anticyclone was centred over eastern Europe, with maximum pressures of 1025 to 1030 Hpa. The event was caused by the wind turning from westerly to southeasterly in the upper and middle troposphere due to a cyclonic circulation. Warm air then invaded the Iberian Peninsula at all tropospheric levels and on the 7th there was convergence at the 700-Hpa level over Catalonia (Figures 12.6 and 12.7). Sea surface temperatures were also higher than continental temperatures (28.5°C between Algeria and the eastern Spanish coast).

The result was heavy rainfall events on 6th and 7th November. The highest rainfall volumes were recorded in: Barcelona with 347 mm (220 mm in 24 hours); Pobla de Lillet, 340 mm; La Molina, 556 mm; Valcebollera, 576 mm; Cabdella, 323 mm; Vilaller, 237 mm; and Senet, 267 mm. More than 100 mm were also recorded in Gerona province (values of 236 mm were reached). The most severe events occurred at river outlets and not in the headwaters. These events were the cause of strong floods in the Llobregat, Besos, Fluvià and Ter rivers.

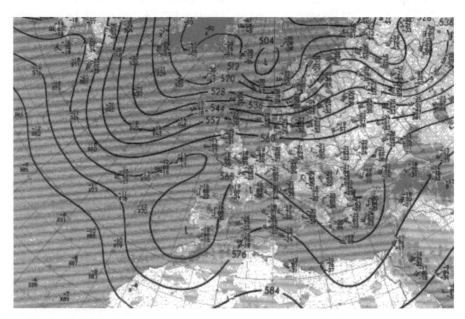

Figure 12.6   Situation at 500-Hpa level on 6 November 1983 at 00:00 GMT

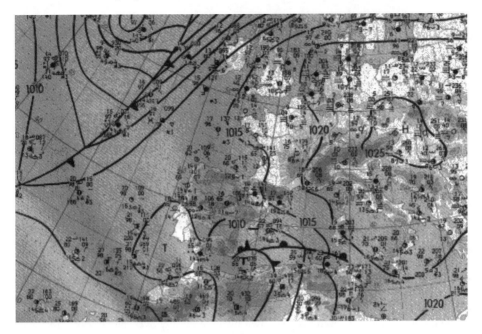

Figure 12.7   Surface situation on 6 November 1983 at 00:00 GMT

12.5.7 July 1986

Summer is a period of abundance of cut-off lows in Spain, however, they are usually weaker than those developed in other seasons and surface conditions are not suitable to induce instability. Yet sometimes conditions are different and convective processes can develop during summer. On 20th and 23rd July a deep polar trough was located between the Azores and Portugal, arriving at the latitude of the Canary Islands. At the same time, a warm air wave came from North Africa that affected most of the Iberian Peninsula promoting thermal lows. Southeast Spain had the maximum temperature values of the summer. The Azores high-pressure cell moved to the north driving a polar air mass to the Iberian Peninsula in the upper troposphere. On 24th to 26th July a V-shaped polar trough evolved to a Z-shaped trough, and a cut-off low from the polar trough crossed the southern sector of the Peninsula. This resulted in great instability in the southeast because of the temperature gradient and divergent winds at the eastern sector of the low. Temperatures dropped because of the vertical motion of the air.

Rainfall exceeded 100 mm in 24 hours in the south of Albacete province and north of Murcia as a result of continental thermo-convective processes caused by the great heat accumulation during the previous days. Coastal areas were protected because of their lower temperatures. On 27th July the low grew weaker over the western Mediterranean.

12.5.8 November 1987

During this event the interaction between different elements and scales finally resulted in the formation of convective systems, one of which was stationary for 36 hours over the Cape of La Nao (Valencia). The result was extreme rainfalls reaching 817 mm in 24 hours on 7th November

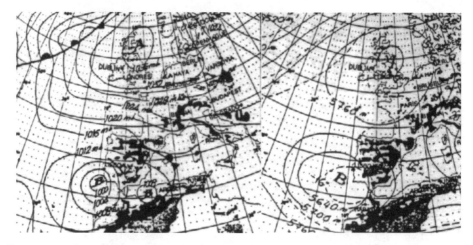

Figure 12.8   Surface situation and 500-Hpa level on 3 November 1987 at 00:00 GMT

at the Oliva raingauge. This is the highest daily rainfall value ever recorded in Spain. Other severe rainfall discharges occurred in San Javier, 304 mm; Sumacarcel, 520 mm; Pobla de Duc, 790 mm; and Denia, 380 mm.

The synoptic conditions were characterised by a cut-off low over the Gulf of Cádiz and a polar trough over eastern Europe with a southwesterly alignment. Both systems produced a very stationary synoptic situation, but between them a high-pressure wave appeared from North Africa. This high pressure resulted in the penetration of warm air from the south-south-west, clearly visible at the 850-Hpa level. During the last days of October, some cold polar troughs arrived to 40°N and on 29th October a cut-off low appeared inside one of these troughs. At the surface a deep low (reaching 996 Hpa in its centre) over the Gulf of Cádiz reflected conditions at higher levels (Figure 12.8). During the 30th the cut-off low became deeper and a branch of the western low crossed northern Africa driving Saharan air to the western Mediterranean, resulting in a strong convective mesoscale system over the sea.

However Riosalido et al. (1989) do not think that these synoptic conditions were enough to account for the volume and intensity of the rainfall event and proposed that the development of mesoscale systems were the reason for such extreme intensities. On 3rd and 4th November several mesoscale convective systems grew and moved along the Mediterranean coast in a northeasterly direction. These were especially concentrated at the Cape of La Nao. On 3rd November the precipitation maximum was located over Valencia and Alicante and the mouth of the River Segura on the 4th.

As a result of the motion of the mesoscale convective systems, one remained almost stationary over the Cape of La Nao. This was caused by the local relief acting as a barrier that allowed the strong thermal gradient between warm air from the Sahara and the colder Peninsular air to produce a stationary front. This system was reinforced by others that moved in a northeasterly direction due to the constant input of warm Saharan air wetted in the Mediterranean.

12.5.9   September 1989

On 2nd September the Azores anticyclone swung strongly to the northeast. This caused a polar air mass to move towards the Cape of San Vicente encouraging the polar air mass to have a

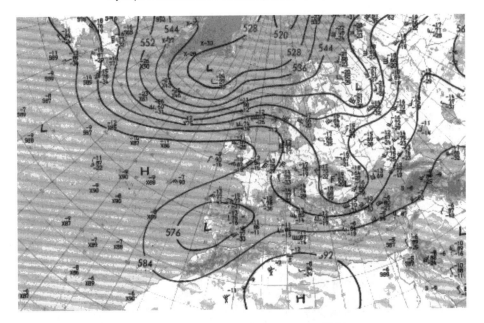

Figure 12.9   500-Hpa level situation on 6 September 1989 at 00:00 GMT

meridian direction and a southward motion, hence encouraging a polar trough to reach
Madeira. In the upper atmosphere, a western circulation showed a very fragmented pattern
(Figure 12.9).

On 4th September a polar perturbation at 500 Hpa appeared over the Cape of San Vicente.
At the surface level, a thermal low over North Africa generated eastern winds towards the
Peninsula (Figure 12.10). On the 5th a warm front was generated by the interruption of a
continental tropical air mass from North Africa. This mass affected the Mediterranean Sea
and the eastern and southeastern Peninsula. At 850 Hpa temperatures of 16°C were reached
over Madrid. On the 6th the situation evolved quickly. The low at 500 Hpa, centred over the
Gulf of Cádiz with a temperature of −14°C, was isolated with respect to the general western
circulation, forming a cut-off low.

The diffluence at 500 Hpa over the Alboran Sea and southeast Spain produced a suction
effect on surface air masses and resulted in the generation of a very deep mesoscale convective
system which generated intense rainfall events all over the area. During 7th September some
relative minima were developed by the suction of air from the lower layers of the troposphere
air. These relative low cells affected North Africa and the Iberian Peninsula, encouraging
continuation of the strong diffluent situation and the generation of new mesoscale convective
systems.

The highest rainfall accumulations were recorded in Tabernas, 256 mm; Verger, 242 mm;
Abarán, 230 mm; Torrevieja, 240 mm. All these data were recorded in less than 24 hours. This
event caused an extreme flash flood in the Rambla de las Moreras in Murcia due to rainfall
intensities higher than 150 mm in 1.5 hours (Capel-Molina, 1989a, 1989b; Rodríguez Estrella
et al., 1992).

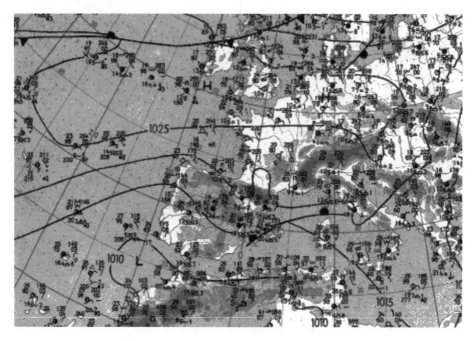

Figure 12.10   Surface situation on 6 September 1989 at 00:00 GMT

12.5.10   August 1996

During the evening of 7th August 1996, a massive storm affected the small catchment of Arás in the central Pyrenees. One hour of high-intensity rainfall produced a strong flash flood with catastrophic consequences for a camp site located on the alluvial fan at the catchment outlet (seven people died). This has been one of the most important events in recent years and is included even though it did not occur over the Mediterranean coast.

The storm was caused by the development of a massive supercell storm over the catchment that orginated due to several synoptic and mesoscalar factors. On the 6th the synoptic situation in the high troposphere showed a low cell reaching Spain from the British Isles with a trough reaching the Iberian Peninsula from the Atlantic with an easterly direction. At low levels both low-pressure cells related to those at higher levels, but the one from the UK drove a cold front that affected the north of Spain.

On the 7th the cold front was stopped by the Pyrenees, producing a massive accumulation of cold air in the troposphere ($-14°C$ at the 500-Hpa level at 12 h). At the same time a low cell at the 850-Hpa level centered over the northern part of the Peninsula and drove a northerly air flow in the eastern Peninsula area that became a northwesterly flow in the Ebro valley. It was an advection of warm and wet air coming from the Mediterranean Sea. All the northeast of the Peninsula had an instability layer at 850 Hpa and higher levels, but a stable layer was present near the ground due to an inversion layer. Hence, some areas were affected by convective processes but others were not. Finally, a reduction of wind regime in high levels of the troposphere caused the limited mobility of the developed convective systems causing the systems to discharge all their water content over the same place.

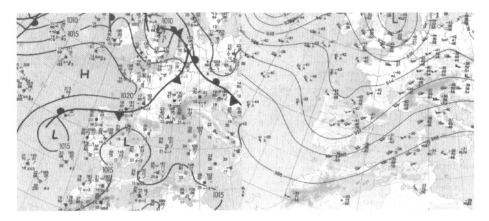

Figure 12.11   Surface situation on 7 August 1996 at 00:00 GMT

The central Ebro valley developed a thermal on the 7th, reinforcing the low-pressure conditions at the 850-Hpa level. This low moved northwards towards the Pyrenees producing mobile convective systems in the central Ebro valley and the Pyrenees. At the same time this thermal low intensified the southeasterly wind, increasing convergence in the Pyrenees and the thermal gradient between the air mass at the north of the cold front and the warm air. The result was the formation of isolated convective cells, one of which caused the Arás catchment flash flood event (Figure 12.11).

Rainfall of 178.4 mm was registered in the Arás basin on 7th August, but it was an extremely isolated event. Most of the surrounding raingauges registered less than 50 mm. Weather radar imagery has shown that most of the rainfall occurred between 16:00 and 18:00, with a maximum of 152 mm between 16:00 and 17:00. The hydrological and geomorphological issues of this massive event have been studied by García-Ruiz et al. (1996) who estimated that the peak discharge was about 500 m^3 s^{-1} at the outlet of the basin.

12.6   CONCLUSIONS AND FORECASTING PROBLEMS

From the analysis of previous work and the case studies it can be concluded that the synoptic situations that produce extreme rainfall events have the following characteristics:

* They usually take place in autumn, but sometimes in late summer.
* A cut-off low appears at 500 Hpa with a core temperature of about −23°C.
* The low usually becomes very deep, extending from 700 to 200 Hpa, and can even extend from 850 to 100 Hpa.
* The position of the Peninsula strongly affects areas receiving precipitation.
* Surface synoptic conditions do not show any particular pattern, although an easterly flow tends to be observed.
* Cyclonic circulation at the 500-Hpa level drives divergent flows.
* Local wind systems such as valley winds or sea breezes can reinforce easterly winds and increase rainfall intensity.
* Heat accumulation during the previous days or months reinforces easterly winds and increases rainfall intensity.

- Topography, especially areas with a south or southeasterly aspect, influences where and how much it will rain.

ACKNOWLEDGEMENTS

The research carried out for this paper was part of the MEDALUS (Mediterranean Desertification and Land Use) collaborative research project. MEDALUS was supported by the European Commission under its Environment and Climate Research Programme (Contract: ENV4-CT95-0118, Climatology and Natural Hazards) and the support is gratefully acknowledged.

REFERENCES

Albentosa, L.M. 1983. Precipitaciones excepcionales e inundaciones durante los días 6 al 8 de noviembre de 1982 en Cataluña. *Estudios Geográficos*, 170-171, 229–273.

Albentosa, L.M. 1991. *Geografía de España. El Clima y las Aguas,* De. Síntesis Madrid, 240pp.

Alonso-Sarría, F. 1995. *Análisis de epiosodios lluviosos y sus consecuencias hidrológicas sobre una cuenca mediterránea semiárida no aforada (Cuenca de la Rambla Salada, Murcia).* Ph.D. Thesis 501pp. Murcia.

Austin, P.M. and Houze, Jr., R.A. 1972. Analysis of the structure of precipitation patterns in New England. *J. Appl. Meteor.*, 11, 926–935.

Bessé, J., Fournie, A. and Ranaudin, M. 1979. *Météorologie, Tome 2. Aerologie-Météorologie Dynamique.* Ecole Nationale de l'Aviation Civile, France.

Browning K.A. 1962. Cellular structure of convective storms. *Meteorological Magazine,* 91, 341–350.

Browning, K.A. and Hill, F.F. 1984. Structure and evolution of a mesoscale convective system near the British Isles, *Quarterly Journal of the Royal Meteorological Society*, 110, 897–914.

Camacho, J.L. 1994. Weather radar and hydrology: precipitation measurement and forecast. *US–Spain Workshop on Natural Hazards, Barcelona, June 1993.* International Decade for Natural Disaster Reduction. Barcelona, 387pp.

Camarasa, A. 1988. El SAIH en la cuenca hidrográfica del Júcar. *Cuadernos de Geografía*, 44, 235–240.

Camarasa, A. 1991. *Génesis de crecidas en pequeñas cuencas semiáridas: Barranc de Carraixet y Rambla del Poyo.* PhD Thesis, University of Valencia.

Capel-Molina, J.J. 1974. Génesis de las inundaciones de octubre de 1973 en el sureste de la Península Ibérica. *Cuadernos Geográficos,* 4, 149–166. University of Granada.

Capel-Molina, J.J. 1977. Los torrenciales aguaceros y crecidas fluviales de los días 25 y 26 de octubre de 1977 en litoral levantino y sur mediterráneo de la Península Ibérica. *Paralelo 37°*, 1, 109–132.

Capel-Molina, J.J. 1981. *Los climas de España.* Ed. Oikos Tau Barcelona, 429pp.

Capel-Molina, J.J. 1989a. Las lluvias torrenciales de Noviembre de 1987 en Levante y Murcia. *Estudios Románicos*, 6, 1551–1562.

Capel-Molina, J.J. 1989b. Convección profunda sobre el Mediterráneo español. Lluvias torrenciales durante los días 4 a 7 de Septiembre de 1989 en Andalucía Oriental, Murcia, Levante, Cataluña y Mallorca. *Paralelo 37°*, 13, 51–79.

Castillo Requena, J.M. 1978. Estudios sobre le comportamiento de la gota de aire frío y la distribución de sus consecuencias pluviométricas en la España peninsular. *Paralelo 37°*, 2.

Chalker, W.R. 1949. Vertical stability in regions of air mass showers. *Bulletin of the American Meteorological Society*, 30, 145–147.

Chisholm, A.J. and Renick, J.H. 1972. The kinematics of multicell and supercell Alberta hailstorms. Alberta Hail studies, Research Council, Alberta. *Hail studies Report*, 72(2), 24–31.

Coantic, L. 1979. Les échanges oceáns-atmosphére. *La Météorologie*, 12, 384–390.

Elias Castillo, F. and Ruiz Beltrán, L. 1979. *Precipitaciones máximas en España.* Ministerio de Agricultura, Madrid, 545pp.

Estrela, M.J. and Millán, M.M. 1994. *Manual práctico de Introducción a la Meteorología*. Ed. CEAM, 351pp.

Fontaine, I. and Portela, P. 1959. Causes méteorologiques des grandes crues cévenales du début de l'automne 1958. *La Méteorologie*, I, 53ff.

Font Tullot, I. 1983a. Algunas observaciones sobre las lluvias excepcionales en la vertiente mediterránea española. *Estudios Geográficos*, 170–171, 55–60.

Font Tullot, I. 1983b. *Climatología de España y Portugal*. Instituto Nacional de Meteorología, Madrid, 286pp.

Gallego, F. 1995. *Situaciones de flujo mediterráneo y precipitaciones asociadas. Aplicación a la predicción cuantitativa en la Cuenca del Segura*. Ph.D. Thesis, Univ. Murcia. Servicio de publicaciones, 412pp.

García de Pedraza, L., 1983. Situaciones atmosféricas tipo que provocan aguaceros torrenciales en comarcas del Mediterráneo español. *Estudios Geográficos*, 170–171, 61–73.

García de Pedraza, L. 1990. Contrastes climáticos en la Región de Murcia. *XVIII Jornadas científicas de la AME, Madrid*, 37–48.

García Miralles, V. and Carrasco Andreu, Q. 1958. *Lluvias de intensidad y extensión extraordinarias causantes de las inundaciones de los días 13 y 14 de Octubre de 1957 en las provincias de Valencia, Castellón y Alicante*. Servicio Meteorológico Nacional, Madrid, 67pp.

García-Ruíz, J.M., White, S.M., Martí, C., Valero, B., Errea, M.P. and Gómez Villar, A., 1996. *La catástrofe del Barranco de Arás (biescas, Pirineo Aragonés) y su contexto espacio-temporal*. CSIC Instituto Pirenaico de Ecología, Zaragoza, 54pp.

Gil Olcina, A. 1983a. *Lluvias torrenciales e inundaciones en Alicante*. Instituto Universitario de Geografía, Universidad de Alicante, Alicante, 128pp.

Gil Olcina, A. 1983b. Inundaciones de Octubre de 1982 en el Campo de Alicante. *Estudios Geográficos*, 170–171.

Gil Olcina, A. 1988. Precipitaciones y regímenes pluviales en la vertiente mediterránea española. *Boletín de la Asociación de Geógrafos Españoles*, 7, 1–12.

Gil Olcina, A. 1989. Causas climáticas de las riadas. A. Gil Olcina and A. Morales (Eds) *Avenidas fluviales e inundaciones en la Cuenca del Mediterráneo*. Instituto Universitario de Geografía de la Universidad de Alicante, Alicante, 15–30.

Haltiner, G.J. and Martin, F.L. 1990. *Meteorología dinámica y física*. Madrid INM, 449pp.

Hardman, M.E. 1983. Cold pools. *Weather*, 38(5), 152–153.

Heras, R. 1973. *Estudio de las máximas crecidas de la zona Alicante–Almería–Málaga y de las lluvias torrenciales de Octubre de 1973*. Centro de Estudios Hidrográficos, Madrid, 28pp.

Jansá-Guardiola, J.M. 1966. Meteorología del Mediterráneo occidental. *SMN Serie A*, 43. Madrid.

Jansá-Clar, A. 1988. *Inestabilidad baroclina y ciclogénesis en el Mediterráneo occidental*. INM, Barcelona, 80pp.

Jansá-Clar, A. 1989. *Notas sobre análisis meteorológico mesoscalar en niveles atmosféricos bajos*. INM, Madrid, 70pp.

Llasat, M.C. 1987. La gota fría, las lluvias y las inundaciones. *Cuadernos de protección civil*, 21, 4–6.

Llasat, M.C. 1991. *Gota fría*. Ed. Boixareu, Barcelona, 164pp.

López-Bermúdez, F. 1973. La Vega Alta del Segura. Clima, Hidrología y Geomorfología, Universidad de Murcia, 288pp.

López-Bermúdez, F. and Gutiérrez Escudero, J.D. 1983. Descripción y experiencias de la avenida e inundación de octubre de 1982 en la Cuenca del Segura. *Estudios Geográficos*, 170–171, 87–120.

López-Bermúdez, F. and Navarro Hervás, F. 1979. Inundaciones catastróficas, precipitaciones torrenciales y erosión en la provincia de Murcia, *Papeles de Geografía*, 8, 49–91.

López-Bermúdez, F. and Romero Díaz, A. 1993. Génesis y consecuencias erosivas de las lluvias de alta intensidad en la región mediterránea. *Cuadernos de Investigación Geográfica*, 18–19, 7–28, Logroño.

López Gómez, A. 1983. Las lluvias catastróficas mediterráneas. *Estudios Geográficos*, 10–171, 11–29.

Maddox, R.A. 1980. Mesoscale convective complexes. *Bulletin of the American Meteorological Society*, 61(11), 1374–1387.

Martin Vide, J. 1987. Precipitaciones torrenciales en España. *Norba*, 1, 63–72.

Martin Vide, J. 1989. *Caracteristiques climatologiques de la precipitació en la franja costera mediterrània de la Península Ibèrica*. Institut Cartogràfic de Catalunya, Barcelona, 245pp.

Medina, M. 1976. *Meteorología básica sinóptica*. Ed. Paraninfo, Madrid, 320pp.

Medina, M. 1984. Métodos para el pronóstico de lluvias torrenciales. *Anales de la Universidad de Murcia*, XLIII (3–4), 1–12.

Meteorological Office 1962. *Weather in the Mediterranean*, 391(1), 391.

Millán, M.M., Estrela, M.J. and Caselles, V. 1994. Torrential precipitations on the Spanish east coast: the role of the Mediterranean Sea surface temperature. *Atmospheric Research*.

Miró-Granada, J. 1976. Avenidas catastróficas en el Mediterráneo occidental. *Revista de Hidrología*, Abril-Junio, 117–133.

Miró-Granada, J. 1983. Consideraciones generales sobre la meteorología de las Riadas del Levante español. *Estudios Geográficos*, 170–171, 31–53.

Morales Gil, A., Bru Ronda, C. and Box Amorós, M. 1983. Las crecidas de los Barrancos de las Ovejas y de Agua Amarga, Octubre de 1982. *Estudios Geográficos*, 170–171, 143–170.

Olcina-Cantos, J. 1994. *Riesgos climáticos en la Península Ibérica*. Ed. Penthalon Madrid, 440pp.

Palmen, E. 1949. On the origin and structure of high level cyclones south of the maximum westerlies. *Tellus*, 1, 22–31.

Palmen, E. and Newton, C.W. 1969. *Atmospheric Circulation Systems*. Academic Press, New York.

Quereda, J. 1982. Las interacciones atmósfera-oceano en la climatología del Mediterraneo occidental. In A. Gil Olcina and A. Morales Gil (Eds) *Avenidas fluviales e inundaciones en la Cuenca del Mediterráneo*. Caja de Ahorros del Mediterráneo and Instituto Universitario de Geografía, 67–87.

Quereda, J. 1983. Los excepcionales temporales de octubre y su relación con las temperaturas del mar. *Cuadernos de Geografía*, 32–33.

Radinovic, D. 1976. Numerical model requirements for the Mediterranean area. *International School of Atmospheric Physics 3rd Course: Meteorology of the Mediterranean*. Erice, Sicilia.

Reuter, H. 1954. *Weather Forecasting Methods and Problems*. Vienna, cited in Jansá-Clar, A. (1988).

Riosalido, R. 1994. Heavy rain in the Spanish Mediterranean area and PREVIMET plan. In Corominas and Georgakakos (Eds) *US–Spain Workshop on Natural Hazards*. International Decade for Natural Disaster Reduction, Barcelona 387pp.

Riosalido, R. 1998. *Estudio Meteorológico de la situación del 7 de Agosto de 1996 (Biescas)*. Dirección General del Instituto Nacional de Meteorología, 90pp.

Riosalido, R., Rivera, A. and Martín León, F. 1989. Desarrollo de un sistema convectivo de mesoescala durante la campaña PREVIMET Mediterráneo-87. *Primer simposio nacional de predictores del INM*. Madrid, 67–83.

Rivera, A. and Martínez, C. 1983. Tratamiento digital de imágenes Meteosat de alta resolución. Aplicación al caso de las inundaciones de Levante en Octubre de 1982. *Revista de la Asociación Meteorológica Española*, 2, 67–79.

Rodríguez Estrella, T., López-Bermúdez, F., Navarro Hervás, F. and Albacete, M. 1992. El riesgo de inundabilidad y zonación para diferentes usos del llano de inundación de la Rambla litoral de las Moreras. La avenida de Septiembre de 1989. In F. López-Bermúdez, F. Conesa-García and M.A. Romero (Eds) *Estudios de Geomorfología en España*, Sociedad Española de Geomorfología, Murcia, 353–364.

Sartor, J.D. 1962. *Essential Factor of Thunderstorm Forecasting Memorandum*. RM-3049-PR USAF Project Rand, Rand Corp.

Segura, F. 1987. Les inundacions de novembre de 1987 al Pais Valencià. *Cuadernos de Geografía*, 24, 205–211.

Showalter, A.K. 1953. A stability index for thunderstorm forecasting. *Bulletin of the American Meteorological Society*, 34, 250–252.

Sumner, G. 1988. *Precipitation Process and Analysis*. John Wiley & Sons, 455pp.

Wainwright, J. 1996. Hillslope response to extreme storm events: the example of the Vaison-La-Romaine event. In M.G. Anderson and S.M. Brooks (Eds) *Advances in Hillslope Processes*, Vol. 2, 997–1026.

Zimmerschied, W. 1954. *La topografia relativa como medio auxiliar indispensable para el análisis del mapa meteorológico, especialmente en su aplicación aeronaútica. Anotaciones acerca de Irrupciones de aire frío de las regiones de Groenlandia y Spitzberg hacia Europa central y occidental a finales de otoño o a principios de invierno*. SMN A.24 40, Madrid.

Zipser, E.J. 1982. Use of a conceptual model of the life-cycle of mesoscale convective systems to improve very-short-range forecasts. In K.A. Browning (Ed.) *Nowcasting*. Academic Press.

Index

Note: Page numbers in *italic* refer to illustrations; those in **bold** type refer to tables.